Fundamentals of Occupational Safety and Health

Sixth Edition

P. Kohn

Published by Bernan Press
An imprint of The Rowman & Littlefield Publishing Group, Inc.
4501 Forbes Boulevard, Suite 200, Lanham, Maryland 20706
www.rowman.com
800-865-3457; info@bernan.com

16 Carlisle Street, London W1D 3BT, United Kingdom

British Library Cataloguing in Publication Information Available

Library of Congress Cataloging-in-Publication Data

Friend, Mark A., author.
  Fundamentals of occupational safety and health / Mark A. Friend and
James P. Kohn. — Sixth edition.
       p. ; cm.
  Includes bibliographical references and index.
  ISBN 978-1-59888-723-5 (pbk. : alk. paper) — ISBN 978-1-59888-724-2
(electronic)
  I. Kohn, James P., author. II. Title.
  [DNLM: 1. Occupational Health—United States. 2. Safety Management—
methods—United States. 3. Accidents, Occupational—prevention & control—
United States. WA 485]
  T55
  363.110973—dc23   2014009928

™ The paper used in this publication meets the minimum requirements of
American National Standard for Information Sciences—Permanence of
Paper for Printed Library Materials, ANSI/NISO Z39.48-1992.

Printed in the United States of America

To the Eternal Memory of

James P. Kohn

Colleague, Teacher, Father, Husband, Friend

His legacy is not only the decades of service he gave to the safety and health profession, the many current and future safety professionals who have and will benefit from his work, and the workers whose lives and health he has helped protect, but also the many hearts he touched.

Summary of Contents

Contents

6 Introduction to Industrial Hygiene   103
Burton R. Ogle, PhD, CIH, CSP, and
Tracy L. Zontek, PhD, CIH, CSP

16 Construction Safety and the Multiemployer Worksite Doctrine 337
Mark A. Friend and Dan Nelson, with additions by William Walker

Figures and Tables

Preface

The first edition of this book was written as a response to the need for a comprehensive, introductory-level text covering the major facets of occupational safety and health. The sixth edition of this book was written with the same objectives as previous editions:

- It should be easily comprehensible by those who lack experience or prior exposure to concepts in occupational safety and health, industrial hygiene, or occupational medicine.
- It should be on a writing level easily understood by a majority of the population, and yet be useful as a text for college courses.
- It should cover the major topics of concern to safety and health professionals and students to provide both with a philosophical base.

The sixth edition incorporates material from the earlier editions and contains necessary updates.

Dr. Burton Ogle, CSP, CIH, and Dr. Tracy Zontek, PhD, CIH, CSP, of Western Carolina University, have rewritten the chapter on industrial hygiene and updated the chapter on hazardous materials. Dr. Celeste Winterberger of Research Triangle Park in North Carolina was a significant contributor to the first edition and her legacy continues in this one. Dr. Scotty Dunlap, CSP, of Eastern Kentucky University updated the chapter on transportation safety, while Dr. Randell Barry updated the chapter on extreme weather, and Brett Carruthers updated the chapter on workplace violence. The rest of the chapters have material added as needed. As the title implies, *Fundamentals of Occupational Safety and Health* covers the basics for the professional's need to understand before accepting responsibility for promoting an environment conducive to reducing hazards and protecting lives. Unlike books exploring only the engineering aspects of the subject, this text attempts to balance the management of safety with the relevant science and the practical aspects of regulatory compliance.

This edition was revised without the assistance of *Dr. James P. Kohn*, who passed away in 1999. He was a friend, a professional, and an innovator in the fields of safety and safety education. All who knew him miss his comradeship and willingness to share knowledge and enthusiasm.

Special thanks go to my wife Kathy for all of her help and inspiration. She is a true friend.

Mark Allen Friend, EdD, CSP
Professor of Doctoral Studies
College of Aviation
Embry-Riddle Aeronautical University
Daytona Beach, Florida

Acknowledgments

The authors wish to thank Glory Mizzille, Daria Hinnant, Doug Gaylord, David McDaniel, Bill Benfield, Mark Thompson, and Nathan Szejniuk for their contributions to the development and editing of the manuscript. Thanks also go to the following individuals for their gracious assistance in providing information, guidance, and text for specific chapters as listed:

Chapter 2   Tom Wagner, Craig MacMurry, Sam Christy, Kelly Nelson, and Wes Heinold

Chapter 3   Tom Wagner, Susan Wilson, and Craig MacMurry

Chapter 4   George Nichols Jr.

Chapter 5   Eddie Allen, Eddie Anderson, Barbara Dail, Ralph Dodge, Doug Gaylord, Eddie Johnson, Barry Maxwell, Amy Tomlin, and Celeste Winterberger

Chapter 6   Burton R. Ogle, PhD, CIH, CSP, and Tracy L. Zontek, PhD, CIH, CSP

Chapter 7   Eddie Anderson, Barbara Dail, Ralph Dodge, Doug Gaylord, Eddie Johnson, Katherine Kohn (ergonomic illustrations), Barry Maxwell, Amy Tomlin, and Celeste Winterberger

Chapter 8   Amber Perry

Chapter 9   Celeste Winterberger (who authored this chapter)

Chapter 11   Brett Carruthers

Chapter 12   Earl Blair (who authored this chapter)

Chapter 13   Brett Carruthers (who authored this chapter)

Chapter 14	Brett Carruthers, Tom Shodoan, Antoine Slaughter, Julie Mitchell, Jeff Porter, James Frey, and Joseph Johnson
Chapter 15	Tracy L. Zontek, PhD, CIH, CSP, and Burton R. Ogle, PhD, CIH, CSP
Chapter 16	William Walker and Dan Nelson (who wrote the section on multiemployer worksite policy)
Chapter 17	Scotty Dunlap (who wrote the chapter)
Chapter 18	Randell J. Barry, PhD
Chapter 19	Robert Getchell, Keith McCullough, Mike Ban Derven, and Neil Brown
Appendix A	Eddie Anderson, Mike Baker, Barbara Dail, Craig Fulcher, Doug Gaylord, Barry Maxwell, David McDaniel, and Ron Skinner
Appendix B	Mike Baker, Craig Fulcher, Doug Gaylord, Darla Hinnant, David McDaniel, Glory Mizzelle, Ron Skinner, Ward Taylor, Chip Tillett, and Jason Whichard

In addition, the authors want to recognize Kathy Friend, MSOS, and Carrie Kohn, MA, for their professional encouragement.

Introduction to Occupational Safety and Health

CHAPTER OBJECTIVES

After completing this chapter, you will be able to

- Explain the importance of occupational safety and health
- Identify key historical figures that have contributed to the profession
- Define basic terminology used in occupational safety and health
- List job titles of individuals performing occupational safety and health activities
- Identify roles and responsibilities of safety and health professionals
- Identify basic guideposts for judging ethical behaviors

CASE STUDY

As a 22-year-old construction worker with 11 months of experience on the job, Bob had finally made it. Since graduating high school, Bob had tried a lot of things, but they just never seemed right for him. He attended a community college for one year and then dropped out. There was too much theory that didn't relate to how he saw the world. Bob tried a number of jobs, but minimum-wage salaries forced him to live at home with his parents. His parents were good people, but he was ready to move on with his life. With this new job, everything was turning out great. Bob was bringing home a good paycheck. He had just moved into a new apartment, which he shared with his high school buddy Tim, and he was going the next day to sign the papers for a brand-new pickup truck.

Bob never made it to the dealership to sign those papers. Maybe he was distracted thinking about that "killer" pickup that he was about to purchase. Perhaps he never realized how dangerous it actually was to work on that scaffolding. After all, it was only 20 feet off the ground and it looked safe. Bob had worked on wet scaffolding before, and although it was wet from the rains the previous night, nothing had ever happened to make him concerned about working at those heights. Bob's world changed when he fell to the ground. His fall put him in a wheelchair, paralyzed from the waist down.

OCCUPATIONAL SAFETY AND HEALTH

The field of occupational safety and health is concerned with minimizing loss by aiding in the preservation and protection of both human and other physical assets in the workplace. The discipline is far reaching in both scope and practice. It primarily involves *monitoring* the workplace and *advising* employers or management on the best ways to prevent and minimize losses. Final responsibility for action always rests on the shoulders of the management, as they are ultimately accountable for workplace behaviors. Management is held accountable by stockholders or owners of the company, the Occupational Safety and Health Administration (OSHA), the courts, and even public opinion. The job of the safety and health professional is to assist management by observing the workplace and providing guidance.

In practice, occupational safety and health addresses moral and economic issues—typically within a framework required by law. The United States government, and governments worldwide, require protection of employees from hazards that may result in injury, illness, or death. Under the Occupational Safety and Health Act of 1970, known as the OSHAct, employers in the United States are required to provide safe and healthy workplaces. The safety and health professionals help employers to do that.

Unfortunately, for some employers the responsibility to protect human life is not as important as other goals or priorities. A company may focus on productivity and profits to the exclusion of safety and health. Its managers may view the occurrence of illnesses and injuries as a routine part of the job. In reality the amount of production required to cover costs associated with accidents in the workplace can be substantial and may far outweigh the expense of providing a safe and healthy working environment. The role of the safety professional requires monitoring workplace conditions and advising management on the importance of making critical corrections for moral, legal, and economic reasons. The effective safety professional will work with the management team and help demonstrate that providing a safe and healthy working environment is the right thing to do for both the employees and the company. The safety and health professional must be able to make a convincing argument, based on sound business practices; otherwise, management may choose to allow safety and health to become low priorities. Neither can exist without management support, and that support is only forthcoming when profits can be made. The organization cannot exist without profit and the job of the safety professional cannot exist without production. Safety does not and never will come first. Safety cannot and will not exist without profitable production, but profitable production is not likely to exist without safety. Safe and profitable production is the ultimate goal of the safety professional.

IMPORTANCE OF OCCUPATIONAL SAFETY AND HEALTH

Morally, legally, and economically, occupational safety and health have become important issues. Companies are attempting to remain profitable in an ever more competitive, global economy. For these companies, addressing safety, health, and environmental issues may mean more than good business practice. For many companies, strong safety, health, and environmental programs may actually mean survival.

Thousands of employees are reportedly killed annually in the United States as a result of on-the-job incidents and many more are injured. The costs associated with these losses are in the billions. Behind the numbers of deaths and injuries are real people—mothers, fathers, sisters, brothers, spouses, sons, or daughters. They are people like Bob whose lives may never be the same again.

The Occupational Safety and Health Administration (OSHA) is the federal agency responsible for workplace safety and health; it attempts to address the safety and health concerns faced by American workers. OSHA may not only levy fines but also seek criminal prosecution of business owners and managers who willfully neglect the safety and health of their employees. In addition, employers may find themselves the target of civil suits levied by the victims and survivors of workplace accidents. Employers with poor safety and health records must also deal with rising medical insurance costs as well as unfavorable workers' compensation premiums. Unfortunately, many employers have not had to bear the full cost of injuring and killing members of their workforces. Workers, uninformed and unaware of their legal rights, have often shouldered the costs of the business not operating safely. The regulations and mechanisms for enforcement are in place in the United States, but the agencies charged with administering safety regulations are generally understaffed. With well over one hundred million workers at millions of worksites of covered employers and not even three thousand OSHA inspectors, the task is clearly a challenge for the agency. In addition, certain categories of workers, such as some federal, state, and municipal employees, do their jobs without protection from OSHA or any government agency. Most federal employees are excluded from OSHA regulations. Many state and municipal employees in states covered by federal OSHA also work without OSHA protection. Continued reduction of accidents in the United States will require an increase in initiative on the parts of all parties involved to include employers, employees, and federal and state governments. An understanding of the issues of today will be enhanced by a review of the past.

EARLY HISTORICAL EXAMINATION OF OCCUPATIONAL SAFETY AND HEALTH

Ancient Greek and Roman Physicians

Concern for occupational safety and health is not a recent issue. Many of today's health and safety problems were first observed over 2,000 years ago. An early account is associated with the Code of Hammurabi that dates back to approximately 2100 BC. In an attempt to recompense victims, it primarily addressed personal injury and losses and prescribed a schedule of punishments and payments for wrongdoers.

Greek and Roman physicians, practicing between 400 BC and 300 AD, expressed concern for the health of individuals exposed to the metals commonly used during this period. These included Hippocrates, the Father of Medicine, and Pliny the Elder, a Roman physician and scientist, both discussed in chapter 6.

Galen, a Roman physician who lived during the second century, wrote about occupational diseases and the dangers of acid mists to copper miners. He was also concerned with the mining, tanning, and chemical occupations, noting several diseases contracted by individuals working in those professions.

The European Renaissance and the Industrial Revolution

Prior to the Renaissance, little information is available on European injury, illness, and property damage-prevention activities. Reports of medieval scribes suffering lead poisoning, while performing the common practice of tipping their quills with their tongues between dips into metallic ink solutions, were repeatedly noted before the fifteenth century. Unfortunately, little else was recorded regarding safety and health in that period. During the European Renaissance, physicians and chemists began noticing the relationship between occupational activities and worker health and safety.

Ulrich Ellenborg, for example, recognized, identified, and reported on "the poisonous and noxious vapors and fumes of metals." In 1437, he recognized that the vapors of some metals, including lead and mercury, were dangerous and described the symptoms of industrial poisoning from these sources. He also became aware of asbestos and lung diseases among miners.

Bernardo Ramazzini, an Italian physician, around 1700 published *De morbis artificum diatribe*, or *The Diseases of Workers*, the first treatise on occupational disease. Considered by some to be the father of occupational medicine, and by others as the father of industrial hygiene, he recommended physicians ask their patients, "What is your trade?" He urged students to learn the nature of occupational diseases in shops, mills, mines, or wherever men toil.

In 1666, fire swept through London, England, and raged for several days. At that time, London was a city of half-timbered, pitch-covered, medieval buildings that ignited at the touch of a spark. From an inn on Pudding Lane the fire spread into Thames Street, where riverfront warehouses were bursting with oil, tallow, and other combustible goods. The customary recourse during a fire of such magnitude was to demolish every building in the path of the flames, in order to deprive the fire of fuel, but the city's mayor hesitated, fearing the high cost of rebuilding. With no building codes at the time, houses were frequently built to the edge of the street. Second and subsequent stories were often cantilevered, with the top floors nearly touching houses across the street. The fire raged and destroyed a large part of London. Early examples of building and fire codes resulted from this disaster.

During the period between 1760 and 1840, history witnessed dramatic advances in technology. *Dr. Percival Pott* (circa 1775) identified the first form of cancer, scrotal cancer in chimney sweeps, and determined its relationship to soot and coal tar exposure. This observation resulted in numerous regulations called the Chimney Sweep Acts that were promulgated between 1788 and 1875. About the same time, several industrialists also became concerned with the welfare of their workers. *Sir Robert Peel*, a mill owner, made the English Parliament aware of the deplorable working conditions often existing in the mills. He reported that orphan labor was frequently used to perform demanding tasks in less than sanitary conditions. His study of these deplorable conditions revealed the mean life expectancy of the working class, under these terrible conditions, was only 22 years, while the mean age of the wealthier class was 44 years.

With advancing technology and the Industrial Revolution came an increase in safety and health hazards. The innovations of mechanical textile machinery, foundry furnaces, steam engines, and numerous other inventions created a new and more dangerous workplace environment. Factories and other workplaces were mazes of moving belts, pulleys, and gears. Human senses were assaulted with fumes, toxic vapors, noise, and heat. The health and safety problem was compounded by the introduction of increasing numbers of women and children into the workforce. Long workdays, unsanitary conditions, and demanding physical labor amplified the likelihood of injury and illness for this new workforce.

At the dawn of the Industrial Revolution in England, *Charles Thackrah* became concerned with occupational safety and health and studied the effects of arts, trades, life habits, civic states, and professions upon health and longevity. By employing basic principles of occupational medicine, he became the first physician in the English-speaking world to establish the practice of industrial medicine. His writings also led to a raised public awareness of the plight of many of the new working class. In 1842, *Edwin Chadwick*, a British lawyer and sanitarian, described the deplorable conditions of factory workers in his "Report into the Sanitary Conditions of the Labouring Population of Great Britain." He stated that life expectancy was much lower in towns than in the countryside and attributed his findings to air pollution.

In the United States, the Industrial Revolution began in the early nineteenth century as factories and mills in New England sprang to life. In

Lowell, Massachusetts, women and girls, as young as six to ten years of age, worked long hours, often from five in the morning until seven in the evening. Their work required their hands to be placed very close to the (running) gears of spinning machines. Many were injured or maimed in the moving gears and pulleys of the textile machinery. Their fingers were cut off or mangled with such frequency that machine-guarding laws were eventually passed. Fatalities in the mining and steel industries were as common as those in the textile industry. In 1877, Massachusetts passed a regulation requiring safeguards on hazardous machinery. This law also tied liability to the actions of employers.

Exposure to toxic metals such as mercury and lead has been an occupational health problem for hundreds of years. Technological advances introduced new and unique hazards, typically overlooked by untrained observers, and only recognized after numerous cases were reported. Dramatic changes in technology and workplace design along with other advancements have not necessarily resulted in healthier or safer workplaces. Employees today may be exposed to as many workplace hazards as were their ancestors in years past. Factory machinery with unguarded gears, mangling fingers and hands, has been replaced by electronic office equipment causing wrist and arm injuries. Many companies have addressed chemical and toxic hazards, but problems still occur, as evidenced by the Connecticut Power Plant explosion and the Deepwater Horizon oil spill in the Gulf of Mexico. The complexity of current safety and health conditions mirrors the complexities associated with modern workplace technology. Practitioners in this profession must develop the broad range of knowledge and skills necessary to ensure the protection of people and company resources. This knowledge base must include a well-grounded understanding of the terms and concepts used in the profession. In addition, occupational safety and health professionals must possess the skills required to effectively perform their roles and responsibilities in safety, health, and the environmental arenas. Specialized knowledge may also be required to address the hazards associated with work in specific industries or locations.

TERMS AND CONCEPTS IN THE SAFETY PROFESSION

Safety professionals are concerned with the preservation of people and company resources. They need the knowledge and skills often acquired through formal education and/or experience in the safety field. Many safety professionals are certified after successfully completing the requirements for the designations "Certified Safety Professional" (CSP), "Certified Industrial Hygienist" (CIH), or others related to occupational safety and health. In order to achieve these designations, safety professionals must have a number of years of experience in the field and undergo a rigorous examination process for each.

Safety professionals attempt to achieve their loss prevention goals through the systematic application of principles taken from a variety of disciplines, including engineering, education, psychology, physiology, industrial hygiene, health physics, and management. Safety professionals are concerned with the elimination or control of hazards that may result in injury, illness, and property damage. They will often use techniques referred to as *loss prevention* and *loss control* to accomplish that goal.

Loss prevention describes a program designed to identify and correct potential accident problems before they result in financial loss or injury. *Loss control*, however, is a program designed to minimize incident-based financial losses. An example of the difference between loss prevention and loss control can be seen in the various activities associated with fire protection and fire prevention programs.

In a fire prevention program, employees can be trained to inspect their areas and remove combustible materials like oily rags or cardboard. These inspection activities would be an example of loss prevention. A fire protection program, on the other hand, might include employee training in the use of fire extinguishers. Employees would then possess the skills necessary to fight a fire. The fire might ignite, but, following training, employees would be prepared to extinguish it and thus minimize the damage. Fire extinguisher training is an example of a loss control technique.

Loss prevention and loss control are important to the safety professional who attempts to recognize, evaluate, and control hazards in the workplace. This is part of the process referred to as "safety management." *Safety management* encompasses the responsibilities of planning, organizing, leading, and controlling activities necessary to achieve an organization's loss-prevention and loss-control goals. Continuing the fire example, the safety professional might wish to establish a safety management program to address this danger. Safety professionals would determine the problems that exist at their facility. They would then establish the details of the fire-training program with goals and objectives of what is to be accomplished (planning). Next, they would determine the trainers and materials necessary to implement the program by establishing a schedule to ensure all activities are accomplished (organizing). Safety professionals must then ensure that the required resources are available and that the people involved in this project coordinate their efforts when required (leading). Finally, the professionals monitor and evaluate the progress of the project (controlling). All of this would be done with the endorsement and engagement of the management team. A more detailed examination of the management of the safety function will be presented in chapter 10.

One of the most important terms used in the safety and health profession is "safety." It is probably the most misinterpreted term by individuals outside of the safety profession. For the layperson, safety means not getting injured. "*Safety*," to the professional, implies reference to the likelihood or risk that a loss event will occur. It can be defined as "operating within an acceptable or low probability of risk associated with conditions or activities having the potential to cause harm to people, equipment, facilities or the enterprise." Although the goal of a company may be to eliminate all loss events and have zero losses in the workplace, this can only be done through the elimination of all activities and all work. Nearly any activity carries some level of risk and exposure to that risk is inherent in engaging in the activity. The safety professional attempts to minimize the risk, but the law of large numbers dictates that exposure to enough risks enough times will eventually and necessarily lead to loss. The role of the safety professional is to minimize loss, but it must be recognized that, in general, losses cannot be completely eliminated.

Risk can be defined as the measure of the probability and severity of a loss event taking place. A *hazard* is a workplace condition or worker action that can or has the potential to result in injury, illness, property damage, or interruption of a process or an activity. As revealed in these definitions, determining occupational risk requires an examination of both the probability of occurrence of a hazard that may cause injury and/or property damage and the severity of the resulting injury or loss.

The evaluation of risks in the workplace starts with the identification of the types of hazards existing at the facility. Establishing a process to ensure hazards are identified is a primary goal of a progressive organization with a strong, safety-management program. The organization then works to eliminate or reduce the risks associated with those hazards to the lowest achievable and reasonable level. Nothing is risk free. Safety professionals identify the tasks and activities having the greatest inherent risk. They then attempt to systematically eliminate or reduce the level of risk as much as feasibly possible given time, personnel, and budget restraints. A detailed discussion about risk assessment will be presented in chapter 5, "Accident Causation and Investigation."

The control of risk ultimately leads to the reduction of losses. There are several different types of losses safety professionals attempt to eliminate or control, including worker-related health and safety losses—injuries, illnesses, and fatalities. Workplace losses can consist of damaged equipment, damaged raw materials or finished products, damaged or destroyed facilities, downtime, service/production interruption, or loss of reputation. Most of this loss occurs as a result of accidents.

While many consider accidents to be events occurring beyond an individual's control, safety professionals look at accidents in a more systematic and determined manner. *Accidents* are unplanned events, often resulting in injuries or damage that interrupt routine operations. They are nearly always preceded by unsafe acts of employees, hazardous conditions in the workplace, or both. When appropriate action is taken, *most* accidents can be eliminated.

JOB TITLES OF INDIVIDUALS PERFORMING OCCUPATIONAL SAFETY AND HEALTH ACTIVITIES

There are many titles given to individuals who perform occupational safety and health activities. The following list describes just a few of those titles:

Industrial Hygienist: Although basically trained in engineering, physics, chemistry, or biology, this individual has acquired, through study and experience, knowledge of the effects on health of chemical and physical agents under various levels of exposure. The industrial hygienist is involved in the monitoring and analytical methods required to detect the extent of exposure and the engineering and other methods used for hazard control.

Risk Manager: The risk manager in an organization is typically responsible for insurance programs and other activities that minimize losses resulting from fire, accidents, and other natural and man-made losses.

Safety Professional: An individual who, by virtue of specialized knowledge and skill and/or educational accomplishments, has achieved professional status in the safety field. This individual may also have earned the status of Certified Safety Professional (CSP) from the Board of Certified Safety Professionals.

Safety Engineer: An individual who, through education, licensing, and/or experience, devotes most or all of employment time to the application of scientific principles and methods for the control and modification of the workplace and other environments to achieve optimum protection for both people and property.

Safety Manager: The individual responsible for establishing and maintaining the safety organization and its activities in an enterprise. Typically, the safety manager administers the safety program and manages subordinates, including the fire prevention coordinator, industrial hygienist, safety specialists, and security personnel.

THE SAFETY AND HEALTH PROFESSIONAL'S ROLE AND RESPONSIBILITY

The specific roles and responsibilities of safety professionals depend upon the jobs in which they are employed or the types of hazards present where they work. Research (Kohn, Timmons, and Besesi, 1991) examining the roles and responsibilities of safety professionals identified the following activities as those most frequently performed:

Accident Investigation: determining the facts and causes related to an accident based on witness interviews and site inspections.

Work with Emergency Response Teams: organizing, training, and coordinating skilled employees to react to emergencies such as fires, accidents, or other disasters.

Environmental Protection: recognizing, evaluating, and controlling hazards that can lead to undesirable releases of harmful substances into air, water, or the soil.

Ergonomic Analysis and Modification: designing or modifying the workplace based on an understanding of human physiological/psychological characteristics, abilities, and limitations.

Fire Protection: eliminating or minimizing fire hazards by inspection, layout of facilities and design of fire suppression systems.

Hazard Recognition: identifying conditions or actions that may cause injury, illness, or property damage.

Hazardous Materials Management: ensuring dangerous chemicals and other products are stored and used in such a manner as to prevent accidents, fires, and the exposure of people to these substances.

Health Hazard Control: recognizing, evaluating, and controlling hazards that can create undesirable health effects, including noise, chemical exposures, radiation, or biological hazards.

Inspection/Audit: evaluating/assessing safety and health risks associated with equipment, materials, processes, or activities. Inspections and audits differ. A more detailed explanation is provided in chapter 10.

Recordkeeping: maintaining safety and health information to meet government requirements, as well as provide data for problem solving and decision making.

Regulatory Compliance: ensuring all mandatory safety and health standards are satisfied.

Training: providing employees with the knowledge and skills necessary to recognize hazards and perform their jobs safely and effectively.

In today's safety environment, safety professionals may also find themselves heavily involved in such issues as security in the workplace, as was the case on 9/11, and natural disaster management to include weather-related events and earthquakes. While these lists provide examples of specific activities performed by the safety professional, the American Society of Safety Engineers (ASSE) has published a document titled "Scope and Functions of the Professional Safety Position." This ASSE publication presents a broad picture of the safety professional's roles and responsibilities (see figure 1-1).

Scope and Functions of the Professional Safety Position
American Society of Safety Engineers

To perform their professional functions, safety professionals must have education, training and experience in a common body of knowledge. Safety professionals need to have a fundamental knowledge of physics, chemistry, biology, physiology, statistics, mathematics, computer science, engineering mechanics, industrial processes, business, communication, and psychology. Professional safety studies include industrial hygiene and toxicology; design of engineering hazard controls; fire protection; ergonomics; system and process safety; safety and health program management; accident investigation and analysis; product safety; construction safety; education and training methods; measurement of safety performance; human behavior; environmental safety and health; and safety, health, and environmental laws, regulations, and standards. Many safety professionals have backgrounds or advanced study in other disciplines, such as management and business administration, engineering, education, physical and social sciences, and other fields. Others have advanced study in safety. This extends their expertise beyond the basics of the safety profession.

Because safety is an element in all human endeavors, safety professionals perform their functions in a variety of contexts in both public and private sectors, often employing specialized knowledge and skills. Typical settings are manufacturing, insurance, risk management, government, education, consulting, construction, health care, engineering and design, waste management, petroleum, facilitates management, retail, transportation, and utilities. Within these contexts, safety professionals must adapt their functions to fit the mission, operations, and climate of their employer.

Not only must safety professionals acquire the knowledge and skill to perform their functions effectively in their employment context, but also through continuing education and training they stay current with new technologies; changes in laws and regulations, and changes in the workforce, workplace, and world business, political, and social climate.

(continued next page)

Figure 1-1.   ASSE scope and functions of the professional safety position

Figure 1-1 *(continued)*

As part of their positions, safety professionals must plan for and manage resources and funds related to their functions. They may be responsible for supervising a diverse staff of professionals.

By acquiring the knowledge and skills of the profession, developing the mind-set and wisdom to act responsibly in the employment context, and keeping up with changes that affect the safety profession, the safety professional is able to perform required safety professional functions with confidence, competence, and respected authority.

Functions of the Professional Safety Position

The major areas relating to the protection of people, property, and the environment are:

A. Anticipate, identify, and evaluate hazardous conditions and practices.

  1. Developing methods for

    a. anticipating and predicting hazards from experience, historical data, and other information sources.

    b. identifying and recognizing hazards in existing or future systems, equipment, products, software, facilities, processes, operations, and procedures during their expected life.

    c. evaluating and assessing the probability and severity of loss events and accidents which may result from actual or potential hazards.

  2. Applying these methods and conducting hazard analyses and interpreting results.

  3. Reviewing, with the assistance of specialists where needed, entire systems, processes, and operations for failure modes; causes and effects of the entire system, process, or operation; and any subsystems or components due to

    a. system, subsystem, or component failures.

    b. human error.

    c. incomplete or faulty decision making, judgments, or administrative actions.

    d. weaknesses in proposed or existing policies, directives, objectives, or practices.

4. Reviewing, compiling, analyzing, and interpreting data from accident and loss event reports, and other sources regarding injuries, illnesses, property damage, environmental effects, or public impacts to

    a. identify causes, trends, and relationship.

    b. ensure completeness, accuracy and validity, or required information.

    c. evaluate the effectiveness of classification schemes and data collection methods.

    d. initiate investigations.

5. Providing advice and counsel about compliance with safety, health, and environmental laws, codes, regulations, and standards.

6. Conducting research studies of existing or potential safety and health problems and issues.

7. Determining the need for surveys and appraisals that help identify conditions or practices affecting safety and health, including those which require the services of specialists, such as physicians, health physicists, industrial hygienists, fire protection engineers, design and process engineers, ergonomists, risk managers, environmental professionals, psychologists, and others.

8. Assessing environments, tasks, and other elements to ensure that physiological and psychological capabilities, capacities, and limits of humans are not exceeded.

B. Develop hazard control designs, methods, procedures, and programs.

1. Formulating and prescribing engineering or administrative controls, preferably before exposures, accidents, and loss events occur to

(continued next page)

Figure 1-1 *(continued)*

    a. eliminate hazards and causes of exposures, accidents, and loss event.

    b. reduce the probability or severity of injuries, illnesses, losses, or environmental damage from potential exposures, accidents, and loss events when hazards cannot be eliminated.

2. Developing methods which integrate safety performance into the goals, operations, and productivity of organizations and their management and into systems, processes, and operations or their components.

3. Developing safety, health, and environmental policies, procedures, codes, and standards for integration into operational policies of organizations, unit operations, purchasing, and contracting.

4. Consulting with and advising individuals and participating on teams

    a. engaged in planning, design, development, and installation or implementation of systems or programs involving hazard controls.

    b. engaged in planning, design, development, fabrication, testing, packaging, and distribution of products or services regarding safety requirements and application of safety principles that will maximize product safety.

5. Advising and assisting human resources specialists when applying hazard analysis results or dealing with the capabilities and limitations of personnel.

6. Staying current with technological developments, laws, regulations, standards, codes, products, methods, and practices related to hazard controls.

C. Implement, administer, and advise others on hazard controls and hazard control programs.

1. Preparing reports that communicate valid and comprehensive recommendations for hazard controls which are based on analy-

sis and interpretation of accident, exposure, loss event, and other data.

2. Using written and graphic materials, presentations, and other communication media to recommend hazard controls and hazard control policies, procedures, and programs to decision-making personnel.

3. Directing or assisting in planning and developing educational and training materials or courses. Conducting or assisting with courses related to designs, policies, procedures, and programs involving hazard recognition and control.

4. Advising others about hazards, hazard controls, relative risk, and related safety matters when they are communicating with the media, community, and public.

5. Managing and implementing hazard controls and hazard control programs that are within the duties of the individual's professional safety position.

D. Measure, audit, and evaluate the effectiveness of hazard controls and hazard control programs.

1. Establishing and implementing techniques, which involve risk analysis, cost, cost-benefit analysis, work sampling, loss rate, and similar methodologies; for periodic and systematic evaluation of hazard control and hazard control program effectiveness.

2. Developing methods to evaluate the costs and effectiveness of hazard controls and programs and measure the contribution of components of systems, organizations, processes, and operations toward the overall effectiveness.

3. Providing results of evaluation assessments, including recommended adjustments and changes to hazard controls or hazard control programs, to individuals or organizations responsible for their management and implementation.

4. Directing, developing, or helping to develop management accountability and audit programs which assess safety performance of entire systems, organizations, processes, and operations or their components and involve both deterrents and incentives.

SAFETY AND ETHICS: DO YOU? DON'T YOU?

Dr. Michael O'Toole
Embry-Riddle Aeronautical University

Background

Acme Engineering specializes in the design and manufacture of custom equipment and facilities. Over the years, the company has gravitated toward designing and making equipment and facilities used in the manufacture of building products. The company has an excellent reputation and is considered the best in its field. Due to the global recession, however, business over the past several years has been slow. Profits have stagnated and talks of potential layoffs are now taking place.

Recently, the sales department received a request for proposal (RFP) from a foreign company based in a third-world country. The RFP calls for design of engineering drawings of a high-speed facility specializing in residential housing building products. Only drawings and specifications are required; another business will actually build and operate the facility. Once complete, the design project will generate approximately $50 million in needed revenues for Acme. Due to its strong world-wide reputation, Acme is the only company receiving the RFP. A team is assembled to ensure requirements of the proposal can be met in the specified time.

In discussions, one of the team members (Tom) questions the fact that, due to the geology of the region where the plant will be built, there are no sources of raw materials readily available. The only option is to rely on recycled sources from a local salvage company. Unfortunately, the recycled material is slightly radioactive. The amount of radiation generated is just below EPA legal standards for similar products found in the United States. The Acme legal department begins a review of applicable laws and regulations and finds there are no parallel regulations in the country where the materials are to be produced and used, and since Acme is not going to build or operate the plant, there is no apparent company liability.

Tom's concern is that Acme would not even consider engaging in the design of this plant if it were to be built in the United States. Others on the team sympathize with Tom's position but point out several key issues:

- Acme is not going to build or operate the plant.
- There are no laws or regulations prohibiting the manufacture and use of these products in the foreign country.
- If Acme turns down the proposal, a competitor will step up to the plate and gladly take the $50 million.
- Acme desperately needs the influx of funds.

Tom seeks the assistance of an epidemiologist to estimate what, if any, adverse health effects might be generated by houses built with products made from the waste material. After careful research and calculation, the epidemiologist estimates there will likely be 28–30 excess lung cancer deaths occurring 35–40 years after construction.

The committee cannot come to consensus, but all notes are sent to the company's top executives. After careful consideration of the facts and arguments, the executive group decides not to go forward with submission of a proposal. The decision is based on several key issues, including the fact that the company is not willing to provide the designs if the plant is to be built in the United States. Additionally, none of the executive group will consider having any of their own family members living in a house constructed with the recycled material.

When contacted by the foreign company, none of the company's competitors steps up to submit a proposal. The plant is never built, so the necessary building products are imported from another location with safe sources for the raw materials.

Business Ethics

There are those who define *ethics* as a moral thermometer of right and wrong; that is, there are no absolutes but the extremes are clearly understood. In most cases it is really only the actions at the "wrong" end of the thermometer that are truly of concern. The real challenge for societies, companies, and individuals is that there is often no one clear definition of right and wrong. On an individual level, decisions about right and wrong derive from family, church, schools, community, and government. All of these impact the development of the individual belief systems and moral guideposts, and they ultimately influence behaviors.

These same factors influence how organizations address or fail to address ethical practices. Organizations are the sum of the parts; that is, the individuals who are members of any given organization influence the ethical standards of that organization. Most societies have certain expectations for the way its members conduct themselves in given circumstances. These social behaviors or expectations are based on the shared belief systems of the group as a whole.

Because of the diversity in any community or organization, there are also differences in expected social norms or ethical behavior. This often results in conflicts and contrary interests that generate ethical dilemmas. Within this context, management operates and attempts to effectively utilize resources (i.e., time, money and personnel) to generate a profit and/or meet shareholder expectations. Since these resources are not unlimited, managers are forced to make difficult decisions on how and where to expend them.

In the case study, management considered maximizing profits while insulating themselves from legal liability. The issue, when examined under the legal microscope, gave a clear signal that no laws were being violated; however, when placed under the ethical microscope, the impact of their decision became clearly a moral one. Outside observers would likely agree that the managers ultimately made the right decision, even though it was not easy to make in considering potential financial profits to be reaped by the company.

So how does one apply the ethical thermometer as a guide for decisions?

- Consider the saying, "If it sounds like the deal is too good to be true, it probably is." One needs to carefully check the facts to ensure that everything is in order and above board.
- Look at how the decision or choice will appear in the "light of day." The decision may eventually become front-page news in the local paper or on the national news. How will it be perceived by the other professionals and the public at large?
- Attempt to remove the emotion and make the decision as objectively as possible. Weigh the pros and cons in light of both the legal and moral consequences. How will the results of the decision affect not only you and your company but also the company constituents, the general public, the environment, and anyone else who may be impacted?
- Lastly, review the guidelines of such organizations as the Board of Certified Safety Professionals. Compare your decision to those guidelines and see if it stands up against careful scrutiny.

Assume you are one of six purchasing agents for a manufacturing company. You all deal with many vendors to order raw materials and supplies at the lowest cost and highest quality for your company. One day you receive a call from a vendor you and several other agents deal with on a regular basis. The vendor invites you and another person of your choice to their company's exclusive hunting lodge, all expenses paid. The contact suggests you keep this "customer goodwill" trip to yourself; none of your peers have been invited.

In this case, several questions need to be asked: Does this seem too good to be true? Why are they offering it to you and no one else? How will you feel if others find out about this? Others include your co-workers, your boss, your subordinate, and even your competitors. What are the pros and cons of making this decision and how will it affect others, including your employer? Is this ethical in light of corporate purchasing guidelines and other professional guidance that may be available?

It should become clear in this simple example that the vendor is trying to curry favor with the purchasing agent to ensure continued and future sales. The issue of customer goodwill is nothing more than a bribe to keep and solidify business. How would it make others feel about you? Would this objectively be considered ethical or unethical?

So, what is a businesses expected to do to ensure ethical behavior by and to their employees? In the global economy and multi-national corporations, executives are responsible for a very diverse workforce with differing values and beliefs as to what is considered right and wrong. It is critical that each business establish an ethics policy to help guide and clarify the expected behaviors of managers and employees in business dealings. The policy needs to address not only areas of general business practices but also expectations for dealings with customers, vendors and its employees. The ethics policies will establish broad boundaries and provide important guideposts so employees can act within the expectations of their organization. Early approaches to establishing ethics policies simply stated that all employees are expected to obey the laws that apply to their business dealings. This can be an important beginning but the example presented earlier demonstrates that meeting the "legal" standards may not be sufficient to address ethical standards for the same activity or behavior.

Below are samples of ethics policies from two corporations to further clarify this issue.

Borealis Corporate Ethics Statement

Borealis Exploration Limited and all subsidiary companies within the Borealis Group abide as a matter of policy by the following principles:

1. Borealis values integrity and honesty in its business dealings. Borealis personnel are encouraged to be truthful and trustworthy in their dealings with customers, clients and associates. Borealis honors contractual obligations and does not knowingly break agreements or seek to manipulate their sense to the detriment of other parties.
2. Borealis is an equal opportunity employer.
3. Borealis regards its employees and consultants as valued members of the extended family of "Borealis people."
4. As such, employees and consultants are encouraged to participate fully in all aspects of company activity, and are encouraged to bring any and all concerns about their work to senior management without fear of reprisal.
5. Borealis regards its shareholders and customers as valued members of the extended family of "Borealis people."
6. As such, Borealis welcomes the advice and input of shareholders and customers in all relevant business activities.
7. Borealis operates in many different countries of the world, and under many different jurisdictions. Borealis respects the laws and customs of the regions in which it operates, and does not knowingly offend against them.
8. Borealis makes extensive use of the Internet to conduct business. Borealis is opposed to the use of censorship or other forms of Government regulation to control the Internet, and supports efforts to create a viable self-regulatory structure which protects the vulnerable from abusive material while retaining the ability to freely exchange information.
9. Borealis respects and values the natural environment and does not knowingly damage or pollute those natural resources which are the heritage of all. In all activities where environmental issues are of concern, Borealis seeks at all times to minimize environmental damage and to work closely with Governmental agencies and local sensibilities to ensure this.
10. Borealis believes in making as much information about the company available to the public as is consistent with good business practice and obligations to third parties. Information so published is, to the best knowledge of the company, truthful and an honest representation of the opinions of Borealis management.

TO: Levi Strauss & Co. Employees Worldwide
FROM: Chief Executive Officer
SUBJECT: Worldwide Code of Business Conduct

LS&CO. has a long and distinguished history of corporate citizenship, including our unwavering commitment to responsible business practices. Our values and strong belief in "doing the right thing" are the foundation of our success.

To ensure we provide our employees with a clear set of standards and guidance for conducting our business with integrity and the highest degree of compliance with the law, we established LS&CO.'s **Worldwide Code of Business Conduct**.

This code certainly does not cover every ethical or legal situation we may encounter in our business operations, but it does provide an excellent summary of important guidelines that define the way we choose to do business.

Our Worldwide Code of Business Conduct applies to all employees around the world; however, with a global footprint of more than 110 countries, we and our affiliates apply the code as appropriate in individual countries, consistent with local laws.

If you have any questions about the Worldwide Code of Business Conduct or how it should be applied in your location, please consult with your manager, Human Resources representative, the Legal department or the Chief Compliance Officer in San Francisco.

The integrity of our employees makes LS&CO. a great place to work. I encourage you to familiarize yourself with our Worldwide Code of Business Conduct and apply these standards to your work every day.

Chip Bergh
President and
Chief Executive Officer
Levi Strauss & Co.

The full text of Levi Strauss & Co.'s Worldwide Code of Business Conduct can be found at http://lsco.s3.amazonaws.com/wp-content/uploads/2014/01/WORLDWIDE-CODE-of-business-conduct.pdf. The Board of Certified Safety Professionals' *Code of Ethics* is available at www.bcsp.org/pdf/BCSPcodeofethics.pdf. And the American Board of Industrial Hygiene's *Code of Ethics* can be found at www.abih.org/sites/default/files/downloads/ABIHCodeofEthics.pdf.

CONCLUSION

The current occupational safety and health profession has its roots in the beginnings of industrial society. Concerns regarding the protection of human health, safety and property form the basis for the profession. The following chapters of this textbook will address, in greater detail, the basic knowledge and skills critical for the successful implementation of a sound occupational safety management program. In every job the application of knowledge and skills must always be guided by ethics and integrity.

QUESTIONS

1. Why does it make good business sense to have a good safety program? List four reasons.
2. Do you think most working individuals are concerned with occupational safety and health issues? Why?
3. Why is it useful to study historical occupational safety and health events?
4. What is your definition of the term *safety*? How does it differ from the professional definition of this term?
5. Why are zero accidents believed by many to be an unrealistic goal—particularly in a large workplace?
6. Explain the criticality of ethics in the safety and health professions. Why are they essential to those professions?

REFERENCES

American Society of Safety Engineers. 2002. *Scope and Functions of the Professional Safety Position.* Des Plaines, IL: Author.

"Brief History." Safety Line Institute. 1998.

Kohn, J. P., Timmons, D. L., and Besesi, M. 1991. "Occupational Health and Safety Professionals: Who Are We? What Do We Do?" *Professional Safety 36*, 1.

BIBLIOGRAPHY

Abercrombie, S. A. 1981. *Dictionary of Terms Used in the Safety Profession.* Park Ridge, IL: American Society of Safety Engineers.

Borealis Exploration Limited. *Corporate Ethics Statement.* Retrieved from Borealis website on January 20, 2010, http://www.borealis.com/investor/ethics.shtml.

DeReamer, R. 1980. *Modern Safety and Health Technology.* New York: Wiley.

Grimaldi, J. V., and Simonds, R. H. 1975. *Safety Management,* 3rd ed. Homewood, IL: Richard D. Irwin.

Levi Strauss & Co. *Worldwide Code of Business Conduct.* Retrieved from Levi Strauss website on January 20, 2010, http://lsco.s3.amazonaws.com/wp-content/uploads/2014/01/WORLDWIDE-CODE-of-business-conduct.pdf.

National Safety Council. *Injury Facts.* 2001 edition; 1998 edition. Itasca, IL: Author.

Samuel Pepys Diary Home Page. 2002. Retrieved from http://www.pepys.info/fire.html.

2

Safety Legislation

CHAPTER OBJECTIVES

After completing this chapter, you will be able to

- Explain the history of safety and health legislation
- Understand the Occupational Safety and Health Act of 1970
- Identify the origins of OSHA standards
- Know the specific requirements of the Act
- Understand the OSHA inspections and resultant actions

CASE STUDY

When his machine shop was inspected, Eric Smith was shocked. With only five employees, Eric thought he was exempt from OSHA inspections. He was even more surprised to learn that one of his employees had complained. After a careful investigation, Eric learned the name of the complainant and nearly fired him. However, Eric's attorney informed him that retaliatory action against an employee for exercising his rights under OSHA regulations is illegal.

LEGISLATIVE HISTORY

The history of occupational safety and health has been dominated by legislation. Governments have observed problems and have attempted to solve those problems through the enactment of laws. In this country, the regulations resulted in the passage of the Occupational Safety and Health Act in 1970. Since its passage, numerous modifications to the Act have increased its applicability.

The eye-for-an-eye principle dominated early attempts to legislate safety. The government effort to encourage safer workplaces first revolved around punishing the wrongdoer. This concept pervaded Babylonian law over 4,000 years ago and was a forerunner to the famous Code of Hammurabi, which was written in 2100 BC, during the thirtieth year of his reign. The code required shipbuilders to repair defects of construction and damage caused by those defects for one year following delivery. Ship captains were required to replace goods lost at sea and to pay a fine equal to half the value of any lost ships that were refloated. If someone other than the master injured a slave, the master received compensation for the loss. Carelessness and neglect were considered unacceptable for skilled workers and professionals. Losses caused by errors on the part of these early professionals were punished using the eye-for-an-eye concept that is also found in the Old Testament. A physician whose mistakes led to the loss of a citizen's life could find his hands cut off, or a builder could have his own child killed if his shoddy work led to the loss of another's child.

From the time of Hammurabi and his contemporaries, little is known about attempts to legislate safety until after the Dark Ages. Early industrial plants were little more than a series of traps consisting of open machinery and unguarded moving equipment capable of maiming or killing a worker in seconds. The British developed their laws out of concern for children (Grimaldi and Simonds, 1989). In the United States the 1800s brought about a number of tragedies that met with little government response. As media coverage became widespread and information became more accessible through books, newspapers, and magazines, American citizens began to expect and demand that their government provide protection from employers. This reactive pattern continued as concern for occupational safety and health grew. When public outcries about disasters and outrageous actions of employers became strong enough to overcome the influence exerted by employers on early legislatures, laws were enacted to protect the workers. Much of the early legislation was reactive; laws were enacted in response to specific tragedies. The following is a review of significant events in the development of safety and health in the 1900s:

In 1903, a fire roared through Chicago's Iroquois Theater, killing 602 people. The fire engulfed the stage almost immediately. As the audience panicked, flames swept across the perimeter of the auditorium and finally to the seats themselves. Fire exits were few and poorly marked. The iron pillars became red hot and eventually melted.

In Monongah, a sleepy West Virginia town in the hills of Appalachia, 362 coal miners were killed in 1907. Nearly every family in the town was affected by the disaster, but little was done for the survivors.

The Pittsburgh Survey (1907–1908) was a 12-month study in Allegheny County, Pennsylvania, sponsored by the Russell Sage Foundation. It found there were 526 occupational fatalities in one year. Survivors and the workers themselves were forced to carry the cost of losses. When the breadwinners were gone, there was no further compensation of any kind. Employer incentives to reduce safety and health risks for workers were called for.

In 1910, the U.S. Bureau of Mines was created by the Department of the Interior to investigate the causes of mine accidents, study health hazards, and find means for taking corrective action.

A fire gutted a new structure in the Triangle Shirtwaist Factory in New York City. This 1911 tragedy killed 145 workers, whose escape was prevented due to the employer's locking of the exits against theft.

In 1911, Wisconsin passed the first successful workers' compensation plan in the United States and by 1948 every state had followed suit.

The first National Safety Congress convened in 1912. It led to the formation of the National Safety Council in 1913.

The Safety to Life Committee, formed by the National Fire Protection Association in 1913, eventually led to the development of the Life Safety Code.

In 1935, Roosevelt's New Deal included the passage of legislation that mandated a 40-hour workweek.

The Walsh-Healey Public Contracts Act of 1936 banned hazardous work done under federal contracts larger than $10,000. This act was a forerunner of the Occupational Safety and Health Act.

In 1968, 78 coal miners were killed in Farmington, West Virginia, when an explosion ripped through the Consolidated Coal Company Mine. Only a few miles from the infamous Monongah site, this disaster devastated the whole community. A town, a state, and a nation were

(continued)

outraged and called for federal intervention into the conditions killing citizens in the workplace.

The Coal Mine Health and Safety Act of 1969 was passed 72–0 in the U.S. Senate, establishing the Mine Enforcement Safety Administration (MESA), later known as the Mine Safety and Health Administration (MSHA). MSHA governs safety within the coal mines as OSHA does for general industry and construction (discussed below).

In 1970, the Williams-Steiger (Occupational Safety and Health) Act was passed, establishing the Occupational Safety and Health Administration (OSHA). Prior to the establishment of OSHA, the responsibility for occupational safety and health rested primarily with state governments.

OCCUPATIONAL SAFETY AND HEALTH ACT

Although more than 90 million Americans were working in 1970, no uniform and comprehensive provisions existed for their protection against workplace safety and health hazards. In 1970, Congress considered annual figures such as these:

- Job-related accidents accounted for more than 14,000 worker deaths.
- Nearly 2.2 million workers were disabled.
- Ten times as many person-days were lost from job-related disabilities.
- Estimated new cases of occupational diseases totaled 300,000.

The Occupational Safety and Health Act (OSHAct) of 1970 was passed by Congress "to assure so far as possible every working man and woman in the Nation safe and healthful working conditions and to preserve our human resources" (United States Department of Labor [US DOL], n.d.). Under the Act, the Occupational Safety and Health Administration (OSHA) was created within the Department of Labor to

- Encourage employers and employees to reduce workplace hazards and to implement new or improve existing safety and health programs
- Provide for research in occupational safety and health to develop innovative ways of dealing with occupational safety and health problems

- Establish separate but dependent responsibilities and rights for employers and employees for the achievement of better safety and health conditions (US DOL, 1991); maintain a reporting and recordkeeping system to monitor job-related injuries and illnesses
- Establish training programs to increase the number and competence of occupational safety and health personnel
- Develop mandatory job safety and health standards and enforce them effectively
- Provide for the development, analysis, evaluation, and approval of state occupational safety and health programs (OSHA Training Institute [OSHATI], 1994, chap. 1)

WHO IS COVERED?

In general, the coverage of the Act extends to all employers and their employees in the 50 states, the District of Columbia, Puerto Rico, and all other territories under federal government jurisdiction. Coverage is provided either directly by federal OSHA or through an OSHA-approved state program. As defined by the Act, an employer is any "person engaged in a business affecting commerce who has employees, but does not include the United States or any State or political subdivision of a State" (US DOL, n.d., p. 3).
*The following are **not** covered under the Act:*

- self-employed persons
- farms at which only immediate members of the farm employer's family are employed
- workplaces already protected by other federal agencies under other federal statutes (US DOL, n.d.)

Under the Act, federal agency heads are also responsible for providing safe and healthful working conditions for their employees. The Act requires agencies to comply with standards consistent with those OSHA issues for private sector employers. OSHA conducts federal workplace inspections in response to employees' reports of hazards and as part of a special program identifying federal workplaces with higher-than-average rates of injuries and illnesses (OSHATI, 1994, chap. 1).

OSHA cannot fine another federal agency for failure to comply with OSHA standards, and it does not have authority to protect federal employee whistleblowers. Federal employee whistleblowers are protected under the Whistleblower Protection Act of 1989 (OSHATI, 1994, chap. 1).

OSHA provisions do not apply to state or local government employees. The Act requires states desiring approval to maintain their own programs to provide safety and health coverage for their own state and local government workers that is at least as effective as their programs for private employees. State plans may also cover *only* public sector employees. Unfortunately for state and municipal employees in states not having their own programs, unless the state has provided otherwise, no governmental protection exists for the employees. For example, at the time of this writing Florida has no state plan of its own. Florida employees are covered by Federal OSHA, but Florida state, county, and municipal employees have no coverage by the Occupational Safety and Health Administration. Since no other provisions have been made for the employees, they effectively have no coverage.

Although OSHA applies to federal government employees, and requires federal agencies to provide safe and healthful places of employment, OSHA does not have the authority to fine another federal government agency. Compliance issues raised at a lower organizational level are raised to a higher level until resolved.

Only employers of 11 or more individuals are required to maintain records of occupational injuries and illnesses. OSHA recordkeeping is not required for certain retail trades and some service industries. Even though employers of ten or fewer employees are exempt from recordkeeping, they must still comply with the other OSHA standards, including the requirement to display the OSHA poster.

OSHA STANDARDS

OSHA standards fall into four categories: General Industry, Maritime, Construction, and Agriculture. The standards are available in the following volumes:

Volume I   General Industry Standards and Interpretations (includes Agriculture)

Volume II   Maritime Standards

Volume III   Construction Standards
Volume IV   Other Regulations and Procedures
Volume V   Field Operations Manual
Volume VI   OSHA Technical Manual

These are available from the Superintendent of Documents, U.S. Government Printing Office, Washington, DC 20402 and commercial publishers. Because some states adopt and enforce their own standards, copies of those may be obtained from the individual states.

ORIGIN OF OSHA STANDARDS

Initially, the OSHA standards were taken from three sources: consensus standards, proprietary standards, and federal laws in effect when the Act became law (OSHATI, 1994, chap. 2).

Consensus standards are developed by industry-wide, standard-developing organizations discussed and substantially agreed on through consensus by industry. OSHA has incorporated standards of the two primary standards groups, the American National Standards Institute (ANSI) and the National Fire Protection Association (NFPA). For example, ANSI Standard B56.1-1969, Standard for Powered Industrial Trucks, covers the safety requirements relating to the elements of design, operation, and maintenance of powered industrial trucks.

Proprietary standards are prepared by professional experts within specific industries, professional societies, and associations. These standards are determined by membership vote, as opposed to consensus. An example of these would be the Compressed Gas Association's Pamphlet P-1, Safe Handling of Compressed Gases. This proprietary standard covers requirements for the handling, storage, and use of compressed gas cylinders (OSHATI, 1994, chap. 2).

Some preexisting federal laws are also enforced by OSHA, including the Federal Supply Contracts Act (Walsh Healey) and the Contract Work Hours and Safety Standards Act (Construction Safety Act). Standards issued under these acts are now enforced in all industries where they apply (OSHATI, 1994, chap. 2).

When the OSHAct was first passed, much criticism stemmed from the fact that the legislation was a hodgepodge of rules of thumb and guidelines never intended to be made into laws. Since the passage of the Act, many of the trivial regulations have been changed or eliminated in an attempt to make it a more reasonable standard for workplace performance.

Horizontal and Vertical Standards

Standards are referred to as being either *horizontal* or *vertical* in their application. Most standards are horizontal or general in that they apply to any employer. Standards relating to fire protection or first aid are examples of horizontal standards (OSHATI, 1994, chap. 2).

Some standards are relevant only to a particular industry, and are called *vertical* or *particular* standards. Examples are standards that apply to longshoring or the construction industry.

FINDING THE OSHA ACT

The Occupational Safety and Health Act appears in the Code of Federal Regulations (CFR) 29. It is divided into the subparts noted in appendixes A and B. Each subpart addresses a different major topical area. These areas include such items as Subpart D, Walking-Working Surfaces. This subpart appears in 1910.21–1910.32. Subpart E, Means of Egress, follows in 1910.35–1910.70. Each subpart is further broken down so that when one looks under Subpart E, one will find the subsection 1910.37, Means of Egress, general. The reader can turn to that section to see an explanation of means of egress and find the section broken down even further. References to specific standards are typically found as follows:

29 CFR 1910.110(b)(13)(ii)(b)(7)(*iii*) indicates that the specific standard in question appears in:

- Title 29
- Code of Federal Regulations (CFR)
- Part 1910
- Section 110

Subsections appear first as lowercase letters, numbers, and Roman numerals. Subsections of subsections appear as italicized, lowercase letters, numbers, and Roman numerals (OSHATI, 1994, chap. 2).

Most universities and many public libraries carry copies of the Code of Federal Regulations (CFRs) and they are available online at www.ecfr.gov. CFRs are also available from the Superintendent of Documents in Washington and from Government Institutes, Inc. OSHA standards are also available online at www.osha.gov. This also contains other useful information, including standards interpretations, downloadable OSHA posters, recordkeeping instructions, and compliance assistance.

SPECIFIC REQUIREMENTS OF THE ACT

Employers are responsible for knowing the standards applicable to their establishments. When an OSHA inspection is performed, the assumption made is that the employer is aware of the law and has already attempted to comply with it. Any violations are subject to corrective legal action, typically consisting of fines. Employees must also comply with all rules and regulations that are applicable to their own actions and conduct, but it is the employer's responsibility to ensure employee compliance. Citations for noncompliance are issued to the employer. For cases involving criminal intent, OSHA may turn the files over to the Department of Justice for prosecution, but this course of action is very rarely followed.

Where OSHA does not have specific, applicable standards, employers are responsible for following the Act's general duty clause (Section 5(a)(1)). The *general duty clause* requires that every working person be provided with a safe and healthful workplace. It specifically states, "Each employer shall furnish to each of his employees employment and a place of employment which is free from recognized hazards that are causing or are likely to cause death or serious physical harm to his employees" (USDOL, 2006). A *recognized hazard* is one that is detectable by the senses or instrumentation, of which there is common knowledge, that is discoverable under usual inspection practices, or of which the employer has knowledge. Incidentally, if the employer has knowledge, an inspector will consider the hazard to be a "willful violation" and it carries an increased penalty.

The general duty clause extends OSHA's authority beyond the specific requirements of the standards when a recognized workplace hazard actually or potentially exists. It is often used when no specific standard applies to a recognized hazard in the workplace. The general duty clause has been used as the basis for numerous OSHA citation, and millions of dollars are collected as a result of these citations. Few other sections of the OSHA standards (Parts 1910 and 1926) cost more in penalties for OSHA violations than Section 5(a)(1). OSHA may also use the general duty clause when a standard exists, but it is clear that the hazards involved warrant additional precautions beyond what the current safety standards require.

EMPLOYER RESPONSIBILITIES AND RIGHTS

Aside from providing a workplace free from recognized hazards likely to cause death or serious physical harm, the employer has other responsibilities. An employer must:

- Examine workplace conditions to make sure they comply with applicable standards
- Minimize or reduce hazards
- Use color codes, posters, labels, or signs when needed to warn employees of potential hazards
- Provide training required by OSHA standards
- Keep OSHA-required records
- Provide access to employee medical records and exposure records to employees or their authorized representatives

Employers have the right to seek advice and off-site consultation as needed by writing, calling, or visiting the nearest OSHA office (OSHATI, 1994, chap. 2).

When an inspection visit occurs, the employer must:

- Be advised by the compliance officer of the reason for the inspection
- Accompany the compliance officer on the inspection
- Be assured of the confidentiality of any trade secrets observed by an OSHA compliance officer during an inspection

Although OSHA does not cite employees for violations of their responsibilities, each worker is required to comply with all applicable occupational safety and health standards. Employees also have the right to ask for safety and health on the job without fear of punishment. If employees are discriminated against for exercising their rights under the Act, OSHA may take the employer to court with no expense to the employees. Again, this is an OSHA action not often taken.

INSPECTIONS

To enforce its standards, OSHA is authorized under the Act to conduct workplace inspections. Every establishment covered by the Act is subject to inspection by OSHA compliance safety and health officers who are chosen for their knowledge and experience in the occupational safety and health field. Compliance officers are vigorously trained in OSHA standards and in the recognition of safety and health hazards. Inspections occur as a result of priorities established by OSHA.

Imminent danger situations are inspected first. Where there is reasonable certainty that an employee is exposed to a hazard likely to cause death or immediate serious physical harm, OSHA will try to respond as soon as possible. OSHA may become alerted to adverse conditions by a complaint or other means. Inspectors may see a story on the evening news or may even be driving by a job site where they see an imminent danger situation.

Catastrophes and *fatal accidents* are investigated after imminent danger situations. If three or more employees are hospitalized (catastrophe) or if an employee is killed, OSHA must be notified within eight hours.

Employee complaints, alleging violation of standards or unsafe or un-healthy working conditions, are investigated after catastrophes and fatal ac-cidents. The employee has the right to remain anonymous to his employer.

Programmed high-hazard inspections are given the next priority. These are aimed at specific high-hazard industries, occupations, or health sub-stances. Selection is based on factors such as death, injury, and illness inci-dence rates and employee exposure to hazardous substances.

Follow-up inspections are given last priority. These are used to determine if previously cited violations have been corrected (OSHATI, 1994, chap. 1).

Inspection Process

Under the Act, "upon presenting appropriate credentials to the owner, op-erator or agent in charge" (US DOL, 1991), an OSHA compliance officer is authorized to:

- Enter without delay and at reasonable times any factory, plant, establishment, construction site or other areas, workplace, or environment where work is performed by an employee of an employer; and
- Inspect and investigate during regular working hours, and at other reasonable times, and within reasonable limits, and in a reasonable manner, any such place of employment and all pertinent conditions, structures, machines, apparatus, devices, equipment, and materials therein, and to question privately any such employer, owner, operator, agent, or employee (US DOL, 1991).

Inspections are generally conducted without advance notice. In fact, alerting an employer in advance of an OSHA inspection can bring a criminal and/ or a jail term. If an employer refuses to admit an OSHA compliance officer or if an employer attempts to interfere with the inspection, the Act permits appropriate legal action. Typically, a compliance officer refused entrance to a workplace or access to any part of it returns with a warrant and conducts the inspection as planned. Workplaces with a history of refusing OSHA access can expect the OSHA compliance officer to initiate the inspection with a warrant in hand.

Once credentials are presented, the compliance officer will explain in an *opening conference* why the inspection is being performed, the scope of the inspection, and the standards that apply. An authorized employee representative is given the opportunity to attend and to accompany the compliance officer on the inspection.

Following the conference, the compliance officer usually proceeds to inspect the OSHA records, including the OSHA 300 log. The inspector may also request copies of other required records such as the hazard communication or lockout/tagout programs.

An *inspection tour* then takes place. The officer may talk to employees in private about safety and health conditions, as well as practices in their workplace. An employer's representative should accompany the officer and keep a careful record of everything inspected. Any comments made by the inspector should be recorded. Split samples should be requested anytime samples are taken. Copies of any photographs taken should be requested. The compliance officer may inspect the premises before reviewing records if he or she desires to do so.

A *closing conference* occurs near the end of the visit. During this time, the employer or an employer's representative should ask questions in order to clearly understand any violations recorded by the compliance officer. The employees' representative may be present during this conference. No penalties will be assigned at this time.

CITATIONS AND PENALTIES

After the compliance officer reports findings, the area director determines what citations, if any, will be issued, and what penalties, if any, will be proposed. Citations inform the employer and employees of the regulations and standards alleged to have been violated and of the proposed length of time set for their abatement. The employer receives the citations and notices of proposed penalties by certified mail within six months of the inspection. The employer must post a copy of each citation at or near the place a violation occurred for three days or until the violation is abated, whichever is longer.

OSHA may issue any of four types of citations:

1. *Citations for willful violations* are issued when the employer disobeys, with an intentional disregard of, or plain indifference to, the requirements of the OSHAct and regulations. These can be assessed if the employer was aware that a hazardous condition existed and made no reasonable effort to eliminate the condition. The employer need not be guilty of malicious intent to be considered in willful violation. Under OSHA's penalty structure, the maximum penalty for a willful violation is now $70,000. The minimum penalty is $5,000. Criminal charges can be brought against an employer if an employee fatality is caused by such negligence. In criminal cases the evidence is turned over to the Department of Justice for disposition. This rarely occurs.

2. *Citations for serious violations* are issued when there is a substantial probability that death or serious physical harm could result and that the employer knew or should have known of the hazard. Violations of the general duty clause are considered serious. The maximum penalty is now $7,000.

3. *Citations for other than serious violations* are issued when a situation would affect safety or health but there is a small probability of the hazard resulting in death or serious physical harm. There is often no penalty assessed, but the hazard must still be corrected. If there is a high probability of the hazard resulting in an injury or illness, then

the maximum penalty is $1,000. The OSHA regional administrators have the authority to impose a penalty of up to $7,000 if the circumstances warrant.

Regulatory citations are issued for:

- No OSHA poster: $1,000
- No OSHA 300 log: $1,000
- Failure to post citations: $3,000
- Failure to report within eight hours a fatality or accident which hospitalizes three or more employees: $5,000

4. ***Repeat violation citations*** are issued when the original violation has been abated, but upon reinspection, another violation of the previously cited section of a standard is noted. They may be inadvertent, but if they are found to be willful, both a willful and a repeat citation may be issued. For a first repeat violation, penalties assessed are multiplied by a factor of 2 for employers with less than 251 employees and by 5 for larger employers. The multiplier goes to 5 for a second repeat offense for small employers and 10 for large. OSHA regional administrators have the authority to use a multiplication factor of up to 10 for small employers in order to achieve the necessary deterrent effect. Failure to abate within the prescribed period can result in a penalty for each day of the violation beyond the abatement date.

It is important to note that the Act also authorizes criminal penalties for certain violations, but as previously mentioned, OSHA rarely pursues this course of action. Penalties may be reduced for employers with less than 251 employees, good-faith efforts, or a good safety record in the last three years. Penalties for employers governed by state programs may differ.

APPEALS PROCESS

Within 15 working days of the receipt of the citation, an employer may submit a written objection to OSHA. Once an inspection takes place, those assigned to receive the citation in the mail for a company need to watch for and turn over any correspondence from OSHA. Companies ignoring the 15-day deadline, or for some other reason failing to meet it, have also missed their right to appeal. Within the 15-day period the employer may contest a citation, a penalty, and/or an abatement date. An OSHA area director has the authority to make adjustments based on objections. If no agreement can be reached with the director, the objection may be forwarded to the Occupational Safety and Health Review Commission, which operates independently of OSHA. Appeals beyond the Commission go through the appeals courts.

In 1992, a chemical reactor operated by Arcadian Corporation failed. OSHA issued a citation alleging that Arcadian's operation and maintenance of the reactor violated the general duty clause. It also proposed 87 penalties of $50,000 each, one for each exposed employee, for a total penalty of $4,350,000.

In using the number of exposed employees as a multiplier, OSHA was following its so-called "egregious" policy, for violations that are not merely "willful" but "egregious" (a term nowhere used in the Occupational Safety and Health Act). Under the "egregious" approach, a penalty will be proposed as to each exposed employee. OSHA claimed that this was lawful under the General Duty Clause, which states that an employer must furnish safe employment to "each" employee.

Arcadian, represented by McDermott, Will, and Emory's OSHA Practice Group, moved to restrict OSHA to only one proposed penalty on the ground that per-employee penalties are illegal. It argued that various provisions of the OSH Act make it clear that a violation is a violative "condition." An administrative law judge agreed, and the full Commission upheld his decision. OSHA appealed to the Fifth Circuit, whose decision was widely anticipated. Four amici curiae filed briefs, including the U.S. Chamber of Commerce in support of Arcadian and the AFL-CIO in support of OSHA.

The Fifth Circuit unanimously upheld the Commission's decision, holding that OSHA could not multiply penalties by the number of exposed employees under the General Duty Clause. The court held that violations of the General Duty Clause are based on the hazardous condition, not the number of employees affected.

The court agreed with Arcadian that the term "each" signifies only "that an employer's duty extends to all employees, regardless of their individual susceptibilities (i.e., age or pregnancy)" (McDermott, Will, and Emory: OSHA Update, June 1997).

OSHA-APPROVED STATE PROGRAMS

The Act encourages states to develop and operate, under OSHA guidance, state job safety and health plans. Once a state plan is approved, OSHA funds up to 50 percent of the operating costs of the program. State plans are required to provide standards and enforcement programs, as well as voluntary compliance activities at least as effective as the federal program. They must also provide coverage for state and local government employees. In addition, OSHA permits states to develop plans limited to public sector coverage (state and local government). In such cases, private sector employment remains under federal jurisdiction.

STANDARDS DEVELOPMENT

Once OSHA has developed plans to propose, amend, or delete a standard, it publishes its intentions in the *Federal Register* as a notice of proposed rulemaking, or often as an earlier advance notice of proposed rulemaking (US DOL, 1991). The advance notice is to solicit information for use in drafting a proposal. A notice of proposed rulemaking will include the terms of the new rule and provide a specific time (at least 30 days from the date of publication, usually 60 days or more) for the public to respond.

Interested parties who submit written arguments and pertinent evidence may request a public hearing on the proposal when none has been announced in the notice. When such a hearing is requested, OSHA will schedule one, and will publish, in advance, the time and location in the *Federal Register*.

After the close of the comment period and public hearing, if one is heard, OSHA must publish in the *Federal Register* the full, final text of any standard amended or adopted and the date it becomes effective, along with an explanation of the standard and the reasons for implementing it. OSHA may also publish a determination that no standard or amendment needs to be issued.

OTHER CONSIDERATIONS

Since the inception of OSHA, the construction industry has been underrepresented. According to one OSHA administrator, construction has always been an afterthought at OSHA. Much of what has been legislated in construction has been after the fact. Once rules were passed for general industry, the question was asked, "Now what do we do with construction?" In recent years, this policy has begun to change. Construction is a hazardous industry and OSHA has recognized those hazards. Construction standards appear in 29 CFR 1926. OSHA has programs giving a high priority to inspection of certain hazards. Refer to chapter 16 on construction safety for more information on focused inspections.

NIOSH AND OSHRC

The OSHAct also created the National Institute of Occupational Safety and Health (NIOSH) and the Occupational Safety and Health Review Commission (OSHRC). NIOSH operates within the Department of Health and Human Services (HHS) under the Centers for Disease Control (CDC) to develop occupational safety and health standards for recommendation to the Secretary of Labor and the Secretary of HHS, and to fulfill the research and training functions of the Secretary of HHS. It is headquartered in Washington, DC, but carries out many of its functions at its facilities in other locations. It also works through contracts with more than multiple Education and Research Centers and many Training Project Grantees around the country. These institutions provide research, education, and training in safety and industrial hygiene on behalf of NIOSH.

The Occupational Safety and Health Review Commission is an independent and autonomous quasi-judicial board charged with hearing cases on appeal from OSHA. The president, with the advice and consent of the Senate, appoints its three members. It meets periodically to review cases on appeal. Once the commission hears a case, if either party is dissatisfied, it may appeal through the federal appeals court system.

FUTURE TRENDS

The agricultural industry has not fared as well as other industries. Although agricultural problems are comparable to those in the construction or mining industries, many agricultural operations are small, family-owned businesses that manage to escape OSHA's attention. If no employees outside the family work in the operation, it is automatically exempt. However, many farmers are unaware of the regulations permitting reporting of incidents and illnesses, so their operations go largely unnoticed. As a result of better recordkeeping and farmers who are becoming more aware of the steps they can take to reduce workplace incidents, OSHA may play a more important role in the agricultural industry.

As the workplace continues to become more complex and OSHA finds that higher numbers of workers are killed or injured in different ways than in the past, the emphasis may shift. Homicide is a primary cause of death for younger workers and for females. NIOSH has carefully studied homicide in the workplace and has produced a number of publications responding to the problem. An issue left largely untouched by OSHA has been that of highway accidents. Employers continue to lose more employees due to vehicular incidents than any other reason.

Although there has been discussion on changing the way OSHA enforces compliance through regulations, companies will continue to be required to comply with applicable standards and codes. Focused inspections keying in on a few critical areas may be used to trigger deeper inspections or permit the compliance officer to get in and get out quickly if no problems are found. Related rates such as those reported for workers' compensation may also be used to target high-hazard companies and industries.

CONCLUSION

The Occupational Safety and Health Act was the culmination of centuries of governmental response to occupational safety and health problems. Since 1970, a number of additional regulations designed to protect workers have been passed, and these will be addressed in other chapters. As was noted above, incident rates and numbers of fatalities have dramatically declined since 1970 and this decline appears to be a result of legislation. It is important to recognize, however, that conformance to legislation is not enough to create a safe and healthful working environment. It merely creates a baseline for companies that might otherwise have no safety and health program at all. Generally, compliance requirements have encouraged a safer and healthier workplace and helped assist otherwise unsafe companies toward preserving the health and safety of their workers.

QUESTIONS

1. Can safety and health be legislated? What are the limitations of safety legislation? Where would we be without it?
2. How do you think the future of safety legislation would be affected by another major occupational disaster of the magnitude of the Monongah mine disaster? Do you think passage of new safety and health legislation is still of a reactive nature?
3. Although safety and health fines have gone up significantly in recent years, do you think it matters how high they are when the likelihood of being inspected is so low? Discuss this with people in industry and note their opinions.

REFERENCES

Grimaldi, J. V., and Simonds, R. H. 1989. *Safety Management*. Homewood, IL: Richard D. Irwin.

McDermott, E., Will, H., and Emory, W. OSHA Update, June 1997. http://www.mwe.com/news/osha0797.htm#arcadian.

OSHA Training Institute. 1994. *A Guide to Voluntary Compliance in Safety and Health*. Atlanta, GA: Author.

United States Department of Labor. 1991. *All about OSHA*. Washington, DC: Government Printing Office.

United States Department of Labor, Occupational Safety and Health Administration. n.d. *SEC. 2. Congressional Findings and Purpose*. https://www.osha.gov/pls/oshaweb/owadisp.show_document?p_id=3356&p_table=OSHACT.

United States Department of Labor, Occupational Safety and Health Administration. 2006. *Occupational Safety and Health Standards*. Washington, DC: Government Printing Office.

BIBLIOGRAPHY

Centers for Disease Control, http://www.cdc.gov/niosh/ (accessed December 12, 2002).

Georgia Tech Research Institute. 1994, December 12–16. *OSHA 501 Training*. Atlanta, GA: Author.

3

Workers' Compensation and Recordkeeping

CHAPTER OBJECTIVES

After completing this chapter, you will be able to

- Explain the concept of workers' compensation
- Describe the evolution of workers' compensation
- Describe the different types of workers' compensation claims
- Explain the basis for workers' compensation rates
- Identify the basic recordkeeping requirements

CASE STUDY

Joe Derek often takes work home with him to complete some of the details that he is unable to finish at the office. Although his employer has never encouraged him to work at home, he is aware of the practice and thankful that the work gets done. One evening on the way home from work, Joe is injured in an automobile accident. His employer is surprised when Joe files for and is awarded workers' compensation.

EARLY WORKERS' COMPENSATION LAWS

Early efforts to compensate victims of workplace accidents date back to the Code of Hammurabi. In modern times, efforts have been made to force employers to pay for injuries suffered by employees on the job. Modern efforts can be traced to Prussia, where in 1838 legislation was passed permitting railroad workers to collect damages for injuries they suffered on the job. In 1884, Otto von Bismarck enacted the first workers' compensation law for all German workers. Wisconsin passed the first successful workers' compensation law in the United States in 1911.

The workers' compensation laws are designed to compensate victims of workplace accidents by forcing the employer to pay the workers money for injuries and time lost. Workers' compensation is considered "no-fault" insurance. No fault is admitted on the part of either the employer or the employee when a claim is made. Prior to passage of these laws, the only recourse a worker had was to sue the employer under civil law, a form of common law. *Common law* is a body of unwritten laws based on judicial decisions of the past, as opposed to *statutory law*, which is prepared and enacted by legislative bodies. In early common law suits, employers successfully utilized the following defenses:

- ***Assumption of risk***. Employees accept risks associated with the job and, by doing so, forfeit any right to collect compensation for injuries.
- ***Contributory negligence***. Since the employees contributed to their own injuries, regardless of how little, they are not permitted to recover compensation.
- ***Fellow-servant rule***. The employer is not at fault because the accident was the fault of another employee or other employees.

These three common law defenses were eliminated after workers' compensation laws were enacted by each state. Employers accepted this legislation because, in order to collect compensation, employees had to waive their rights to sue. Employees accepted the legislation because it eliminated the requirement to sue the employer in order to collect compensation. By 1948, every state had passed workers' compensation legislation, but that legislation may vary from state to state. When those laws were passed, they were referred to as "work*men*'s compensation" laws. Today, they are referred to as "*workers'* compensation" laws.

Prior to the passage of the workers' compensation laws, the full cost of production was not borne by the employer, but instead was passed on to the employee. Rather than engineer a solution to a simple problem, such as providing a guard on an open saw blade, the employer permitted the employee to run the risk of being injured. It was cheaper to replace the employee than guard the machines.

Most states prohibit the suing of businesses complying with the Act except for deliberate assaults and conditions so flagrantly unsafe as to make injury virtually certain. This has generally worked as a win/win situation. In some states, however, employees who have collected workers' compensation benefits have been successfully following up with lawsuits against the employer by using loopholes in the laws. Following a successful suit, the employee simply repays the workers' compensation fund for benefits received. This relatively recent practice is a result of sympathetic juries who often perceive employers and their insurance companies as bottomless sources of cash. Lawyers sometimes engage employees on a no-fee basis where legal expenses are only paid if the suit is successful. This is an attractive proposition for some employees since there are no upfront expenses and they may profit substantially if the suit is successful. This relationship violates the spirit of the workers' compensation legislation but is a result of one of the inherent flaws in some state regulations.

MODERN WORKERS' COMPENSATION LAWS

When a worker is injured and stands to collect benefits, that worker must first typically undergo a waiting period, usually lasting from three to fourteen days. Benefits may or may not be paid retroactively to the first day of lost work and wages. Once the benefits begin, they are limited to two-thirds of the worker's wages or capped at a given dollar amount per month depending on the state. Benefits extend for a limited period of time and amount to a maximum amount of money. Because employees off work for a period of time often suffer a decrease in their standard of living, litigation may appear to be an attractive option. If workers were permitted to receive benefits equal to their wages, however, there would be a strong incentive not to return to work. Even at two-thirds wages, some employees may malinger in order to continue to receive benefits without working.

Injuries are categorized in one of the following ways:

- *Partial:* when the employee can still work but is unable to perform all duties of the job due to the injury, as is often the case with a broken finger or a severed toe.

- *Total:* when the employee is unable to work or perform substantial duties on the job, as is often the case with a severe back injury or blindness.
- *Temporary:* when the employee is expected to fully recover, as is the case with a broken limb or a sprain.
- *Permanent:* when the employee will suffer the effects of the injury for life, as is the case with a severed limb, blindness, or permanent hearing loss.

Specific cases are categorized as temporary partial, temporary total, permanent partial, or permanent total. Other categories of benefits include *retraining incentive benefits* for employees who may have specific injuries. These benefits are paid for a limited period of time to aid the injured in pursuing additional education or training. Vocational rehabilitation services may be offered to employees who are eligible for permanent total disability benefits and actively participate in a vocational rehabilitation program. *Survivors* of employees killed in industrial accidents may be entitled to benefits as well.

Even businesses with only a few employees are required to carry workers' compensation insurance. Some states require companies to provide benefits even for one part-time employee. A large company having the appropriate financial resources to cover the anticipated losses of self-insurance is one of the few exceptions to the requirement of purchasing workers' compensation insurance from an outside carrier.

Who is covered by workers' compensation? The following is an example of the coverage that exists in one state:

- Every person, including a minor, whether lawfully or unlawfully employed, in the service of an employer
- Every executive officer of a corporation
- Every person in the service of the state, county, or city
- Every person who is a member of a volunteer ambulance service, fire department, or police department
- Every person who is a regularly enrolled volunteer member or trainee of the civil defense corps
- Every person who is an active member of the National Guard
- Every person performing service in the course of the trade, business, profession, or occupation of an employer at the time of injury
- Every person regularly selling or distributing newspapers on the street or to customers at their homes or places of business

- Owner(s) of a business, whether or not employing any other person to perform a service for hire

EXEMPTIONS

The following employees are typically exempt from the coverage of workers' compensation:

- Any person employed as a domestic servant in a private home
- Any person employed for not exceeding 20 consecutive workdays in or about the private home of the employer
- Any person performing services in return for aid or sustenance only, received from any religious or charitable organization
- Any person employed in agriculture
- Any person participating as a driver or passenger in a voluntary vanpool or carpool program while that person is on the way to or from his place of employment
- Any person who would otherwise be covered but elects not to be covered

If an employee is injured during a work assignment outside the state where he is employed, he is still eligible to collect benefits. In some states, benefits are reduced if the employee is not using the safety equipment required by the employer at the time of the injury.

PREMIUM CALCULATION

Rates employers pay were historically based on the annual payroll in a given Standard Industrial Classification (SIC) code. SIC codes are the classifications used by the federal government for businesses. Companies are categorized by major industry in divisions, such as Agriculture, Forestry, and Fishing; Mining; Construction; Manufacturing; or Transportation. Within those divisions are major groups. In Division D (Manufacturing), for example, there is Major Group 25, "Furniture and Fixtures"; Industry number 2511, "Wood Household Furniture, Except Upholstered"; Industry number 2512, "Wood Household Furniture, Upholstered"; and so forth. States still use a similar method of categorizing employers. Workers' compensation

premiums, or the amounts paid for the insurance, are expressed in dollars per $100 of payroll for companies in a given four-digit category or classification. Industries or companies in classifications that are considered to be higher risk because of higher numbers of injuries or deaths pay more than those in lower classifications or categories. The rates are published in a manual and are calculated as follows:

Manual Premium = Payroll/100 × Rate (based on the SIC code)

Experience Modification

In addition to the classification rate, companies are assigned an **experience modification factor** or **mod** based upon the claims activity of their specific business. It begins at one and then changes either upward or downward as the claims activity of the business changes. It is possible for the experience modification to change yearly, but it typically takes into account the past three years. The standard premium incorporates the mod and is calculated as follows:

Standard Premium = Manual Premium ×
Experience Modification Factor

Retrospective Rating

If the mod rate is significantly higher than the manual rating, a company may request the premium be paid on a retrospective-rating basis. Companies with strongly improving safety and health programs sometimes request this rating. The insurer will charge its actual costs for claims plus an appropriate amount for overhead.

SELF-INSURED

Employers may be self-insured if they are able to furnish to the workers' compensation board satisfactory proof of their fiscal ability to directly pay the compensation and if they deposit an acceptable security, indemnity, or bond (Champa, 1982). Self-insurance is no insurance. Companies simply pay the claims themselves.

Regardless of the program employers use to cover their workers' compensation insurance, they ultimately pay the cost for that insurance in addition to overhead. Workers' compensation is a cost of doing business that can be lowered through the administration of effective safety and health programs.

RECORDKEEPING

Case Study

Jamie Smith is discouraged by the recordkeeping requirements of the company. She keeps forms for OSHA and an additional form for workers' compensation requirements. When a compliance officer arrives for a routine inspection, she is surprised to learn that she does not have to keep all the forms. She is also surprised to learn a form must be posted during the months of February through April. She has not posted the form in the past, but now assumes she should after the comment made by the compliance officer. The biggest surprise arrives three months later when the company receives a citation because the appropriate company official did not sign the form.

Background

Before the Occupational Safety and Health Act of 1970, there was no centralized and systematic method for monitoring occupational safety and health problems. Statistics regarding job injuries and illnesses were collected by some states and by some private organizations. National figures were based on less than reliable projections. With the passage of OSHA came the first basis for consistent nationwide procedures, a vital requirement for gauging problems and solving them.

Accurate injury and illness records are essential in providing information for the safety and health program. *Injury and illness recording is a requirement under law (29 CFR 1904). The information maintained on forms OSHA 300, 300a, and 301 can be used to the advantage of the safety professional to:*

- *Reveal* which operations are most hazardous
- *Determine* weaknesses in the safety and health program
- *Judge* the effectiveness of the program by comparing it with past records or records of other similar plans

- *Aid* in accident analysis and investigation
- *Identify* the causes of occupational diseases by relating them to particular exposures or processes, or both
- *Satisfy* legal and insurance requirements

Accurate records can be used to analyze illnesses and injuries so problem areas can be identified and corrected (NIOSH, 1979, p. 20).

Who Must Keep Records

Employers with 11 or more employees are subject to OSHA recordkeeping requirements as stated by 29 CFR 1904. They must maintain a record of occupational injuries and illnesses as they occur. Employers with ten or fewer employees are exempt from keeping such records unless selected by the Bureau of Labor Statistics (BLS) to participate in the Annual Survey of Occupational Injuries and Illnesses. Certain low-hazard industries are also exempt. They are, however, subject to the rest of the OSHA regulations applicable to their industry and jobs, and must display the OSHA poster. The purpose of keeping records is to permit the BLS survey material to be compiled, to help define high-hazard industries, and to inform employees of the status of their employer's record (www.osha.gov).

Forms

OSHA 300, 300A, and 301 Forms

Employers are required by the rule to keep injury and illness records on three forms (see figures 3-1, 3-2, and 3-3):

1. the OSHA 300 Log of Work-Related Injuries and Illnesses
2. the annual OSHA 300A Summary of Work-Related Injuries and Illnesses
3. the OSHA 301 Injury and Illness Incident Report

Employers are required to keep separate 300 Logs for each establishment expected to be in operation for one year or longer. They must use these or equivalent forms for recordable injuries and illnesses.

OSHA's Form 300 (Rev. 01/2004)

Log of Work-Related Injuries and Illnesses

Attention: This form contains information relating to employee health and must be used in a manner that protects the confidentiality of employees to the extent possible while the information is being used for occupational safety and health purposes.

Year 20___

U.S. Department of Labor
Occupational Safety and Health Administration

Form approved OMB no. 1218-0176

You must record information about every work-related death and about every work-related injury or illness that involves loss of consciousness, restricted work activity or job transfer, days away from work, or medical treatment beyond first aid. You must also record significant work-related injuries and illnesses that are diagnosed by a physician or licensed health care professional. You must also record work-related injuries and illnesses that meet any of the specific recording criteria listed in 29 CFR Part 1904.8 through 1904.12. Feel free to use two lines for a single case if you need to. You must complete an injury and illness incident report (OSHA Form 301) or equivalent form for each injury or illness recorded on this form. If you're not sure whether a case is recordable, call your local OSHA office for help.

Establishment name _____

City _____ State _____

Identify the person

(A) Case no.	(B) Employee's name	(C) Job title (e.g., Welder)	(D) Date of injury or onset of illness
			month/day

Describe the case

(E) Where the event occurred (e.g. Loading dock north end)	(F) Describe injury or illness, parts of body affected, and object/substance that directly injured or made person ill (e.g. Second degree burns on right forearm from acetylene torch)

Classify the case

CHECK ONLY ONE box for each case based on the most serious outcome for that case:

Death (G)	Days away from work (H)	Job transfer or restriction (I)	Other recordable cases (J)

Remained at Work

Enter the number of days the injured or ill worker was:

Away from work (K)	On job transfer or restriction (L)
___ days	___ days

Check the "Injury" column or choose one type of illness:

(M)

Injury (1)	Skin disorder (2)	Respiratory condition (3)	Poisoning (4)	Hearing loss (5)	All other illnesses (6)

Page totals

Be sure to transfer these totals to the Summary page (Form 300A) before you post it.

Public reporting burden for this collection of information is estimated to average 14 minutes per response, including time to review the instructions, search and gather the data needed, and complete and review the collection of information. Persons are not required to respond to the collection of information unless it displays a currently valid OMB control number. If you have any comments about these estimates or any other aspects of this data collection, contact: US Department of Labor, OSHA Office of Statistical Analysis, Room N-3644, 200 Constitution Avenue, NW, Washington, DC 20210. Do not send the completed forms to this office.

Page ___ of ___

Figure 3-1. OSHA Form 300

OSHA's Form 300A (Rev. 01/2004)

Summary of Work-Related Injuries and Illnesses

Year 20 ____

U.S. Department of Labor
Occupational Safety and Health Administration

Form approved OMB no. 1218-0176

All establishments covered by Part 1904 must complete this Summary page, even if no work-related injuries or illnesses occurred during the year. Remember to review the Log to verify that the entries are complete and accurate before completing this summary.

Using the Log, count the individual entries you made for each category. Then write the totals below, making sure you've added the entries from every page of the Log. If you had no cases, write "0."

Employees, former employees, and their representatives have the right to review the OSHA Form 300 in its entirety. They also have limited access to the OSHA Form 301 or its equivalent. See 29 CFR Part 1904.35, in OSHA's recordkeeping rule, for further details on the access provisions for these forms.

Number of Cases

Total number of deaths	Total number of cases with days away from work	Total number of cases with job transfer or restriction	Total number of other recordable cases
____	____	____	____
(G)	(H)	(I)	(J)

Number of Days

Total number of days away from work	Total number of days of job transfer or restriction
____	____
(K)	(L)

Injury and Illness Types

Total number of . . .
(M)

(1) Injuries ____	(4) Poisonings ____
(2) Skin disorders ____	(5) Hearing loss ____
(3) Respiratory conditions ____	(6) All other illnesses ____

Post this Summary page from February 1 to April 30 of the year following the year covered by the form.

Establishment information

Your establishment name ____

Street ____

City ____ State ____ ZIP ____

Industry description (e.g., Manufacture of motor truck trailers) ____

Standard Industrial Classification (SIC), if known (e.g., 3715) ____

OR

North American Industrial Classification (NAICS), if known (e.g., 336212) ____

Employment information (If you don't have these figures, see the Worksheet on the back of this page to estimate.)

Annual average number of employees ____

Total hours worked by all employees last year ____

Sign here

Knowingly falsifying this document may result in a fine.

I certify that I have examined this document and that to the best of my knowledge the entries are true, accurate, and complete.

Company executive ____ Title ____

Phone (___) ___ - ____ Date ___ / ___ / ___

Figure 3-2. OSHA Form 300A

OSHA's Form 301

Injury and Illness Incident Report

U.S. Department of Labor
Occupational Safety and Health Administration

Form approved OMB no. 1218-0176

Attention: This form contains information relating to employee health and must be used in a manner that protects the confidentiality of employees to the extent possible while the information is being used for occupational safety and health purposes.

This *Injury and Illness Incident Report* is one of the first forms you must fill out when a recordable work-related injury or illness has occurred. Together with the *Log of Work-Related Injuries and Illnesses* and the accompanying *Summary*, these forms help the employer and OSHA develop a picture of the extent and severity of work-related incidents.

Within 7 calendar days after you receive information that a recordable work-related injury or illness has occurred, you must fill out this form or an equivalent form. Some state workers' compensation, insurance, or other reports may be acceptable substitutes. To be considered an equivalent form, any substitute must contain all the information asked for on this form.

According to Public Law 91-596 and 29 CFR 1904, OSHA's recordkeeping rule, you must keep this form on file for 5 years following the year to which it pertains.

If you need additional copies of this form, you may photocopy and use as many as you need.

Completed by _____

Title _____

Phone (_____) _____ - _____   Date ___/___/___

Information about the employee

1) Full name _____

2) Street _____

City _____ State _____ ZIP _____

3) Date of birth ___/___/___

4) Date hired ___/___/___

5) ☐ Male   ☐ Female

Information about the physician or other health care professional

6) Name of physician or other health care professional _____

7) If treatment was given away from the worksite, where was it given?

Facility _____

Street _____

City _____ State _____ ZIP _____

8) Was employee treated in an emergency room?
☐ Yes
☐ No

9) Was employee hospitalized overnight as an in-patient?
☐ Yes
☐ No

Information about the case

10) Case number from the Log _____ (Transfer the case number from the Log after you record the case.)

11) Date of injury or illness ___/___/___

12) Time employee began work _____ AM / PM

13) Time of event _____ AM / PM   ☐ Check if time cannot be determined

14) **What was the employee doing just before the incident occurred?** Describe the activity, as well as the tools, equipment, or material the employee was using. Be specific. *Examples:* "climbing a ladder while carrying roofing materials"; "spraying chlorine from hand sprayer"; "daily computer key-entry."

15) **What happened?** Tell us how the injury occurred. *Examples:* "When ladder slipped on wet floor, worker fell 20 feet"; "Worker was sprayed with chlorine when gasket broke during replacement"; "Worker developed soreness in wrist over time."

16) **What was the injury or illness?** Tell us the part of the body that was affected and how it was affected; be more specific than "hurt," "pain," or sore." *Examples:* "strained back"; "chemical burn, hand"; "carpal tunnel syndrome."

17) **What object or substance directly harmed the employee?** *Examples:* "concrete floor"; "chlorine"; "radial arm saw." *If this question does not apply to the incident, leave it blank.*

18) **If the employee died, when did death occur?** Date of death ___/___/___

Figure 3-3. OSHA Form 301

Employers are permitted to use an insurance form or any form that keeps equivalent information instead of the OSHA 301 Incident Report. If their insurance form does not have all the information found on OSHA 301, they may add to it as needed.

Recordable Occupational Injuries and Illnesses

The employer must enter each recordable injury or illness on the OSHA 300 Log and 301 Incident Report within seven (7) calendar days of receiving information that a recordable injury or illness has occurred.

For recordkeeping purposes, an injury or illness is an abnormal condition or disorder. Injuries include cases such as, but not limited to, a cut, fracture, sprain, or amputation. Illnesses include those that are both acute and chronic, such as, but not limited to, a skin disease, respiratory disorder, or poisoning. Injuries and illnesses are recordable only if they are new, work-related cases that meet one or more of the Part 1904 recording criteria. The distinction between injury and illness is no longer a factor for determining which cases are recordable.

Any work-related injuries and illnesses that result in one or more of the following must be recorded:

- Death
- Days away from work
- Restricted work
- Transfer to another job
- Medical treatment beyond first aid
- Loss of consciousness
- Diagnosis of a significant injury or illness

First Aid Cases

If a case is only considered to be *first aid*, it does not have to be recorded. *For recordkeeping purposes, "first aid" means only the following treatments (any treatment not included in this list is not considered first aid for recordkeeping purposes):*

- Using a nonprescription medication at nonprescription strength
- Administering tetanus immunizations

- Cleaning, flushing, or soaking wounds on the surface of the skin
- Using wound coverings such as bandages, Band-Aids, gauze pads, and so on, or using butterfly bandages or Steri-Strips
- Using hot or cold therapy
- Using any nonrigid means of support, such as elastic bandages, wraps, nonrigid back belts, and so on
- Using temporary immobilization devices while transporting an accident victim
- Drilling of a fingernail or toenail to relieve pressure, or draining fluid from a blister
- Using eye patches
- Removing foreign bodies from the eye using only irrigation or a cotton swab
- Removing splinters or foreign material from areas other than the eye by irrigation, tweezers, cotton swabs, or other simple means
- Using finger guards
- Using massages
- Drinking fluids for relief of heat stress

For recordkeeping purposes, medical treatment does *not* include visits to a physician or other licensed health care professional solely for observation or counseling; diagnostic procedures such as X rays and blood tests, including the administration of prescription medications used solely for diagnostic purposes (e.g., eyedrops to dilate pupils); or any treatment contained on the list of first aid treatments (www.osha.gov).

Fatalities

If an on-the-job accident occurs, resulting in the death of an employee or the hospitalization of three or more employees, the employer is required to report the accident in detail within eight hours (29 CFR 1904.8) by filing a report by telephone or in person to the nearest OSHA area office or state plan office. The hospitalization of three or more is referred to by OSHA as a "*cat*astrophe," and, of course, a death is a "*fat*ality." An event that must be reported within eight hours is referred to as a **fatcat**. A motor vehicle accident occurring on a public highway, not in a construction zone, does not need to be reported.

Privacy Concern Cases

The rule requires the employer to protect the privacy of the injured or ill employee. The employer must not enter an employee's name on the OSHA 300 Log when recording a privacy concern case. The employer must keep a separate, confidential list of the case numbers and employee names, and provide it to the government on request. If the work-related injury or illness involves an intimate body part or the reproductive system; resulted from a sexual assault; results in mental illness; leads to HIV infection, hepatitis, or tuberculosis; results from needlestick and sharps injuries contaminated with another person's blood or other potentially infectious material; or leads to other illnesses, not injuries, and the employee independently and voluntarily requests that his or her name not be entered on the OSHA 300 Log, it is considered a *privacy concern case*.

Posting Annual Summary Requirements

After the end of the year, employers must review the log to verify its accuracy, summarize the 300 log information on the 300A summary form, and certify the summary. A company executive must sign the certification. The information must then be posted from February 1 to April 30. The employer must keep the records for five years following the calendar year covered. If the business is sold, records must be transferred to the new owner.

An other-than-serious citation will normally be issued if an employer fails to post the OSHA 300A Summary by February or fails to certify the Summary or keep it posted for three months (until May). The unadjusted penalty for this violation is $1,000. A citation won't be issued if the Summary would have reflected no injuries or illnesses anyway.

A copy of the OSHA forms can be obtained from the U.S. Department of Labor or the state department of labor—OSH Division. They are also available online at www.osha.gov.

In summary, OSHA places special importance on posting and recordkeeping. The compliance officer will inspect records the employer is required to keep, and ensure that a copy of OSHA No. 300A is or has been posted and that the OSHA workplace poster is prominently displayed. Information is critical to the effective operation of any safety program, so nothing should occur to inhibit the flow of that information upward.

CONCLUSION

Employers need to be careful in the application of laws in particular states. Although the types of awards vary, employers are subject to workers' compensation liability any time an employee is injured in the course of work. This could occur in a number of ways. In the case of an employee taking work home from the office, that employee is subjecting the employer to a possible claim while en route, as well as while working at home. Employees required to travel in the course of their jobs may also be covered by workers' compensation even during their off-hours.

Managers should make every effort to ensure the safety of their employees at all times. When things go wrong, they must be prepared to track injuries and illnesses. These records must be kept in a timely manner and include all required information.

QUESTIONS

1. What is the value of records to the employer and to OSHA? Recordkeeping requirements by OSHA are heavily emphasized. Prior to walking around in an inspection, the compliance officer will first look at records. Why do you think he usually reviews the records *before* the inspection?

2. Is workers' compensation beneficial to the employer? Although it costs the employer money, why is it desirable from the employer's perspective?

3. How do complete company records provide protection to the employer? If the law didn't require them, should the employer still keep detailed occupational safety and health records?

4. What is a "fatcat" and what are the reporting requirements when one occurs?

REFERENCES

Champa, Shirley A. 1982. *Kentucky Workers' Compensation.* Norcross, GA: Harrison.

National Institute for Occupational Safety and Health. 1979. *Self-Evaluation of Occupational Safety and Health Programs.* Washington, DC: NIOSH.

BIBLIOGRAPHY

Kentucky Labor Cabinet. 1991. *Facts about Occupational Safety and Health.* Frankfort, KY: Kentucky Labor Cabinet.

United States Department of Labor. 1991. *All about OSHA.* Washington, DC: U.S. Government Printing Office.

United States Department of Labor. 2000. *All about OSHA.* Washington, DC: U.S. Government Printing Office.

United States Department of Labor. 2002. *Inspections.* Washington, DC: U.S. Government Printing Office.

United States Department of Labor. 2003. *Recordkeeping Guidelines.* Washington, DC: U.S. Government Printing Office.

United States Department of Labor. OSHA website. Washington, DC, http://www.osha.gov (accessed December 2, 2002).

4

Safety-Related Business Laws

CHAPTER OBJECTIVES

After completing this chapter, you will be able to

- Understand the important legal terminology related to civil law
- Identify the different types of torts
- Understand product liability and associated risks
- Understand product safety and the Product Safety Act
- Identify the types of warranties
- Understand contracts

CASE STUDY

Larry Wilson's son was badly burned in a motorcycle accident. Larry decided to sue the manufacturer to see if he could be compensated for uninsured medical bills, pain and suffering on his son's behalf, and future expected care costs. In a criminal case, when an individual violates a law, a party is presumed innocent until proven guilty. In a product liability case, like this one, a party wins or loses based on the preponderance of evidence; guilt is not necessarily involved. The party with the most convincing argument wins. It is not unusual for a case of this nature to be settled before trial if parties are willing and can reach an agreement on terms. Additional legal expenses can then be saved. However, each is gambling that the company or individual will be financially better off with a settlement. Incidentally, this case was to be tried before a jury, as many such cases are.

The really interesting and ironic aspects were the circumstances surrounding the accident. Wilson had purchased the motorcycle from a neighbor for $15 earlier that day. He pushed it to his house, parked it on the lower level, and closed the door. The next time he thought about the motorcycle was when he heard shouts coming from the area, where he found his son lying on the floor, engulfed in flames.

A subsequent investigation by the defendant's attorneys found the motorcycle gas tank had no cap when Wilson bought it; yet he stored it next to the water heater. He took no steps to warn his son, nor did he give the boy permission to play with the motorcycle. When his son tried to kick start the motorcycle, he dropped it. Defense attorneys speculated that gas spilled and was ignited by the nearby water heater. The boy was pinned by the weight of the cycle and suffered burns as a result. Additionally, no doors had been opened to permit the exhaust from the motorcycle to leave the house had it, in fact, started.

Many involved in the case, including the motorcycle manufacturer's legal counsel, believed that the manufacturer had no liability. The person most likely to have been sued was the neighbor who sold the motorcycle without a gas cap, but he was dismissed early, because he lacked "deep pockets." The company managers and defense attorneys believed that once the case went to trial, a jury might award a large cash settlement to the Wilson family. Even though preponderance of evidence may not have favored the Wilsons, juries are often sympathetic to the plights of injured people regardless of fault. They may perceive that manufacturers have unlimited funds and the injured people suing have few financial resources. The company settled out of court for $750,000 before the case ever went to trial.

IMPORTANT TERMINOLOGY

Criminal law is addressed in the courts through prosecution, and the penalty suffered is either a fine or imprisonment. *Civil law* is addressed in the courts through litigation, and penalties include monetary damages. For example, while driving the company truck John Smith hits the brakes, they fail to operate properly, and the truck runs off the road and into a ravine. As a result of the accident, John suffers severe back pain and is unable to work for several months. John perceives the benefits he receives are insufficient to compensate him for the inconvenience, pain, and suffering he has undergone. John contacts a local attorney and, after discussing the accident with her, decides to sue under civil law. John is seeking recovery beyond the workers' compensation benefits.

After meeting with his counsel, John decides to sue his employer, as owner of the truck; the dealership where the truck was last serviced; the manufacturer of the truck; and the owner of the land where the truck went off the road. John is the *plaintiff* and each of the parties being sued is a *defendant*. His lawyer has advised John to follow the *deep pockets theory* and to sue everyone who might have money in the case. As in the Wilsons' case, John will only go after individuals who are most likely to have the money to pay. After *depositions* are given by each of the parties in the suit, the landowner is removed from the list because John's legal counsel does not believe this party shared fault. Aside from the small piece of property, the landowner has few assets and the defense counsel believes it is in the plaintiff's best interest to excuse him from the suit. The policy of some companies is to never settle out of court if they do not absolutely believe they are at fault. Their perception is that settlements encourage additional lawsuits. Other companies settle when possible to avoid the legal expenses associated with a long court case, as well as the adverse publicity often accompanying a trial.

Lawsuits, like the ones mentioned above, are typically initiated following a *tort*. A *tort* is an act, or absence of an act, causing a person to be injured, a reputation to be marred, or property to be damaged. A *lawsuit* differs from a criminal case in that the purpose of a lawsuit is to obtain compensation for damage suffered. A *criminal case* seeks to punish the wrongdoer without compensation. A tort may give rise to a criminal action or a lawsuit.

Today, any accident resulting in death, injury, or property damage can bring about a lawsuit, criminal prosecution, or both. Anyone can sue another party anytime for practically any reason. Lawsuits range from the justified to the ridiculous. Most are a result of someone who perceives he or she has been wronged and seeks compensation.

From a safety perspective, one of the major concerns is *liability*. Liability is either voluntarily assumed, as by contract, or imposed, as is the case following a lawsuit. Liability resulting from torts is of the latter type. It is important to understand tort law because managers can sometimes, unwittingly find themselves on the losing end of a lawsuit. *Activities that can act as catalysts for lawsuits and that are difficult, if not impossible, to defend include the following:*

- *Intentional torts* occur when the defendants intended for their actions to cause the consequences of their act; they are knowingly committed. Examples of intentional torts are assault, battery, false imprisonment, and defamation. They are defined below.

    —*Assault* occurs when words or actions cause another person to fear personal harm or offensive contact. If one employee threatens to hit another employee with a shovel, the first employee may be guilty of assault. If a person is made to feel apprehensive or afraid that he will be hurt or offensively contacted, then the other party may be guilty of assault. A key element of apprehension is the knowledge the assault occurred. If a worker shakes his fist at another worker or threatens another worker in such a way that the other worker is unaware of the motions, then no assault has occurred.

    —*Battery* is harmful or offensive bodily contact. If an employee hits another employee or even grabs the other employee by the shirt collar, the offending employee may be guilty of battery. The act may have caused little or no injury, but if it offends a reasonable person, then it may be considered battery.

    —*False imprisonment* occurs when a person is restrained within a fixed area against their will. If, however, the restrained individual has a means of exit available, then no false imprisonment has occurred. For example, merchants who restrain suspected shoplifters might be sued for false imprisonment. If, however, the merchant merely stands between the shoplifter and the door to discuss the issue, and the shoplifter has another means of exit, no imprisonment has occurred.

    —*Defamation* is the act of injuring another individual's reputation by disgracing or diminishing that person in the eyes of others. Defamation can occur through *libel* when the communication is written or broadcast through media such as radio or television. It can also occur through *slander* where the communication is spoken. Defamation is grounds for a lawsuit. The best defense against a defamation suit is the truth, but employees should be warned about the consequences of defamation lawsuits. A negative reference letter or an adverse report by telephone about someone given may trigger a defamation suit.

All of the above are examples of intentional torts. In order to successfully sue someone for an intentional tort, the plaintiff has to present a *preponderance of evidence* as opposed to proving the defendant guilty, as a prosecutor tries to do in a criminal case. In an intentional tort lawsuit, the verdict is awarded to the individual who presents enough evidence to convince the judge or jury he should receive the verdict in his favor.

Individuals need to also be aware of ***unintentional torts*** when parties are sued due to their unintentional actions. Typically, the parties being sued are negligent. ***Negligence*** involves creating an unreasonable risk of harm due to failure to exercise reasonable care.

"Reasonable care" is a relative term dependent on the circumstances and the party involved. The legal standard for reasonable care is the reasonable person who always acts prudently under the circumstances and is never careless or negligent. If the party happens to be a professional, the standard is raised. For example, a person who is qualified to work in a profession or trade requiring special skills is expected to act in a manner consistent with that of a prudent and careful surgeon, carpenter, accountant, or safety professional. The skill level and expectations are higher for a professional than they are for the nonprofessional in a given area.

If standards of performance exist, the professional should be aware of those standards and make every effort to comply with them. For example, industrial hygienists have certain standards that are maintained and published by the American Board of Industrial Hygiene. Safety professionals have standards published by the Board of Certified Safety Professionals. When industrial hygienists, safety practitioners, or other professionals remain ignorant of the standards of their trades or choose to ignore those standards, they are increasing their risk of being sued and of being unsuccessful in the defense of a suit. Professionals are held to a higher standard of performance by the courts.

Even though a plaintiff can successfully present a case for negligence and a preponderance of evidence, the defendant may not receive a favorable award because of a successful defense. Successful defenses that have prevented awards include contributory negligence, comparative negligence, and assumption of risk.

Contributory negligence occurs when the plaintiff helped cause or actually contributed to the loss. If an employee mops the floor in an area where pedestrians have the option to go around but choose not to, and a person slips, falls, and is injured on the wet floor, the person who fell contributed to the accident through negligence and may be denied an award.

In some states, the defendant and the plaintiff may simply share the above award. In a ***comparative negligence award*** the amount is apportioned based on the percentage of fault. Therefore, in a comparative negligence defense, if the defendant is found to be 40 percent at fault and the plaintiff 60 percent at fault, and the award is for $100,000, the actual cash paid by the defendant is $60,000. If the plaintiff's share of fault is found to be greater than 50 percent, then the plaintiff may receive no award.

Assumption of risk occurs when the injured party knowingly accepts the risk involved in the action that leads to injury(s). For example, a delivery person chooses to deliver a package by walking up ice-covered steps, falls, and is injured and takes legal action against the owner of the steps. The defendant's argument could be that the delivery person could see the steps were icy, yet chose to deliver the package and, therefore, assumed the risk of falling.

ROLE OF THE SAFETY PROFESSIONAL

The safety professional should be aware of the various pitfalls open to management and other employees relative to rules of doing business. Employees unaware of the consequences of their actions may engage in activities detrimental to the well-being of the organization. In a lawsuit, even in situations where no dollar amount is awarded to a plaintiff, the company may have to expend large amounts of money in defending itself. The safety professional should look for exposures to litigation due to torts and help the company avoid those exposures through advice to and education of managers and other employees.

PRODUCT LIABILITY

Within the last few decades, litigation costs surrounding product safety have evolved, with some lawsuits reaching the million- or billion-dollar award level in damages. According to Seiden, products should "prevent and/or mitigate personal injury and illness under reasonably foreseeable conditions of service, including reasonably foreseeable use and misuse" (1984, p. 15). Much of the litigation evolved from the consumer movement of the late 1960s and early 1970s. One result of this movement was the passage of the Product Safety Act.

Product Safety Act

The Product Safety Act was officially enacted on October 27, 1972, with four main purposes:

- To protect the public against unreasonable risks of injury associated with consumer products
- To assist consumers in evaluating the comparative safety of consumer products
- To develop uniform safety standards for consumer products and to minimize conflicting state and local regulations
- To promote research and investigation into the causes and prevention of product-related deaths, illnesses, and injuries (Bureau of National Affairs [BNA], 1973, app. 3)

This Act is administered by the Consumer Product Safety Commission, which promulgates consumer product safety standards. These standards are "reasonably necessary to prevent or reduce an unreasonable risk of injury associated with such product" (BNA, 1973, app. 7). *The text of the Product Safety Act gives two categories of requirements for product safety:* as to performance, composition, contents, design, construction, finish, or packaging of a consumer product,

- that a consumer product be marked with or accompanied by clear and adequate warnings and instructions,
- or requirements respecting the form of warnings and instructions (Colangelo and Thornton, 1981, p. 1).

Companies manufacturing products and not considering the above requirements place themselves at increased risk for product liability. *Product liability* "describes an action, such as a lawsuit, in which an injured party (the plaintiff) seeks to recover damages for personal injury or loss of property from a seller or manufacturer (the defendant) when it is alleged that the injuries or economic loss resulted from a defective product" (Colangelo and Thornton, 1981). Manufacturers, in turn, can effectively minimize product liability by developing products with safeguards against predictable types of defects, deficiencies, abuse, or misuse. However, there are many unforeseeable factors that can hamper achievement of this goal. For example, injuries to consumers using (or misusing) products can facilitate product liability litigation, thus increasing the manufacturer's responsibility for producing a safe product and the cost of that product to consumers.

Theories of Product Liability

There are four important theories that form the basis of product liability and provide a fundamental understanding of product liability litigation:

1. *Strict liability* is the concept that a manufacturer of a product is liable for injuries received due to product defects, without the necessity for a plaintiff to show negligence or fault (Hammer, 1972). In some instances, a party may be liable for damages caused, even though there was no negligence or intentional tort. Some activities, though they are desirable and often necessary, are so inherently dangerous they are a risk to others regardless of how carefully they are conducted. In such cases the purveyor of the activity is held to the concept of *strict liability* or liability without fault. An example of strict liability is when a company stores explosives or even flammable liquids in large quantities, or emits noxious fumes. Once damages are linked to the activity, the manufacturer will generally be held liable, regardless of the circumstances.

2. *Negligence* is the failure of the manufacturer to exercise reasonable care. In addition, it also includes the responsibility to carry out a legal duty so injury or property damage does not occur to another (Hammer, 1972).

3. *Breach of express warranty* occurs when the product does not meet the claims made by the manufacturer and, as a result, damage or injury occurs to another (Colangelo and Thornton, 1981). When a company gives a statement of performance concerning a product, this statement becomes part of the product. If the product fails to live up to the standards outlined by this statement, then the product breaches the express warranty. This warranty may be written, oral, or simply in the form of a demonstration. The assumption made by the buyer, and justifiably so, is that the product purchased will perform as well as was shown in the demonstration.

4. *Breach of implied warranty* occurs when a product fails to fulfill the purpose for which it was intended. An implied warranty is a warranty implied by the law rather than specifically made by the seller. The

basis for this rests primarily on definitions found in the Uniform Commercial Code. In essence, the seller may be held liable if the seller makes an improper recommendation regarding use of the product (Colangelo and Thornton, 1981). Even though no particular warranties are expressed, products are purchased for a particular purpose. When a product fails to fulfill that purpose, it is in violation of the implied warranty. If, for example, a consumer buys a balloon and the balloon will not hold air, it is in violation of the implied warranty. The assumption made by the purchaser is that all balloons hold air.

Manufacturers or sellers are not liable for all injuries that may result from a product. This would be the concept of absolute liability. However, these four basic theories clearly establish the product responsibilities of the manufacturer or seller in most states (Brauer, 2006).

Liabilities extend to the final user of the product, regardless of how many times it changes hands. The liability extending to wholesalers, retailers, and to the final user is known as *vertical privity*. *Vertical privity* refers to parties up and down the distribution chain. *Horizontal privity* refers to users on a given level of the chain. Injured bystanders are protected under the concept of horizontal privity.

Lawsuits

In a product liability lawsuit, the plaintiff must present certain evidence to support the claim. *With the exception of express warranty cases, the plaintiff must prove (Brauer, 2006):*

- The product was defective
- The defect existed at the time it left the defendant's hands
- The defect caused the injury or harm and was proximate to the injury

In reference to the four theories of product liability, each has its own requirements for evidence presentation (Brauer, 2006):

- *Strict Liability*. No other evidence is required to establish a case.
- *Negligence*. The plaintiff must show the defendant was negligent in some duty toward the plaintiff.

- ***Express Warranty and Implied Warranty***. The plaintiff must show a product failed to meet the implied or express warranty or misrepresented claims for the product.

The manufacturers or sellers, on the other hand, have a number of defenses at their disposal for the four theories (Brauer, 2006). They may try to show that, although the product is dangerous, the danger itself is not a defect. According to Brauer, the defendant could try to show the plaintiff "altered the product or unreasonably misused it," or "the product did not cause the injury or was not the proximate cause." The defendant might also claim that "the product met accepted standards of the government, industry, or self-imposed standards related to the product, to the claimed defects, and to the use of the product."

There are risks manufacturers or sellers face when they put a product on the market. ***There are, however, a number of ways liability can be minimized:***

- Defend the manufacturer in design, manufacturing, packaging, and the marketplace
- Assemble a team for product design review to thoroughly analyze the product for hazards and acceptable controls (Brauer, 2006)
- Remove unreasonable dangers from products and environments; prevent defects from reaching the marketplace; account for the use and misuse of the product; and consider hazards, potential injury, compliance with standards, and claims for products
- Ensure that warnings identify remaining hazards and provide instructions necessary for user protection

At every step in the product's life cycle, beginning with product conception, the safety professional should have input. (See chapter 9 for additional information.) Safety should not be an afterthought in the design and manufacture of products if liability is to be minimized. Safety must be built into each and every product leaving the facility.

CONTRACTS

A ***contract*** is a legally enforceable promise. It may be expressed, implied-in-fact, or quasi-contractual. Contracts can be made in a number of ways.

Express contracts are made when a person engages in an oral or written commitment to another to enter into a binding relationship. *Implied-in-fact contracts* occur when parties indicate through their actions, rather than words, they intend to agree. *Quasi contracts* occur when one party would be enriched unjustly if permitted to enjoy something received from another party. Usually Party A provides work or services to Party B even though an actual contract did not exist. Party B may have to make reasonable restitution to Party A for the services performed.

In order for a contract to be valid, four conditions must be present. First, there must be an agreement; this requires an *offer*. An offer states, in reasonably certain terms, what is promised and what each party will do. The courts consider the intention of the offering party. For example, if the person is joking, asking for details, or simply negotiating, an offer may not take place. On the other hand, a proposal may be an offer even though all of the terms have not been specified.

Once an offer has been extended, *acceptance* may be made by the other party, or the other party may make a counteroffer. In any case, in order for the contract to come into existence, there must be an agreement to the terms of the offer and then acceptance. If the offer specifies it must be accepted in a particular way in terms of time or means, unless the terms are adhered to, there is no contract.

Typically, there is no enforcement of the contract unless there is *consideration*. Consideration is exchanged in terms of goods, services, or restraint from doing something. Both parties must be obliged to provide something, but the value provided by each party does not have to be equal. Sometimes a promise is exchanged for a small sum, but if fraud or improper conduct is evidenced, the contract is unenforceable.

All parties to the contract must have the legal *capacity* to enter into the contract. They must be of legal age, sober, and of sound mind. If, for example, a party enters into a contract with a minor (one who is typically under the age of 18), the contract may be *void*. There are exceptions to the above conditions, but these are generally upheld. If the contract has an illegal intent or calls for some action that violates public policy, then it is typically void.

Enforcement of contracts occurs under civil law. When one party fails to live up to his or her part of the contract, the other party may attempt to enforce the terms of the contract by bringing a lawsuit. The courts may require an award of money or may force a party in the contract to perform the promise specified. This remedy, known as *specific performance*, may be necessary when a monetary remedy would be insufficient to compensate for the unique requirements of the contract. For instance, if a specific piece of real estate or a rare object of art is involved, the court may require that particular property to be delivered.

Certain points of law are sometimes pitfalls for companies. For example, insurance contracts and some leases are considered *contracts of adhesion*. These documents are often offered to the purchaser or lessee. Because greater power rests with the author of the document, ambiguities and language subject to interpretation will typically be construed against the writer. In such cases, the contract writer is considered to be the expert, so interpretations will tend to favor the weaker party, such as the insured or the lessee.

Exculpatory or *hold harmless* clauses are sometimes written to limit or eliminate liability in certain contracts. For example, when a company hires a contractor to perform work in its business, the contractor may require the company to first sign an exculpatory agreement stating the contractor will not be responsible for any damages the contractor causes. Clauses in contracts similar to this do not provide much protection for contractors or others whose intentional acts or negligence causes injury or damage. Companies still use them, hoping to deter another party from actually suing.

Safety professionals should be aware of the *parol evidence* rule, which states that the written contract supersedes any oral promises. When the written terms are agreed upon, they will govern the contract. For example, a salesperson's promises, if not included in the contract, will usually not be enforced. It is, therefore, worthwhile to review all written contracts based on a sales pitch. Insurance contracts, leases, and service or maintenance contracts are typical of the types of documents that might differ from the buyer's expectations.

INSURANCE

Companies can build safety into a product every step of the way. No unusual warranties may be given, and the salespeople may do an excellent and thorough job of explaining the product features. Buyers can even sign an exculpatory clause. However, even with the best products and the best of intentions, companies may still be sued. Sometimes, the only thing that stands between them and potential bankruptcy is insurance. Insurance companies not only pay up to specified amounts when lawsuits are lost, but also provide defense attorneys for the defendant. It is sometimes less expensive to defend against a lawsuit than to simply pay up. Insurance is often the best protection against spurious lawsuits.

CONCLUSION

Safety practitioners need to be aware of the basic legal environment surrounding the workplace. Although they often depend on legal counsel for specific advice, an awareness of the laws likely to affect the profits of the company is critical. Knowledge of potential pitfalls and methods to avoid those traps can only help to increase company profitability.

The concepts discussed in this chapter are complex and hold many exceptions and potential problems for manufacturers and consumers. Complete details are beyond the scope of this text. Additional research or legal consultation is recommended for answers to specific problems or questions.

QUESTIONS

1. Is the job of the safety manager one that requires concern for contracts, liability, and insurance, or should the safety practitioner simply be concerned with "safety"?
2. Do you think that there should be a cap on the amounts of jury awards that may be given beyond actual losses? Why or why not? Who pays for these awards?
3. How does a concern for torts become a part of the safety practitioner's concerns? Are you aware of situations where an employee or an agent of a company has committed a tort against another and the company was held financially responsible? Who pays in this case?
4. Do you ever use exculpatory or hold harmless clauses in your work? Are you ever asked to sign them when you rent, purchase, or use property? What are the circumstances when you are likely to see them?

REFERENCES

Brauer, R. L. 2006. *Safety and Health for Engineers*. Hoboken, NJ: John Wiley.
Bureau of National Affairs. 1973. *The Consumer Product Safety Act: Text, Analysis, Legislative History*. Washington, DC: Author.

Colangelo, V. J., and Thornton, P. A. 1981. *Engineering Aspects of Product Liability*. Metals Park, OH: American Society for Metals.

Hammer, W. 1972. *Handbook of System and Product Safety*. Englewood Cliffs, NJ: Prentice-Hall.

Seiden, R. M. 1984. *Product Safety Engineering for Managers: A Practical Handbook and Guide*. Englewood Cliffs, NJ: Prentice-Hall.

5

Accident Causation and Investigation: Theory and Application

CHAPTER OBJECTIVES

After completing this chapter, you will be able to

- Explain the benefits of understanding accident causation theory
- Define the terminology associated with accident causation theory
- Identify the activities involved in risk assessment
- Compare and contrast the various accident causation theories
- Explain the purpose of accident investigation
- List the activities involved in accident investigation

INTRODUCTION

Safety and health professionals are responsible for aiding management in anticipating, identifying, evaluating, and controlling hazards in the workplace. Responsibilities include advising the management team of the risks facing the facility. The professional then calls upon management to eliminate or minimize the hazards associated with those risks before losses occur. Knowledge of accident causation theories permits the professional to more thoroughly recognize and communicate information regarding organizational safety problems.

Employees with little safety and health experience might walk near a punch press leaking lubricating oil on a shop floor and recognize a slipping hazard. For the more experienced professional, several questions are raised:

- What is the lubricant?
- Is it a fire hazard?
- Is it a health hazard?
- How do the operators clean the spillage?
- Is the solvent used for cleanup a health hazard?
- Are hazard communication programs, training, and personal protective equipment required to use these products?
- Should less hazardous lubricants and solvents be considered as substitutes? Is the punch press guarded?
- Are there repetitive motion or other ergonomic hazards associated with the operation of the press?

It is possible to greatly expand the list of questions one might ask, given the few signals provided in this simple scenario. The more information and experience available, the more safety and health hazards are likely to be anticipated and recognized.

There are many theories associated with the causes of accidents, ranging from simple to complex. Some focus on employees and how their action or lack thereof contributes to accidents. Others focus on management and its responsibilities for preventing conditions leading to accidents. Theories are not facts; they are tools predicting relationships that *may* exist in the future.

Accidents are not events happening by chance; they have specific causes. Nor are they random events; they are usually predictable and preventable. By applying one or more theories, the professional is more likely to predict accidents and initiate activities preventing their occurrence or recurrence. Accidents represent failures in the system or management problems in the organization. Management controls all the variables surrounding accidents. It creates the work environment. It advertises for, hires, places, trains, and supervises workers. Management can reward for work it wants done and punish when procedures are violated. If the worker makes a fatal error, both OSHA and the courts will likely hold management responsible. Public opinion rarely favors management in accident situations.

In an accident investigation, management may blame the worker. This is referred to as the *pilot error syndrome*; in other words, the pilot is blamed because he is dead, injured, or considered a weak scapegoat and cannot defend himself. *It is important to emphasize the safety professional is looking not for a place to assign blame, but for errors in the system.* These errors can then be addressed so problems and accidents will be prevented in the future. Anytime the professional fails to find the error in the system, causing or permitting the incident to occur, the investigation stops short. Once the worker or pilot is blamed, management and the management system are off the hook; therefore, there is no reason to continue to search for why the accident occurred.

Anytime the accident investigator seeks to place blame, employees run for cover. Stories will be concocted and the truth may not be found. A safety professional in a major company was investigating a helicopter accident following a crash involving a pilot and a few employees on a surveillance mission of the plant property. Although no one was injured, the copter was a total loss and the investigation yielded no causes to be addressed. All aboard agreed the crash was due to wind shear. Some months later, the investigator was engaged in informal conversation with one of those onboard and questioned him further regarding the incident with a promise of no reprisals. After encouragement, the employee revealed that during the flight, the pilot yelled, "Watch this!" He then proceeded to chase a deer through the woods with the copter. With the pilot's attention focused on the animal, he failed to notice that the clearing where he was flying became narrower; nor did he see the rotor clip some trees until it was too late to avoid damage. Once on the ground, the pilot coached the other employees to agree on a story to tell the safety investigator. With newly revealed information, the company established a policy regarding flying altitudes, and the likelihood of the problem resurfacing was reduced. The investigator was true to his word and pursued no disciplinary action. To do so would likely have slowed the flow of information in future accident investigations. Even in this case of flagrant employee disregard of safety, management accepted responsibility for the problem and initiated appropriate action.

THE CONCEPTS OF RISK, INCIDENTS, AND ACCIDENTS

"Accident" refers to a loss-producing, unintended event. *"Incident"* is an unintended event possibly not causing a loss. For example, a truck with a flat tire pulls off the side of the road into tall grass, bordering a hill. The

driver does not realize the grass is disguising a slope; he doesn't check the ground before driving onto it and the truck rolls down the hill. If no slope were present—only solid ground next to the road—the truck would not roll down the hill. In either case, if the driver never checks the slope, it is simply a matter of chance as to whether a loss occurs. If the truck does not encounter a slope and no loss occurs, it could be described as an "incident." Incidents include so-called "near misses" and other accident-type events not resulting in loss. If the truck rolls down the hill, causing injury, death, or property damage, the event is now described as an "accident." Whether or not an incident is also an accident, accumulating data about such events is worthwhile. The response to either should be the same in terms of active intervention with engineering or administrative controls or the establishment of other controls.

As defined in chapter 1, *risk* describes both the probability and severity of a loss event. **Probability** refers to the likelihood of the occurrence of an event. When used as part of a risk assessment tool in system safety, probabilities can be categorized as *frequent*, *probable*, *occasional*, *remote*, and *improbable* (Roland and Moriarity, 1990). By studying safety-related data, it is possible to determine statistical trends for a variety of factors, such as the types of accidents (e.g., falling to the same level, being struck by a falling object, or injuring as a result of overexertion) or location where injuries are taking place (e.g., warehouse, assembly, paint shop). Using the data on the frequency of occurrence of these events, it is possible to classify injuries, property damage, or other loss factors in terms of one of the probability categories.

Risk = Probability × Severity

Probability and frequency of loss events are only half of the risk picture. The severity of the loss event must also be considered. *"Severity"* refers to the magnitude of the loss in a given period of time. **When used as part of the risk assessment tool in system safety, the severity of a particular condition is classified in one of four categories:**

1. Catastrophic
2. Critical
3. Marginal
4. Negligible

These four categories correspond to death or loss of a system, severe injury or major damage, minor injury or system damage, and no injury or system damage respectively (Roland and Moriarity, 1990). When viewed on a severity continuum, near misses, scratches not requiring first aid, or brief assembly line stoppage might be considered negligible reactions to an incident. At the opposite end of the severity continuum, multiple fatalities or explosions that destroy an entire building represent catastrophic loss events.

All companies face risk and the resultant potential losses on a daily basis. They willingly accept risk—even welcome it—in the hope of gaining financial return. All new marketing and business ventures are examples of businesses taking risks. These are considered *speculative loss exposures* because they offer the opportunities for gain and loss. Exposures offering potential for loss with no opportunity for gain are referred to as *pure loss exposures*. Examples of pure loss exposures include those to theft, fire, or accident.

Determining exposures a company faces is a formidable task, but it must be undertaken. The best starting point is usually with the line employees. Ask people doing the job what risks they face and what problems must be handled. The last thing you want to hear, after the fact, is "That was an accident waiting to happen." Employees often have more insight into problems and potential problems than anyone else. Companies may also choose to bring in an outside consultant or depend on the advice of a representative such as a loss control expert from their insurance company. This person will review activities, procedures, and processes to determine where exposures occur. Once exposures are identified, the company selects the critical exposures and prioritizes them in order of importance.

A second approach is to follow the process and review each step and the associated hazards. For example, start with the raw materials and consider what could go wrong with each. If more hazardous materials could be replaced by those with less hazardous properties in an economical manner, attempt to do so. Look at how each material is handled and ask and what could go wrong. As the raw materials make their way through the system, look at them every step of the way. Raw materials eventually become partially or completely finished products. Look at each and consider hazards as they are worked and any new interactions take place. This may also require a backward look at given interactions along the way. For example, a manufacturer assembles raw materials and machines them to prepare parts for assembly. At some point, the product may have to be tested. The safety professional now has to question the testing and everything that takes place prior to that process. Are there any new hazards introduced along the way? This must be considered and steps taken to minimize or eliminate associated hazards.

A third approach is to ask employees to notify the safety professional anytime a near miss occurs. These can be done anonymously through "suggestion boxes" and they may also include ideas employees may have to improve the system. Additional approaches to discovering hazards before they result in accidents are suggested in chapter 9 on system safety.

Even large corporations have limited resources and must make decisions as to where to commit them. While controlling losses may seem to the safety manager like the most important way to utilize resources, the marketing director may feel that a new product introduction should have precedence. The safety function competes with every other entity in the enterprise for resources, so it is important for the safety manager to clearly understand what needs are to be addressed, why, how much addressing the needs will cost, and the projected return on investment. The effects of investments in safety are measured by management against the *expected* return on investment as well as the return on investment of other potential ventures.

Establishing priorities allows the safety professional to determine problems to attack first and where to spend the allocated safety budget. Non–first aid injuries occurring with great frequency may appear, at first glance, to be events not ranking very high on the safety professional's things-to-do list. However, occasional confined-space entry fatalities are usually of great concern because employee deaths often result. Multiplying the probability of loss from a fire or any event in a given year's time and the likely amount to be lost from such a fire or event (severity) gives the safety practitioner a relative gauge as to where to place resources for controlling losses. Comparing the cost of controls to the estimated costs of losses saved over time is also useful.

By considering both probability and severity, the risk assessment provides the safety practitioner with a much sounder perspective for judging the significance of hazards. Once the risk assessment is performed, it is possible to determine the types of controls that most effectively eliminate the hazards (see figure 5-1).

There are a number of control techniques available for treating loss exposures. These can be broken into two categories: loss control techniques and financing techniques. Loss control techniques include the approaches mentioned in chapter 6, "Introduction to Industrial Hygiene." A company may choose engineering or administrative controls or personal protective equip-

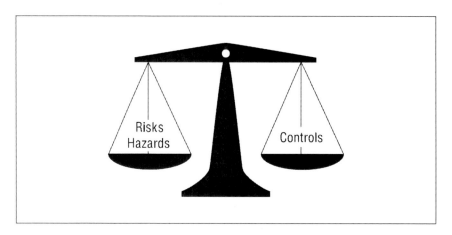

Figure 5-1.   An illustration of the balance between risks and hazards in the workplace and controls necessary to minimize their effects

ment to deal with losses. Engineering controls include building a ventilation system to reduce explosive vapor levels, whereas administrative controls might limit exposures to toxic materials. Issuing personal protective equipment (PPE) such as respirators is the last line of defense against hazards in the workplace. Refer to chapter 6 for a detailed examination of these three types of control methods.

A company might try to avoid the loss altogether. If a company is in the business of producing football helmets, it may choose not to market the product due to the potential liability it would face if a wearer of one of its helmets was injured. Instead, it may choose to produce novelty lamps that look like football helmets to limit its liability exposure.

Sometimes a company can reduce exposure by substitution. Instead of using a strong acid as a solvent to remove excess glue from a finished product, a company could use citric acid for removing glue to reduce the likelihood of a worker being injured.

Occasionally, the company may choose to *transfer* the liability to another party, rather than run the risk of loss itself. If removing glue could not be accomplished safely in the plant, the company may choose to have the product shipped to a contractor who would remove the glue in the contractor's plant. If the contractor's workers are overcome by vapors from the solvent, then the contractor would typically hold the liability.

Another form of risk transfer is insurance. *Insurance* is designed to permit the company to shift the financial consequences of the risk to an insurance company. By paying the insurance company's *premiums*, the organization can expect specified *benefits* to remunerate for a loss. The insurance company enters into similar relationships with a number of other companies by selling them insurance policies too. With large numbers of insureds, it can more accurately estimate its own losses. When an insured company turns in a *claim*, the insurer is able to pay because it receives enough money from similar insureds not having losses to cover the cost of claims and still make a profit.

Some companies simply retain their loss exposures without dealing with them. When companies choose to retain their own exposures, they may ignore them, or attempt to reduce them using one of the methods already mentioned, or they may, in fact, self-insure. Ignoring the risks may make the owners more confident, but dealing with the risks will make them more prepared for loss. Self-insurance is simply no insurance; the company retains the loss exposure. It should only be undertaken by companies with the financial resources necessary to absorb potential losses.

The safety practitioner can use accident causation models to pinpoint hazards in the occupational environment. Systematic, proactive hazard identification will assist the practitioner in establishing loss control strategies and determining the cost-benefits of the controls to be implemented.

The following review briefly examines some of the most popular accident causation theories. The bibliography found at the end of this chapter provides the student with resources to examine this area in greater detail. The intent is to provide examples of accident causation theories representative of past and current thinking. The bottom line benefit is to provide the tools necessary to seek out and eliminate the causes of accidents.

ACCIDENT CAUSATION THEORIES

Single Factor Theory

The *single factor theory* states there is a single and relatively simple cause for all accidents. A good example of this theory would be in determining the cause of worker hand lacerations. Because utility knives are used in the operation, knowing something about the cause of these accidents does not necessarily stop the problem. Other contributing factors, such as the product or the work methods, as well as corresponding corrective actions, are overlooked when a single factor is considered the only cause. This theory is virtually useless for accident and loss prevention.

Domino Theories

There are several domino theories of accident causation. While each domino theory presents a different explanation for the cause of accidents, they all have one thing in common. *All domino theories are divided into three phases:*

1. *Pre-contact phase:* refers to those events or conditions that lead up to the accident
2. *Contact phase:* refers to the phase during which the individual, machinery, or facility comes into contact with the energy forms or forces beyond their physical capability to manage
3. *Post-contact phase:* refers to the results of the accident or energy exposure—physical injury, illness, production downtime, damage to equipment and/or facility, and loss of reputation are just some of the possible results that can occur during the post-contact phase of the domino theory

Domino theories represent accidents as predictable, chronological sequences of events or causal factors. Each causal factor builds on and affects the others. If allowed to exist without any form of intervention, these hazards will interact to produce the accident. In domino games, where the pieces are lined up and the first one is knocked over, the first domino sets into motion a chain reaction of events resulting in the toppling of the remaining dominos. In just that same way, accidents, according to the domino theories, will result if the sequence of pre-contact phase causes is not interrupted.

Heinrich's Domino Theory

H. W. Heinrich developed the original domino theory of accident causation in the late 1920s. Although written decades ago, his work in accident causation is still the basis for several contemporary theories.

According to Heinrich's early theory, the following five factors influence all accidents and are represented by individual dominos:

1. Negative character traits leading a person to behave in an unsafe manner can be inherited or acquired as a result of the social environment
2. Negative character traits are why individuals behave in an unsafe manner and why hazardous conditions exist

3. Unsafe acts committed by individuals and mechanical or physical hazards are the direct causes of accidents
4. Falls and the impact of moving objects typically cause accidents resulting in injury
5. Typical injuries resulting from accidents include lacerations and fractures

The two key points in Heinrich's domino theory are that (1) injuries are caused by the action of preceding factors, and (2) removal of the events leading up to the incident, especially employee unsafe acts or hazardous workplace conditions, prevents accidents and injuries. Heinrich believed that unsafe acts caused more accidents than unsafe conditions. Therefore, his philosophy of accident prevention focused on eliminating unsafe acts and the people-related factors that lead to injuries (Brauer, 1990).

Bird and Loftus's Domino Theory

Bird and Loftus (1976) updated the domino sequence to reflect the management's relationship with the causes and effects of all incidents. ***Bird***

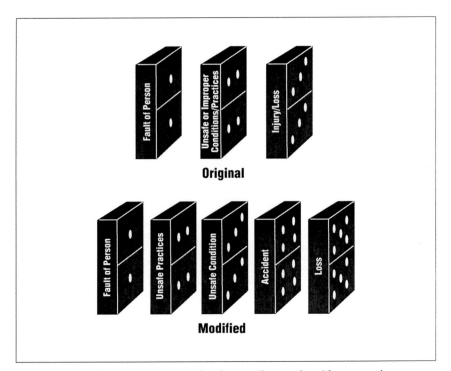

Figure 5-2.   An illustration of Heinrich's domino theory of accident causation

and Loftus's theory uses five dominos that represent the following events involved in all incidents:

1. ***Lack of Control—Management.*** Control in this instance refers to the functions of a manager: planning, organizing, leading, and controlling. Purchasing substandard equipment or tools, not providing adequate training, or failing to install adequate engineering controls are just a few examples represented by this domino.

2. ***Basic Cause(s)—Origin(s).*** The basic causes are frequently classified into two groups: (1) personal factors such as lack of knowledge or skill, improper motivation, and/or physical or mental problems, and (2) job factors including inadequate work standards, inadequate design or maintenance, normal tool or equipment wear and tear, and/or abnormal tool usage such as lifting more weight than the rated capacity of an overhead crane. These basic causes explain why people engage in substandard practices.

3. ***Immediate Causes(s)—Symptoms.*** The primary symptoms of all incidents are unsafe acts and unsafe conditions. "When the basic causes of incidents that could downgrade a business operation exist, they provide the opportunity for the occurrence of substandard practices and conditions (sometimes called errors) that could cause this domino to fall and lead directly to loss" (Bird and Loftus, 1976, p. 44).

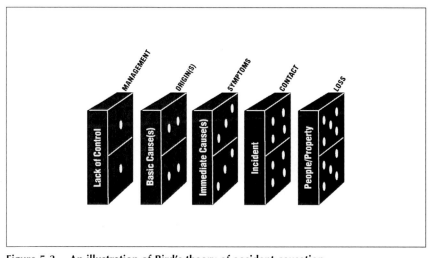

Figure 5-3.   An illustration of Bird's theory of accident causation

4. *Incident—Contact.* "An undesired event that could or does make contact with a source of energy above the threshold limit of body or structure" (Bird and Loftus, 1976). The categories of contact incident events are often represented by the 11 accident types. The 11 accident types include struck-by, struck-against, contact-by, contact-with, caught-in, caught-on, caught-between, foot-level-fall, fall-to-below, overexertion, and exposure (ANSI Z 16.2).

5. *People-Property-Loss. Loss* refers to the adverse results of the accident. It is often evaluated in terms of property damage, as well as the effects upon humans, such as injuries and the working environment. The central point in this theory is that management is responsible for the safety and health of the employees. Like Heinrich's theory, the Bird and Loftus domino theory emphasizes that contact incidents can be avoided if unsafe acts and conditions are prevented. Using the first three dominos to identify conditions permitting incidents to occur, and then ensuring the appropriate management activities are performed, can eliminate accidents and related losses according to this theory.

Marcum's Domino Theory

According to C. E. Marcum's 1978 "Seven Domino Sequence of Misactsidents," a *misactsident* is an identifiable sequence of misacts associated with inadequate task preparation leading to substandard performance and miscompensated risks. The misactsident permits individuals and facilities

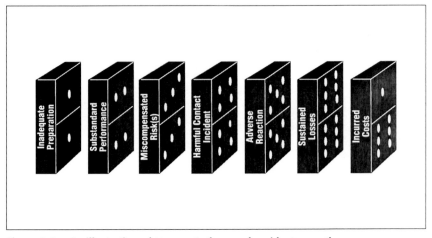

Figure 5-4. An illustration of Marcum's theory of accident causation

to come in contact with harmful agents, energy forms, forces, or substances in ways that initiate adverse reactions sufficiently extensive so that unwarranted losses are sustained and resultant costs incurred.

Like the Bird and Loftus theory of accident causation, Marcum's theory focuses on management responsibility for protecting employee safety as well as preventing the downgrading of an organization. Downgrading of an organization includes incurring losses to equipment and facilities and to intangible assets of the organization such as reputation or corporate goodwill. This theory attempts to examine management accident response protocols to ensure that sustained losses and the subsequent incurred costs were minimized. Marcum keeps these two post-contact-phase components separate to permit closer monitoring of the two variables during accident-analysis activities. Throughout this theory, Marcum focuses on the human element of misacts. This includes misacts of employees who fail to recognize or appreciate risks in the workplace, as well as misacts of organization management who permit risks to go unrecognized, unappreciated, and/or uncorrected.

Marcum uses the term *misactsidents* to emphasize the deterministic aspects of his accident causation theory. Accidents, Marcum believes, are considered by most as events occurring by chance. There is no control over them. They may even be considered acts of God. Marcum emphasizes that accidents have specific causes and can be controlled—that safety can and must be managed.

Multiple Causation Accident Theories

Multiple Factors Theory

Manuele (1997a) believes the domino theories are too simplistic. He proposes the term *unsafe act* also be eliminated. He suggests the chief culprits in accident causation are less than adequate safety policies, standards, and procedures, as well as inadequate implementation accountability systems. Manuele attempts to pull different causation theories together into one working theory. His approach also incorporates some of the following ideas.

Grose's multiple factors theory uses four Ms to represent factors causing an accident: machine, media, man, and management (Brauer, 1990). *Machine* refers to tools, equipment, or vehicles contributing to the cause of an accident. *Media* includes the environmental conditions surrounding an accident, such as the weather conditions or walking surfaces. *Man* deals with the people and human factors contributing to the incident. *Management* also incorporates the other three Ms, looking at the methods used to select equipment, train personnel, or ensure a relatively hazard-free environment.

The multiple factors theory examines characteristics of each of the four Ms:

1. *Machinery:* Examination of machinery characteristics includes the design, shape, size, or specific type of energy used to operate the equipment
2. *Man:* Characteristics of man are psychological state; gender; age; physiological variables (including height, weight, or condition); and cognitive attributes (such as memory, recall, or knowledge level)
3. *Media:* Snow or water on a roadway, temperature of a building, and outdoor temperature can be characteristics of media
4. *Management:* Characteristics of management could include safety rules, organizational structure, or policy and procedures

Multiple factors theories attempt to identify specific workplace characteristics that reveal underlying, and often hidden causes of an accident by pointing to existing hazardous conditions. When viewed as a whole, the characteristics can direct the investigator's attention to the specific causes of an accident.

Systems Theory of Causation

One variation of the multiple causation theory is R. J. Firenzie's theory of accident causation. Firenzie's theory is based on interaction among three components: *person*, *machine*, and *environment*. Human variables of information, decisions, and perception of risks combine with machine hazards and environmental factors affecting the likelihood of an accident.

For example, as a person operates a noisy bulldozer on a hot day, other activities must take place for the operator to safely and effectively perform the job. The person consciously or subconsciously will collect information, weigh risks, and make decisions as to how to perform the task. How close should the bulldozer get to the 20-foot-high spoil bank or the electrical power lines? How fast should it be moving? The operator, based on knowledge and experience, makes countless decisions—all of which affect the probability of an accident.

Psychological/Behavioral Accident Causation Theories

Goals Freedom Alertness Theory

Dr. Willard Kerr's theory of accident causation regards an accident as a low-quality work behavior. He considers it similar to production waste during manufacturing, except that the scrap happens to be human. Raising the level of quality and safety involves raising the level of worker awareness. According to Kerr, alertness can only be obtained within a positive organizational culture and psychological climate. The more positive the workplace climate, the greater the alertness and work quality is. As alertness decreases, the probability of an accident rises.

Motivation Reward Satisfaction Model

This theory of accident causation builds on Dr. Willard Kerr's goals freedom alertness theory and Herzberg's hygienic management theory. Stated simply, the "freedom to set reasonably attainable goals is typically accompanied by higher-quality work performance" (Heinrich, Petersen, and Roos, 1980, p. 44). If an accident occurs, it is due to a lull in alertness. Safety performance depends on degree of motivation and capability to work; factors affecting these variables will either promote or prevent accidents.

According to Petersen, rewards strongly affect performance. They originate from a variety of sources and can be physical and/or psychological. Money or praise is not considered to be the primary motivation factor. Rewards, including doing a good job, learning new skills, expanding personal knowledge, and participating on a successful team, are some of the numerous intrinsic reinforcements associated with enriched jobs. If employees see the rewards from their work as equitable, they are more likely to be motivated and, in turn, produce positive safety results.

Human Factors Theory

The *human factors theory* is based on the concept that accidents are the result of human error. ***Factors that cause human error are:***

- ***Overload:*** *Overload* occurs when a person is burdened with excessive tasks or responsibilities. For example, the employee not only must perform his or her job but also must handle excessive noise, stress, personal problems, and unclear instructions.

- *Inappropriate activities:* "Inappropriate activities" is another term for human error. When individuals undertake a task without proper training, they are acting inappropriately.
- *Inappropriate response:* "Inappropriate response" occurs, for example, when an employee detects a hazardous condition but does not correct it, or removes a safeguard from a machine to increase productivity.

Overload, inappropriate activities, and inappropriate responses are all human factors causing human error and, ultimately, accidents.

Energy-Related Accident Causation Theories

Energy Release Theory

Accidents result when energy, out of control, puts more stress on a person or property than tolerable without damage. Controlling the energy involved or changing the structures that energy could damage can prevent accidents. William Johnson expanded Haddon's ten strategies to twelve when they are applied to the accident investigation strategy referred to as Management Oversight and Risk Tree (MORT). *Johnson's barriers to accident-causing energy include:*

- *Limit the Energy.* Example: Limit the amount of flammable or combustible materials that are stored on the shop floor. Use low-voltage equipment.
- *Substitute a safer energy form.* Example: Use non-asbestos brake pads or select nonflammable or nontoxic solvents.
- *Prevent the build-up.* Example: Utilize fuses, circuit breakers, and gas detectors.
- *Prevent the release of the energy.* Example: Attach toe boards on scaffolds to prevent tools from striking people or objects below.
- *Provide for slow release.* Example: Utilize safety-release valves.
- *Channel the release away; separate it in time or space.* Example: Ground electrical appliances.
- *Place a barrier on the energy source.* Example: Place machine guards or utilize acoustic enclosures.
- *Place a barrier between the energy source and the persons or objects to be protected.* Example: Use rails on elevated surfaces; use fire doors.

- *Place a barrier on the persons or objects to be protected.* Example: Require personal protective equipment (PPE) and respirators.
- *Raise the injury or damage threshold.* Example: Acclimatize to a hot or cold work environment.
- *Ameliorate the effects.* Example: Incorporate administrative controls such as job rotation to reduce the duration of exposure to loud noise.
- *Rehabilitate.* Example: Treat injured employees or repair damaged objects.

Haddon and Johnson focus on energy as the source of the hazard. Identifying energy sources and preventing or minimizing exposures can prevent accidents.

Swiss Cheese

More recently James Reason, a professor emeritus from Manchester University in the United Kingdom, contributed his theories in which he explains (2008) that a widespread myth is that errors occur "out of the blue" and are highly variable in their form. Errors are not random and they tend to be recurrent and predictable. Most organizations build a safety system with productive "planes" later referred to as "Swiss cheese slices" that provide defenses or safeguards separating potential losses from local hazards. Each layer is an additional defense, that as long as it is not penetrated, no accident will occur. These defenses include such components as reliable equipment, skilled operators, effective training, and effective management. Holes may exist or develop over time in those planes or slices. The slices are constantly rotating or moving and the holes or gaps are constantly opening and shutting. According to Reason, when a series of holes "line up," an accident trajectory may pass through them, thus causing harm to people, assets, and the environment. For example, a driver may back his vehicle out of a parking space by checking the rear-view mirrors but not turning around. Checking the rear-view mirrors provides a safeguard between the moving vehicle and an object in its path. It is usually effective because there is rarely anyone standing in the blind spot still not seen by the driver. Even if someone was standing there, they would likely hear the car, see the back-up lights or movement, and move themselves. Repeating this routine dozens or even hundreds of times has never led to an accident. One day the backing occurs near a busy intersection where lots of noise drowns out the sound of this vehicle. Movement all around prevents a person, standing exactly in the blind spot, from noticing the back-up lights and movement of the vehicle. The circumstances are such that the holes in the layers of cheese or defenses exactly line up to create a set of circumstances in which a pedestrian is injured or killed due to the ineffective backing practices of the driver.

All of the above theories provide explanations of why incidents and accidents occur. As much as management tries to avoid them, they still occur because there is always some residual risk. Living and working involve risk. Enough exposure to enough risk yields incidents and accidents. The job of the safety professional is to identify the risk, advise management to reduce or eliminate as much as is feasible, and continue to monitor to help ensure that the assets of the company, especially its employees, are adequately protected from those hazards.

INCIDENT INVESTIGATION

Incidents include near misses and accidents. *Accidents* result in loss. *Near misses* result in no loss. The purpose of any incident investigation is to determine cause(s). As already mentioned, the circumstances differentiating a near miss from a loss-producing incident or accident may be merely a matter of chance; therefore, all incidents can be considered candidates for investigation, since correcting the contributing circumstances may help prevent future incidents.

Incident investigation is concerned with fact-finding, not fault-finding. During the accident investigation, it is important to find out answers to the questions *who*, *what*, *where*, *when*, *why*, and *how*.

Who questions include:

- Who are the victims?
- Who are the witnesses?
- Who has any information that will help determine the actual causes of the incident?

What questions include:

- What events led up to the accident?
- What were victims and witnesses doing prior to and during the incident?

- What did individuals notice that may have a bearing on the incident?
- What were the backgrounds and experiences of all the parties involved?

When questions include:

- When did the incident occur?
- When did you notice important elements associated with the incident? When did you become concerned that a problem existed?

Where questions include:

- Where were victims and witnesses prior to and during the incident?
- Where were equipment and/or machinery?
- Where were the PPE or the locks and tags for energy sources?

Why questions include:

- Why, in your opinion, did the incident take place?
- Why were particular methods used to perform a task?
- Why were these conditions existing at the time of the incident?

How questions include:

- How did the incident take place?
- How did the victims and witnesses react in given situations?
- How did you first learn of the incident?

When an investigation team arrives at the scene, key elements of personnel, tools/equipment, raw materials or finished product, and structure and environment must be searched for possible clues. Systematically consider each as a source of clues and pull those clues together to find likely causes.

Preplanning and preparation are of vital importance. Know roles and responsibilities of each investigator and prepare for prompt arrival at the accident scene so evidence does not disappear and witness recollections are not lost. Ensure all tools and equipment necessary to conduct the investigation are organized and available at all times. Utilize a checklist to confirm all appropriate individuals are contacted—from the plant manager or corporate CEO to OSHA and EPA officials. Periodic mock drills ensure effective responses to emergencies by highlighting weaknesses. An enterprise was checking its building evacuation plan for the first time and ran a mock drill. When the occupants filled the stairwells, the weight forced the door frames at the base of the stairs onto the doors so they would not open. Had this happened in an actual emergency, all occupants would have been trapped inside the building. The drill enabled members of the management team to identify a risk that would have otherwise remained undetected.

Drills also aid in effective decision making. Effective decisions are more likely to be made during practice than in the heat and scrutiny of a critical investigation. Publicity, time constraints, and political pressures inhibit effective decision making. These factors are less intense and less likely to exist during practice. Standard operating procedures should be developed during mock investigations. The following is a brief protocol of some of the steps that can be followed at a facility to prepare for accident investigations.

When an accident occurs:

1. Notify the following individuals immediately (list important individuals here). Be certain next of kin are told before media reports occur.
2. Secure the scene to prevent additional accidents; bring in accident-trained technical personnel to determine:

   - Damage to and hazard from power, gas, and fluids distribution systems
   - The structural integrity of the building and equipment
   - The best way to remove or make harmless explosives and/or hazardous materials

Caution: Make sure the positions of switches, equipment, and materials are recorded before they are moved or removed.

3. Evaluate the condition of any injured personnel; determine:

   - What is the degree of injury?
   - What must be done immediately to save life?
   - What should be done to relieve suffering until the injured individual can be removed to a medical facility?
   - What can be done to remove all danger of increased injury?

Caution: Move an injured person only if there is danger of further injury.

4. Identify the elements at the accident scene:

- People involved
  - injured
  - principals
  - witnesses
- Equipment involved
  - in use
  - standby
  - secured or standing
  - materials involved
  - in use
  - ready for use
  - stored in area
- Environmental factors
  - weather
  - lighting
  - heat
  - noise
  - any additional contributing factors
- Keep detailed notes for reference

5. Secure the accident scene:

- Barricade the area to prevent removal or defacement of possible evidence
- Isolate potential witnesses

6. Collect and preserve the evidence:

- Make drawings of the area
- Pick up, store, and label evidence
- If evidence cannot be removed from the scene or it is too large to bag, take pictures or make drawings, making sure to note the location on the drawing of the area
- Make notes of any observations regarding the accident scene

7. Develop witness questions:

- Form open-ended questions based on initial observations and evidence collected

(continued)

- Include control questions to ensure the accuracy of the statistical data and to permit later evaluation of witness reliability

8. Interview the witnesses:

   - Interview each witness separately
   - Find a suitable location
   - Be prepared to take notes on and/or record the interview
   - Take short notes as a memory device
   - If a recording device is used, ask for permission to record before the interview
   - Watch for witness cues during the interview
   - Take into account personality types:
     - introvert
     - extrovert
     - suspicious
     - prejudiced
     - other personality traits
   - Notice any nonverbal messages:
     - body language
     - voice changes
   - Establish initial communication
   - Assure the individual that the purpose of the interview is for accident prevention, not assigning blame
   - Note whether the witness was:
     - the injured party
     - an eyewitness
     - an "ear witness"
   - Take an initial statement
   - Have the witness describe the incident in his or her own terms
   - Avoid interrupting the witness during the statement
   - Expand the interview for detail
   - Ask the developed open-ended questions
   - Space the control questions throughout this portion of the interview
   - Close the interview
   - Ask the witness for his suggestions on how the accident could have been prevented

- Thank the witness for his or her time
- Evaluate the witness statements
- Develop a witness analysis matrix based on the control questions
- Place the control question numbers on the horizontal axis
- Put the witness statements on the vertical axis
- Place an X in those columns where the witness has accurately answered the control question
- Check credibility of individual testimony based on control questions
- High witness credibility is based on the number of accurate responses to the control questions

Caution: Just because a witness gives inaccurate answers to the control questions, it does not totally invalidate the testimony.

9. If necessary, conduct follow-up witness interviews:

- Let the witness know there is a gap or deficiency in a critical area of the investigation.
- Allow the individual to reconsider or reexamine his or her observations

Caution: Be sure to remain nonjudgmental.

10. Synthesize the information gathered from the witnesses interviewed and evidence collected to determine the accident cause
11. Use the data gathered from the accident to perform an accident trend analysis
12. Make changes to operating procedures, equipment, and/or training based on the accident trend analysis

Caution: Some interviewers prefer to use audio and/or video equipment to record witness interviews. Interviews may include witness speculations, hearsay, and unfounded conclusions on the part of biased or uninformed individuals. Recordings of those interviews may later be subpoenaed during litigation or criminal hearings. When only notes are taken, speculations can be omitted and signatures can be affixed to statements limited to facts.

CONCLUSION

Several accident causation models have been presented in this review. The theories focus on people variables, management aspects, and physical characteristics of hazards. The benefit of understanding accident causation is in recognizing how hazards in the workplace result in losses. Eliminating hazards before they result in losses is the proactive responsibility of everyone in an organization. Professionals recognize it is not always possible to identify and eliminate all hazards. Accidents may still occur in spite of a proactive safety program. It is at this point that an effective accident investigation program is of vital importance for the collection of critical data.

QUESTIONS

1. What are some of the benefits associated with understanding accident causation theory? Explain your answer.
2. What are the advantages of considering accidents as management problems?
3. What are the two factors associated with risk? Explain how these two factors impact on the selection of accident controls.
4. Is it important to conduct accident investigations? Why?
5. What are the six key questions that should be asked during accident investigations? Explain the importance of these six questions to the health and safety professional.

REFERENCES

Bird, F. E., and Loftus, R. G. 1976. *Loss Control Management*. Loganville, GA: Institute Press.

Brauer, R. L. 1990. *Safety and Health for Engineers*. New York: Van Nostrand Reinhold.

Heinrich, H. W., Petersen, D., and Roos, N. 1980. *Industrial Accident Prevention*. New York: McGraw-Hill.

Manuele, F. A. 1997a. *On the Practice of Safety*. New York: John Wiley & Sons.

Reason, James. 2008. *The Human Contribution*. Burlington, VT: Ashgate.

Roland, H. E., and Moriarty, B. 1990. *System Safety Engineering and Management*, 2nd ed. New York: John Wiley & Sons.

BIBLIOGRAPHY

Bird, F. E., Jr., and Germain, G. L. 1992. *Practical Loss Control Leadership.* Loganville, GA: International Loss Control Institute.

Ferry, T. S. 1984. *Safety Program Administration for Engineers and Managers.* Springfield, IL: Charles C. Thomas.

Ferry, T. S. 1988. *Modern Accident Investigation and Analysis*, 2nd ed. New York: John Wiley & Sons.

Goetsch, D. L. 1993. *Industrial Safety and Health: In the Age of High Technology.* New York: Maxwell Macmillan International.

Grose, V. L. 1992, August. System Safety in Rapid Rail Transit. *ASSE Journal, 22,* 18–26.

Hale, A. R., and Glendon, A. I. 1987. *Individual Behavior in the Control of Danger.* New York: Elsevier Science Publishers B.V.

Kuhlmann, R. L. 1977. *Professional Accident Investigation: Loganville Investigative Methods and Techniques.* Loganville, GA: Institute Press.

Marcum, C. E. 1978. *Modern Safety Management Practice.* Morgantown, WV: Worldwide Safety Institute.

Manuele, F. A. 1997b. A Causation Model for Hazardous Incidents. *Occupational Hazards,* 160–65.

Rasmussen, J., Duncan, K., and Leplat, J. (Eds.). 1987. *New Technology and Human Error.* New York: John Wiley & Sons.

6

Introduction to Industrial Hygiene

Burton R. Ogle, PhD, CIH, CSP, and
Tracy L. Zontek, PhD, CIH, CSP

Western Carolina University

CHAPTER OBJECTIVES

After completing this chapter, students will be able to

- Define terms and acronyms associated with industrial hygiene
- Identify the classes of occupational hazards
- Describe basic toxic effects and routes of entry
- Identify various agencies and organizations providing support to industrial hygiene
- Differentiate between occupational and paraoccupational exposures
- Characterize the roles and responsibilities of the industrial hygienist
- Identify the various disciplines that form the foundations of industrial hygiene
- Understand the primary approaches to protecting workers in their work environment
- Describe the primary controls employed by industrial hygienists to control workplace exposures
- Explain the various types of sampling that may be conducted by the industrial hygienist
- Understand the concepts of equipment calibration and accurate laboratory analysis
- Identify how industrial hygiene fits into an overall occupational safety and health program

CASE STUDY

A manufacturing firm recently began a new process of cleaning using xylene. Xylene was chosen because it was less hazardous than the previous solvent (a potential human carcinogen). Employees have been complaining that the xylene had an annoying, sweet odor and has caused dry skin on their hands, eye and throat irritation, and headaches. Xylene is used intermittently to clean out machines and tools, approximately one-half gallon per cleaning. Operators manually spray the xylene, and then wipe down equipment and tools. They always wear safety glasses and occasionally leather gloves when handling sharp objects, along with hearing protection if they will be in the area for more than a few minutes, since the background noise level is typically 85–90 decibels. The industrial hygienist (IH) from their insurance carrier has been asked to evaluate this use of xylene.

Upon arrival, the IH reviews the Safety Data Sheet (SDS) for xylene with the supervisor and employees who are using it. During review, the employees are surprised to find that xylene is very flammable, and decide that spraying the xylene should be discontinued. In addition, they need to ensure that there is no ignition or static discharge source present when the xylene is in use. To determine how much xylene employees are exposed to, the IH conducts air monitoring in the employee breathing zone while they are using the xylene. During the sampling, the IH notes that the odor threshold is about 1 ppm. This is much lower than the regulatory limits; however, the odor threshold should never be used to determine if chemical concentrations are safe. The only way to determine the actual concentration is to take air samples; personal samples for employees are taken both in the short term (15 minutes) using a direct reading instrument and over the entire day (8 hours) using a charcoal tube connected to a sampling pump according to the National Institute for Occupational Safety and Health (NIOSH) Manual of Analytical Methods. The short-term sample is above the ACGIH Short Term Exposure Limit (STEL); the long-term sample is below both the ACGIH TLV-TWA and OSHA PEL limits.

The IH also discusses other hazards of xylene use with management and employees. Xylene can be absorbed through intact skin; therefore the air monitoring results represent only one route of exposure. The actual employee exposure to xylene was likely much higher, since employees did not wear the appropriate chemically resistant gloves and absorbed some xylene through their skin. Xylene has also been shown to act synergistically with noise exposure, causing more hearing loss than if the employee was exposed to noise alone.

Based on these observations, the IH leaves the company with the following recommendations:

1. Find a less hazardous cleaning solvent for this operation to substitute for the xylene.
2. If a less hazardous solvent cannot be found, use the hierarchy of controls to limit employee exposure.

   a. Can the operation be automated or enclosed to reduce employee exposure?
   b. Can local exhaust ventilation be used to remove the chemical and prevent employee exposure?
   c. Can a smaller amount of the chemical be used to clean? Can the xylene just be used for a final cleaning?

3. Until a less hazardous solvent can be found, employees must:

   a. Wear goggles, and a chemically resistant apron and gloves to prevent skin exposure and absorption of xylene. Leather gloves should never be worn when working with solvents.
   b. Ensure that all ignition and static discharge sources are controlled.
   c. Utilize local exhaust ventilation on equipment to remove xylene vapors whenever possible.
   d. If exhaust ventilation cannot be used, employer must institute a respiratory protection program. Since the ACGIH STEL has been exceeded, respiratory protection must be worn during the cleaning operation if a less hazardous product or local exhaust ventilation is not used.

4. Employees using xylene must wear hearing protection due to the synergistic effect of noise and solvent.

WHAT IS INDUSTRIAL HYGIENE?

Industrial hygiene has most often been described as the art and science devoted to the *recognition, evaluation,* and *control* of workplace health hazards. In addition to these three tenets, it has been suggested that we add

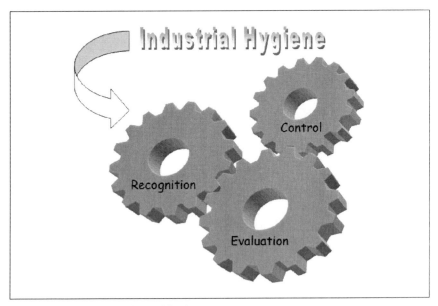

Figure 6-1. Tenets of industrial hygiene

a fourth, *anticipation*. We must be careful here to avoid "expecting the expected." Emphasis on the anticipation of workplace hazards may result in overlooking the *unexpected* health hazard. A well-trained, well-prepared industrial hygienist (IH) is equipped to deal with virtually any situation that arises. We will take a closer look at recognition, evaluation, and control later in this chapter.

What Do We Mean by the "Art" of Industrial Hygiene?

"Art" implies that industrial hygiene cannot be practiced by simply following a recipe. With each unique workplace setting and each new day, an IH must reinvent himself or herself. An IH does not simply report the measurements and levels of workplace hazards, but also interprets those results. Often, industrial hygiene measurements lie within "gray" areas; that is, they are neither safe, nor unsafe. An IH must make professional recommendations based upon a preponderance of the quantitative data gathered as well as intangible threads of information gathered by observation. Experience, preparation, and study of the associated sciences will help enable the IH to become a competent practitioner of the craft.

What About the Science of Industrial Hygiene?

Industrial hygiene is steeped in a variety of sciences, including engineering, epidemiology (the study of the distribution of disease within a population), physics, statistics, biology, microbiology, chemistry, anatomy, physiology, and toxicology. Knowledge of the applied sciences, such as public and environmental health, computer applications, geographic information systems, and industrial psychology, is essential to the IH who wishes to practice comprehensive industrial hygiene.

What Are the Health Hazards That the IH Is Charged with Recognizing, Evaluating, and Controlling?

In theory, the IH is responsible for **chemical**, **biological**, **physical**, and **ergonomic** health hazards within the work environment. In practice, no one individual can be an expert in all of these areas. Control of workplace health hazards has become so complex that in workplaces where most or all of these hazards are present, the management of these hazards is assigned to a team of occupational health professionals. Such a team might include an occupational physician, occupational nurse, industrial hygienist, safety professional, health physicist (radiation protection), and others. Workplaces with high-risk microbiological exposures may also employ biological safety professionals.

Industrial hygiene requires the cunning of a detective, the discipline of a scientist, and the persuasion of a TV evangelist. The IH must uncover the secrets of the often invisible hazardous agents. What are they? Where do they originate? How much is present? What harm might they cause? How can I control or eliminate them? Industrial hygiene investigations must be performed with scientific method. Procedures must be reproducible, accurate, and precise. And in order to be effective, an IH must *sell* the recommendations. The IH must convince the worker that recommended work practices and controls are absolutely necessary to a safe and healthful workplace. The IH must convince the administration that hazard control expenses are worth every dime.

HISTORY OF INDUSTRIAL HYGIENE

While industrial hygiene is a relatively young profession, the historical references to occupational medicine and toxicology date back to early civilization.

Diseases resulting from exposure to hazardous substances have existed as long as people have interacted with their environment. Below is a brief introduction to a select group of distinguished historical characters who made an immeasurable contribution to industrial hygiene.

Hippocrates (470–410 B.C.), a Greek physician, is best known as the *father of medicine*. He believed in the natural healing process of rest, a good diet, fresh air, and cleanliness. Hippocrates wrote of the plight of miners exposed to lead and other work-related contaminants. Hippocrates developed a treatise of medical ideology that modern physicians acknowledge prior to beginning their practice. This is better known as the *Hippocratic Oath*.

Pliny the Elder (23 A.D.–79 A.D.) was a Roman senator, writer, and scientist. In his writings, he described the dangers to workers exposed to zinc and sulfur. He is believed to be the first to recommend *respiratory protection* when he suggested that miners cover their mouth and nose with an animal bladder when working in the mines. Pliny, a scientist to the end, lost his life trying to investigate the eruption of Mount Vesuvius.

Georgius Agricola (1494–1555), a.k.a. Georg Bauer, a Saxon physician, is best known as the *father of geology*. Agricola's most famous work— *De Re Metallica* ("On the Nature of Metals")—describes illnesses experienced by miners, the need for ventilation (an advanced industrial hygiene control measure), and ergonomic issues related to mining. This book was so respected that it remained the standard text on mining for two centuries. President Herbert Hoover (formally educated as a geological engineer) and his wife, L. H. Hoover, translated many of Agricola's works into English.

Paracelsus (1493–1591), a Swiss physician and chemist, is best known as the *Father of Toxicology*. Along with many of his predecessors, he wrote of the toxicity related to mining. In his work *On the Miners' Sickness and Other Diseases of Miners*, he established the important concepts of *acute* and *chronic* toxicity. Unique to Paracelsus was his claim that all substances were poisons—the dose differentiates a poison from a remedy. This concept is the basis for the *dose-response* relationship.

Bernardino Ramazzini (1633–1714), an Italian physician, is best known as the *father of occupational medicine*. His most famous work, *De Morbis Artificum* ("Diseases of Workers"), urged all physicians to ask their patients, "Of what trade are you?" This simple question became a powerful tool to help establish the previously often overlooked relationship between occupation and illness.

Percival Pott (1713–1788), a London surgeon, is best known for being the first to establish the relationship between an *occupation* (chimney sweep), a *toxin* (polyaromatic hydrocarbons [PAH] from soot), and *malignancy* (testicular cancer). Pott was a brilliant physician whose discoveries and theories are in large part still accepted and utilized today.

Alice Hamilton (1869–1970). This petite American physician was a true hero in the field of industrial medicine. She was the first woman to be named to the Harvard Medical School staff; her writings included *Industrial Poisons in the United States* (1925), *Industrial Toxicology* (1934), and *Exploring the Dangerous Trades* (1943). Dr. Hamilton was a pioneer who performed much of her own field research. When she sought to better understand the hazards associated with lead, she introduced herself to leadworkers, went to their modest homes, interviewed their family members, and observed them firsthand performing their dirty, dangerous jobs.

TOXICOLOGY

Toxicology is known simply as the *science of poisons*. We know from Paracelsus that all substances are toxic—it is the *dose* that makes the poison.

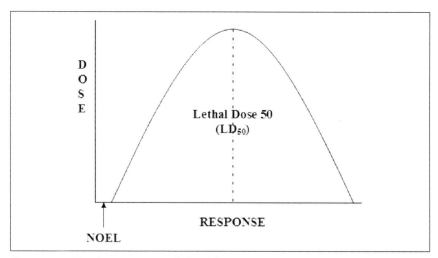

Figure 6-2. The dose-response relationship

Further, we know that for all noncarcinogenic substances there is some level below which there is no adverse effect. This level is known as the **threshold level** or no observed adverse effect level (NOAEL or NOEL). As we increase that level or dose, within any population there will be some measurable response to that dose. We refer to this relationship as **dose-response**. Within any given population there will be **sensitive** and **tolerant** individuals. Those sensitive individuals will typically respond to much lower doses than the majority of the population, while the tolerant individuals must receive a greater dose than the majority of the population in order to elicit a response. If the response being studied is mortality, then the dose that elicits death in exactly one-half of the studied population is called the **lethal dose-50 (LD_{50})**.

Routes of Entry

In order for any substance to cause injury to an individual, it must first be taken up by the body. Knowing the way in which substances are taken up is critical in order to understand and prevent toxicity. The body may be exposed to toxic substances through any one or a combination of four routes:

- **ingestion**—taken into the body orally
- **inhalation**—taken into the body through the lungs
- **absorption**—taken into the body through dermal absorption
- **injection**—taken into the body through broken skin

Occupationally, inhalation is the most common route of entry by toxic substances. It is fairly common for substances to have more than one route of entry, and some may cause injury through all four routes.

Acute and Chronic Exposures

When an individual receives a relatively large dose over a single (or brief) time period, we refer to this type of exposure as **acute**. An example of an acute exposure is when a child accidentally ingests a household chemical. When an individual receives a relatively low dose over a long time period (perhaps over a lifetime), we call this type of exposure **chronic**. An example of a chronic exposure is asbestos exposure—most harm from asbestos fibers takes decades to manifest. Seldom do chronic and acute exposures to the same toxin have much in common. Often the symptoms vary and the health

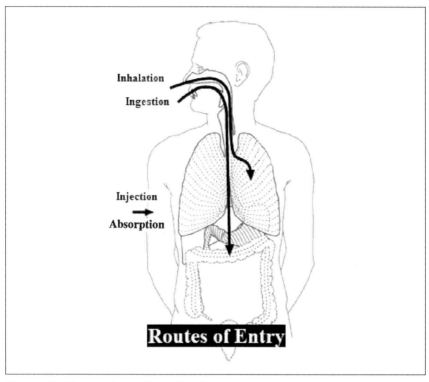

Figure 6-3. Routes of entry for toxic substances

effects are quite dissimilar. For example, passive cigarette smoke may cause an individual to cough and experience symptoms of eye and upper respiratory irritation. Over a lifetime, a person regularly exposed to secondhand cigarette smoke may experience cardiovascular disease, respiratory disease, and even cancer. Often we are tempted to examine acute symptoms to predict chronic exposures; however, we see from this example that the acute symptoms from passive cigarette smoke provide us with no information that would help predict chronic outcomes.

No matter which mode of entry a toxin takes, the health effects can be either *local* or *systemic*, depending upon the physical and chemical characteristics of the substance. **Local effects** are the tissue reactions of the body areas that come in direct contact with the contaminant. **Systemic effects**, on the other hand, occur in the target sites of the body, which are not the initial contact locations, but which, because of their affinity for that particular substance, readily absorb it. This reaction may occur far away from the point of initial contact. If an employee were working with sulfuric acid and inhaled sufficient quantities of it, irritation of the upper respiratory system would be likely (irritation of the nose, bronchi, trachea, etc.). Inhalation of heavy metals, such as lead, would have systemic effects (lead does not have a substantial impact upon the respiratory system, which in this case is the site of contact). However, the primary dangers from lead are to the central nervous system, the blood, and the kidneys.

Chemical Interactions

When two or more substances are taken into the body, there are three possible interactions that may take place:

- Antagonism
- Additive
- Synergism

From toxicological studies, we know and can predict the amount of injury caused by many individual substances. When we combine these substances, and the effects are as predicted, we call this *additive*. We express this as $1 + 1 = 2$. When we combine two or more substances and a lower overall effect is observed—that is, their combined effect is less than the sum of their individual effects—we call this *antagonism*. This is expressed as $1 + 1 < 2$. When we combine two or more substances and the overall

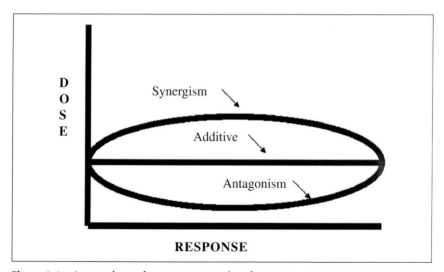

Figure 6-4.   Interactions of two or more toxic substances

effect is greater than the predicted effects—that is, their combined effect is greater than the sum of their individual effects—we call this *synergism*. This may be expressed as $1 + 1 > 2$.

Classification of Toxic Materials

There are numerous ways to classify toxic materials. In general, toxic materials are categorized according to their physical or chemical characteristics or the physiological organs that they target. The following groups of toxic materials are representative of the common methods used to classify toxic substances.

Physical Classification: This method of classification attempts to examine toxic agents according to the form in which they exist in the occupational environment. These classifications include solids, liquids, gases, and vapors.

Chemical Classification: This method of classification uses the chemical structure, nature, and composition that a substance possesses. Examples of chemical classifications include aliphatic compounds, aromatics, acids, alcohols, ketones, esters, and ethers.

Physiological Classification: This extensive method of classification uses potential human injury as a means of categorizing certain agents. Below are several physiological classifications.

Irritants: Chemicals that are not corrosive, but that cause a reversible inflammatory effect on living tissue by chemical action at the site of contact.

Asphyxiants: Substances that prevent the body from taking in or utilizing oxygen, resulting in deprivation and suffocation. *Simple asphyxiants* are those agents that displace oxygen in an atmosphere, resulting in what is called an "oxygen-deficient" atmosphere. We typically consider an oxygen-deficient atmosphere to be one containing less than 19.5 percent oxygen by volume (our normal ambient atmosphere contains about 21 percent oxygen). Methane, carbon dioxide, argon, nitrogen, and hydrogen are examples of gases that can displace oxygen and reduce its concentration below the level necessary to support life. *Chemical asphyxiants* are substances that prevent the uptake, use, and/or transportation of oxygen by the body. Gases such as carbon monoxide, hydrogen cyanide, and hydrogen sulfide are examples of chemical asphyxiants.

Nephrotoxins: Agents that produce kidney damage. Examples of these agents include heavy metals (such as cadmium, chromium, lead, and mercury), halogenated hydrocarbons (such as bromobenzene and carbon tetrachloride), antineoplastic drugs such as cisplatin, and others.

Neurotoxins: Agents that produce damage to the central and/or peripheral nervous system. Examples of neurotoxins include carbon monoxide, lead, methyl mercury, cyanide, alcohols, and others.

Hepatotoxins: Agents that produce liver damage, such as liver necrosis, fatty liver, cirrhosis, and carcinogenesis. Examples of hepatotoxins include carbon tetrachloride, aflatoxins, phosphorus, ethanol, bromobenzene, and nitrosamines.

Respiratory Toxins: Agents that irritate or damage the pulmonary system. These agents may cause coughing, chest tightness, and shortness of breath or no immediate (acute) symptoms at all. Examples of respiratory agents include asbestos, silica, sulfur dioxide, ozone, oxides of nitrogen, chlorine, phosgene, and many others.

Reproductive Toxins: Agents that affect the reproductive capabilities, causing, for example, chromosomal damage (mutations) and effects on fetuses (teratogenesis). Reproductive toxins may affect both females and males. Examples of these agents include lead and DBCP, which can lead to birth defects or sterility.

Hematopoietic Agents: Chemical agents that act on the blood to decrease the hemoglobin function or to deprive the body tissues of oxygen. Symptoms affecting the body include cyanosis and the loss of consciousness. Chemicals that cause effects such as these include carbon monoxide and cyanides.

Cutaneous Hazards: Substances that cause skin injuries, such as defatting, rashes, or irritations. Examples of cutaneous hazards include ketones and chlorinated compounds.

Eye Hazards: Substances that injure the eye or otherwise reduce visual capacity. Examples include organic solvents, alkalis, acids, infrared and ultraviolet radiation, physical hazards, and others.

Carcinogens: Substances that result in uncontrolled cell growth. Examples of carcinogens are cigarette smoke, benzene, carbon tetrachloride, ultraviolet radiation, asbestos, and others.

Mutagens: Substances causing cell injury that results in chromosomal damage. Examples of mutagens are X rays and benzo(α)pyrene.

Teratogens: Substances that lead to a birth defect. Examples of teratogens are cigarette smoke, alcohol, thalidomide, and others.

Toxicity versus Risk

The toxicity of a chemical is not the only factor to consider when evaluating risk. Toxicity is the *ability* of a substance to cause harm or adversely affect an organism. Hazard is simply the *probability* that harm will occur. Risk can be expressed as toxicity times hazard—how toxic is the substance and the probability this type of exposure will occur.

When we fail to examine the actual risk of a substance or activity, humans tend to let emotion take over and cloud their perception of risk. For example, most of us believe that performing in a rodeo or hang gliding is a *risky* activity. Nevertheless, the actual risk of serious injury from these activities is much lower than it is for motorcycle riding or cigarette smoking. When a substance has been labeled by the scientific community as hazardous or carcinogenic, such as saccharin, the public perception is that contact with these substances should be eliminated. We must remember—all substances are toxic; it is the dose that makes the poison. Industrial hygienists minimize risk by choosing less toxic substances and decreasing the probability that employees will come into contact with them.

Precautionary Principle

When a toxin is perceived to be so dangerous that it raises significant threats of harm to the environment or human health, precautionary measures should be taken even if some cause-and-effect relationships are not fully established scientifically—this is the consensus definition of the *precautionary principle.* This is perhaps the only situation where an IH will look to safety first and wait for the science and toxicology to emerge. An application of the precautionary principle is with nanotechnology. Nanotechnology deals with particles that are less than one ten thousandth of the size of the human hair. The existing health and safety information on nanoparticles is limited. Prudent practices with nanotechnology assume that vigorous protective measures are necessary for safe handling. Science

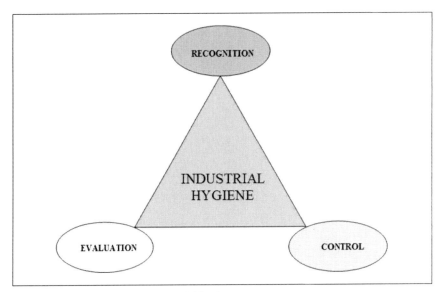

Figure 6-5. The practice of industrial hygiene

may later tell us that our current practices were overkill, but we would rather accept the trouble and expense of excessive protection than find out that our precautions were too meager.

INDUSTRIAL HYGIENE PRACTICE

As we discussed earlier, the tenets of industrial hygiene are the *recognition, evaluation,* and *control* of workplace health hazards. The goal of the IH is to provide a safe and healthful work environment for all workers.

Recognition

The initial tenet, *recognition,* requires the IH to perform what is known as a **"walk-through"** survey or inspection. If an IH is unfamiliar with the workplace or processes, an initial literature review, Internet search of credible sources, and/or review of OSHA 300 logs for the industry may provide insight into the types of hazards commonly found. A walk through survey includes but is not limited to the following: visual reconnaissance of all areas of the workplace; taking pictures; preparing drawings; interviewing employees; review of company safety policies, previous inspection reports, injury

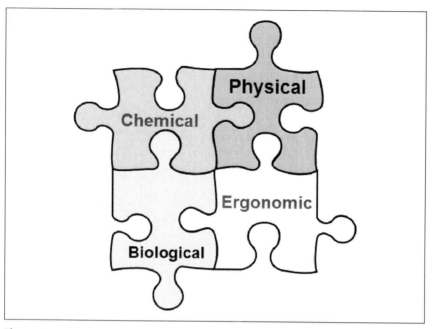

Figure 6-6. Four categories of occupational health hazards

logs, and medical records; and other measures. By thoroughly examining the workplace and observing work practices and hazard controls, the IH will be able to recognize visible hazards and/or identify areas that need additional monitoring for contaminants that cannot be evaluated visually. Inspections may be initiated as a result of a complaint or general inspection schedule, or to determine if existing or new contaminant controls are effective. Review of standard operating procedures and company policies, along with a walk-through inspection, can elucidate whether safety policies are just a piece of paper or an actual day-to-day practice in the field.

Industrial hygienists categorize occupational health hazards into four categories, referred to as environmental stressors: chemical, physical, biological, and ergonomic.

1. **Chemical stressors** include substances such as solvents, acids, caustics, and alcohols.
2. **Physical stressors** include ionizing radiation (alpha, beta, gamma, neutron, X-ray radiation), nonionizing radiation (infrared, ultraviolet, visible light, radio frequency, microwave, and laser radiation), noise, and temperature.

3. **Biological stressors** include hazards such as bacteria, viruses, mold, fungus, and insect-related contaminants.
4. **Ergonomic stressors** are the human psychological and physiological injuries or illnesses associated with repetitive and cumulative trauma, fatigue, and exertion.

Knowledge of work processes and an awareness of environmental stressors allow safety and health professionals to anticipate and recognize potential health hazards. ***Questions that should be answered during preliminary investigations and inspections might include:***

- What environmental stressors are present in the facility?
- Where are the points of origin of those environmental stressors?
- What are the forms that those environmental stressors take? Are they dusts, vapors, gases, mists?
- What are the work processes at the facility?
- Have these work processes changed in any way?
- What are the raw materials, intermediate products, and finished products that are involved in the work process? Do they pose health risks?
- What physical and mental tasks are required? Do these tasks require frequent repetitive motions or excessive forces for long durations?
- What control methods are currently being used?

Once these basic questions are answered, the safety and health professional must then evaluate the potential health hazards by determining the concentration and duration of exposure to the contaminant or stressor. The potential mode of entry or exposure of the stressor must be determined. Based upon the physical and chemical properties of the stressor, the probability of contact, absorption, inhalation, or ingestion must also be addressed. Then the rate of generation of the contaminant must be determined. The industrial hygienist must determine if, based upon the nature, toxicity, concentration, and duration of exposure, a significant health risk exists, thus creating the potential for causing injury or illness.

Evaluation

After the initial walk through and recognition phase, the IH must prioritize evaluations based on risk, the toxicity and the hazard. Prioritization is often necessary due to limited time or resources, making it impossible to address all possible issues at one time.

The majority of industrial hygiene *evaluation* is based upon analytical measurement of workplace hazards, and the most commonly monitored hazards are airborne chemical hazards. A variety of methods are utilized to conduct IH sampling, including bulk, intermittent, or continuous sampling; sampling across media, requiring subsequent laboratory analysis or direct reading (real time) monitoring; and qualitative or quantitative measurements. Prior to evaluation, the IH must understand the manner in which the potential contaminant appears in the workplace.

Contaminants can take a wide variety of forms. The following are some of the key terms associated with contaminant forms:

- **Dusts:** Solid particulates generated by crushing, grinding, chipping, or abrasion. Examples of dusts include coal, wood, and sand.
- **Fumes:** Solid particulates (usually metals) generated by condensation from a gaseous state. An example of a fume is welding emissions.
- **Aerosols:** Liquid droplets or solid particulate dispersed in air. An example of an aerosol is overspray from spray painting.
- **Mists:** Suspended liquid droplets generated by the condensation from gas to liquid state.
- **Gases:** Substances which are in the gaseous state at normal temperature and pressure (NTP). Examples: oxygen, nitrogen, and carbon dioxide.
- **Vapors:** Gaseous phases of a material which is ordinarily a solid or liquid at room temperature. Examples: gasoline, toluene, xylene, and benzene.

For many substances, including dusts, fumes, and mists, measurement is typically expressed in terms of weight of the substance captured on the sampling media per unit volume of air sampled. This is usually represented in milligrams of contaminant per cubic meter of air sampled. The volume of air sampled is determined by the flow rate of air volume passing through the sampling media over a fixed period of time. Fiber density is measured as the ratio of fibers to the volume of air sampled, typically in cubic centimeters. Fibers such as asbestos, cotton, or fiberglass are collected on the surface of the sampling media as a unit volume of air crosses the media surface. Fibers are typically counted via light or electron microscopy.

While particulates are primarily sampled to determine toxic effects upon workers, explosivity may also be of concern. Dusts such as grain may explode when concentrations are sufficiently elevated and an ignition source is present. When monitoring is done to evaluate explosive hazards in the occupational environment, dust-to-air concentrations, as well as particle size and sources of ignition, must be determined.

Gas and vapor toxicity is measured in units of concentration that compare the volume of the contaminant to the volume of air sampled. This volume-to-volume ratio is stated in parts per million or ppm. The percentage of a gas or vapor is used as the unit of measure when monitoring for flammable or oxygen content in the ambient atmosphere. This percentage represents the ratio of the gas or vapor of concern to the total volume of the air sampled.

During the evaluation phase, the IH may collect the sample on a sampling media in accordance with standard methods, as described in the OSHA Technical Manual or NIOSH Manual of Analytical Methods. The IH seeks to measure the amount of the contaminant present in the ambient atmosphere, such as volume of gas, weight of dust, or number of fibers. The volume of air sampled may also be readily determined. This is established by pre- and post-calibrating the sampling equipment used to establish the volume of air crossing the sample collection media during fixed units of time. Once again, the volume of air sampled is the product of the volume of air passing across the media during a given interval of time. The quantitative amount of the substance collected is then divided by the volume of air sampled and expressed in terms of ppm, percentage, mg/M^3, or fibers per cc, depending upon the specific substance being monitored.

Once the concentration of the contaminant in the employees' breathing zone has been established, the health and safety professional or industrial hygienist compares the results to established criteria. Established criteria have been set by three groups: OSHA, NIOSH, and the American Confer-ence of Governmental Industrial Hygienists (ACGIH). As previously dis-cussed in earlier chapters, OSHA requirements are mandatory standards. This means that professionals must ensure that employee exposures fall below the specified limits or risk citations and penalties for failing to com-ply with these standards. NIOSH provides recommended exposure levels; these are voluntary guidelines. The ACGIH standards are also voluntary; however, they are reviewed annually by panels of recognized experts. Companies are not required to comply with the ACGIH standards, which tend to be more conservative (lower recommended concentrations). The OSHA health chemical hazard standards are referred to as Permissible Exposure Limits (PELs). The NIOSH recommendations are referred to as

Recommended Exposure Levels (RELs). The ACGIH standards are referred to as Threshold Limit Values (TLVs). TLVs are reviewed and published annually in the publication *Threshold Limit Values and Biological Exposure Indices*. TLVs are based on the best available, scientific information from industrial experience, human exposures, experimental animal studies, and when possible all three. They are based upon science and not typically influenced by congressional politics. Many OSHA PELs have not been changed since their inception in 1970, despite advances in our understanding of workplace health hazards. Since PELs are law, they must pass through Congress, and are subject to both politics and lobbyists, thus limiting their timelines and effectiveness in promoting a safe workplace. When protecting employees, an IH often compares to the ACGIH TLVs to use the most current and scientific standards.

There are three categories of TLVs:

1. **Threshold limit value-time weighted average (TLV-TWA):** The time-weighted average concentration for a normal 8-hour workday and a 40-hour work week, to which nearly all workers may be repeatedly exposed day after day, without adverse effects.
2. **Threshold limit value-short term exposure limit (TLV-STEL):** The concentration to which workers can be exposed continuously for a short period of time without suffering (1) irritation, (2) chronic or irreversible tissue damage, or (3) narcosis of sufficient degree to increase the likelihood of accidental injury, impaired self-rescue, or materially reduced work efficiency, provided the TLV-TWA is not exceeded. The STEL is equal to a 15-minute TWA exposure, which should not be exceeded at any time during a workday even if the 8-hour TWA is within the TLV. There should be at least 60 minutes between successive exposures for a maximum of four exposures/day or work shift.
3. **Threshold limit value-ceiling (TLV-C):** The concentration which should not be exceeded, even instantaneously.

OSHA's PELs are typically expressed in terms of time-weighted averages or ceilings using the same definition and parameters as ACGIH's TLVs (the first PELs were the 1968 TLVs). OSHA permissible exposure limits are published in the tables found in Subpart Z of 29 CFR 1910. While health and safety professionals know that the OSHA PELs must be met to avoid citations, progressive companies attempt to meet the more conservative ACGIH standards. By meeting the TLV recommendations for maximum worker exposure to contaminants, health risks are reduced and there is a greater margin of compliance in the event that health hazard control methods are compromised.

Control

Controls can be defined as processes, procedures, or method changes that correct existing health problems and prevent or minimize the risk of health hazards in the workplace. *Engineering controls* are the method of choice because of their ability to isolate or eliminate health hazards. By eliminating health hazards at the point of origin, the occupational health and safety professional eliminates the release of the contaminant into the workplace environment, ultimately preventing (or greatly reducing) employee exposure. Examples of engineering controls include the use of ventilation systems to reduce the concentration of contaminants or enclosing and shielding hot work areas. Effective exhaust ventilation removes the hazard "at the source," before it can reach the worker's breathing zone. Enclosures place a barrier between the employee and the health hazard point of origin. In both examples, the employee is separated, and thus protected, from the health hazard in the workplace environment.

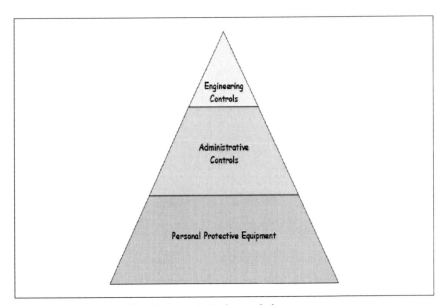

Figure 6-7. Controls of health hazards in the workplace

When it is impossible to implement engineering controls or when fabrication and construction of engineering controls will take time, administrative controls and personal protective equipment options may be required. *Administrative controls* are those health hazard control methods that are employee oriented or process management oriented. In other words, they are the control methods that the management of a facility has influence over through manufacturing method or employee work assignment activities. **Job rotation**, moving employees from one workstation or task to another at regularly assigned intervals, is one example of an administrative control. For example, if noise is the workplace health hazard of concern, it may be possible to rotate the employee to a less noisy area during portions of the shift. In this example, the overall exposure to the noise hazard is reduced, thus reducing the employees' risk of hearing loss. Other administrative control options include substituting less toxic materials in the manufacturing and work processing, and establishing training programs that make the employee aware of the existing health hazard. The use of citric acid–based solvents in place of a cancer-causing agent like carbon tetrachloride or the use of enamel-based paint in place of lead-based paint would be an example of the administrative control method referred to as substitution.

Personal protective equipment (PPE) is the third category of health hazard control available to the industrial hygienist or occupational health and safety professional. It is considered the last line of defense because the barrier separating the employee from the health hazard must be worn correctly and consistently. If the employee does not wear the PPE or it fails, they will be at greater risk for illness or injury. It is an unfortunate fact that for many occupational safety and health programs, PPE is the symbol of the safety program rather than a last resort.

The main reason why administrative controls and PPE are not considered the preferred methods for protecting worker health is that the health hazards still exist in the workplace and can present a risk. In many instances, both administrative controls and PPE can be circumvented. Employees, for example, may use the wrong respirator for a particular hazard. This mistake will result in the employee inhaling toxic vapors or dusts, thus increasing their risk of occupational disease.

One way that industrial hygienists, especially in small to medium sized workplaces, evaluate risks and determine control methods is through a concept termed ***control banding***. The underlying assumption of control banding is that there are only a finite number of methods to control hazards; therefore, the hazards can be grouped into bands or groups. Each hazard band then has a certain set of controls (safe handling and practice, substitution, enclosure/isolation, ventilation). Hazard bands are created by evaluating the chemical toxicity, work process, process duration, quantity used, physical state and other relevant factors. For example, a chemical that is a sensitizer may be placed in a more stringent control band to ensure adequate engineering controls. A chemical that is a skin irritant would be placed in a less stringent control band that only required general safe handling practices. Control banding as a method to control hazards is still being studied to determine its effectiveness across the wide spectrum of workplaces.

CONCLUSION

Occupational health hazards have assumed a significant position for the occupational safety and health professional. With increasing numbers of new chemical products being produced and new manufacturing methods being used, the industrial hygienist and occupational safety and health professional must remain vigilant. Employee health and well-being are dependent upon the industrial hygienist's ability to recognize, evaluate, and control environmental stressors. It is a very complex and demanding area of occupational safety and health, but the rewards more than outweigh the demands.

QUESTIONS

1. What are some of the responsibilities of an industrial hygienist?
2. Why is industrial hygiene an art and a science? Provide an example that requires both art and science to solve a workplace health hazard.
3. Are all substances toxic? Can you explain why a substance may or may not be toxic?
4. What are the routes of entry for substances to be taken up by the body? Provide an example of each type. Which route is the most common occupational route of entry?
5. Differentiate between the terms *acute* and *chronic* using the appropriate dose level and time period.
6. Why should an industrial hygienist consider both local and systemic effects of substances?

7. Differentiate between the additive, antagonistic, and synergistic effects of substances in the body.
8. What are the three classification methods of toxic materials?
9. Discuss the influence of emotion when characterizing toxicity and risk.
10. What is the difference between PELs and TLVs? Explain why knowledge of these exposure measurements is important.
11. What are three categories of health hazard controls? Provide two workplace examples for each category.
12. What type of health hazard control should be instituted first? Why is personal protective equipment a last resort for controlling exposure?

BIBLIOGRAPHY

American Conference of Governmental Industrial Hygienists. 1994a. *TLVs Threshold Limit Values and Biological Exposure Indices for 1994–1995*. Cincinnati, OH: Author.

American Conference of Governmental Industrial Hygienists. 1994b. *Industrial Ventilation— A Manual of Recommended Practice* (21st ed.). Cincinnati, OH: Author.

Canadian Centre for Occupational Health and Safety. 2008. *Control Banding*. Retrieved from http://www.ccohs.ca/oshanswers/chemicals/control_banding.html#_1_1.

Canadian Centre for Occupational Health and Safety. 2009. *OSH Answers*. Retrieved from http://www.ccohs.ca/oshanswers/hsprograms/hazard_risk.html.

Centers for Disease Control and Prevention. 2013. *NIOSH: Control Banding*. Retrieved from http://www.cdc.gov/niosh/topics/ctrlbanding/.

Clayton, F., and Clayton, G. (Eds.). 1981. *Patty's Industrial Hygiene and Toxicology*. New York: John Wiley & Sons.

Confer, R., and Confer, T. 1994. *Occupational Health and Safety: Terms, Definitions, and Abbreviations*. Boca Raton, FL: Lewis Publishers.

Craft, B. F. 1983. Occupational and Environmental Health Standards. In R. W. Rom (Ed.), *Environmental and Occupational Medicine*. Boston: Little, Brown.

Loomis, T. A. 1978. *Essentials of Toxicology*. Philadelphia: Lea and Febiger.

National Institute of Occupational Safety and Health. 1973. *The Industrial Environment—Its Evaluation and Control*. Washington, DC: U.S. Government Printing Office.

National Safety Council. 1994. *Accident Facts*. Itasca, IL: National Safety Council.

Occupational Safety and Health Administration. 1993. *OSHA Industrial Hygiene Technical Manual*. Chicago: Commerce Clearing House.

Occupational Safety and Health Administration. 1993. *Code of Federal Regulations, Title 29, Sections 1910.134, 1910.1000, 1910.1200*. Rockville, MD: Government Institutes.

Ogle, Randall B., MSEH, CIH, CSP, CHMM. 2006. Personal interview.

Plog, B. A. (Ed.). 1988. *Fundamentals of Industrial Hygiene*, 3rd ed. Chicago: National Safety Council.

Ergonomics and Safety Management

CHAPTER OBJECTIVES

After completing this chapter, you will be able to

- Define ergonomic terminology
- List and describe the components of the operator-machine system
- Explain the role of anthropometrics when solving ergonomic problems
- Explain the role of biomechanics when solving ergonomic problems
- List the categories of workstations
- Identify when selected types of workstations should be considered the design of choice

CASE STUDY

Employees at a small facility were experiencing a high frequency of back injuries while manufacturing and finishing air compressor tanks. The management of the plant was at a loss as to what to do about rising workers' compensation premiums. Faced with the prospect of losing compensation insurance coverage, it contacted a certified professional ergonomist (CPE) to conduct an analysis of the workplace.

The consultant first reviewed various safety and health records, including OSHA 300 and 300A forms, nurse's logs, turnover records, and workers' compensation claims. Health interviews were then conducted.

Following basic ergonomic assessment protocol, workplace layouts and critical design measurements were noted. The consultant recorded videos of various work locations in the plant where the majority of ergonomic problems were reported. Using the videos, the consultant analyzed each job observed by breaking it down into major behavioral steps. Each step was further analyzed for:

- body motion
- activity frequency and duration
- pace issues
- lifting-related risk factors

Lifting tasks were studied and compared to NIOSH lifting guidelines. Weight of the objects lifted, distance the objects were held away from the body, and location of the objects at the beginning and end of the lifts were some of the measurements taken. Individual risk factors were studied along with environmental stressors that also influence human performance such as temperature, lighting, and noise.

All analyses pointed to two locations where manual material handling was contributing to a back injury epidemic. Most injuries occurred during storage activities in the warehouse and in the paint shop where tanks were loaded onto an overhead hook conveyor system.

The consultant, working with management and the company safety committee, developed several strategies for reducing the frequency of back injuries. Placing the tank storage pallets on scissor lifts, reducing bending and lifting activities, and using a portable jib for lifting were suggested for the paint shop. In the warehouse, placing the finished product at floor level, instead of on shelves, and eliminating full-arm-extended reaching and lifting postures reduced risk factors. Through implementation of these strategies, the company saw improved productivity and employee morale at the two locations, along with a 63 percent reduction in the frequency of back injuries.

INTRODUCTION TO ERGONOMICS

Due to its significance as a health and safety issue—in terms of medical costs, employee days lost from work, and human suffering, ergonomics has caught the attention of management. Operations requiring lifting, movement of parts or other materials, repetitive motion, unusual postures, or stationary positions are likely to cause sprains, strains, and musculoskeletal problems. Many problems can be minimized with relatively simple and inexpensive interventions. Some require intervention on a higher level, such as through workstation or plant redesign, to reduce the amount of material handling required.

Definition of the Term "Ergonomics"

Ergonomics is based on two Greek words: *ergos*, meaning work, and *nomos*, meaning "the study of" or "the principles of." In other words, "ergonomics" refers to the study of work. ***Ergonomics is the discipline examining the capabilities and limitations of people.*** There are several fields of study contributing to the discipline. Through their application, information about human characteristics and their connection with workplace tools, materials, or facilities is compiled. Ergonomics applies this information to the design of working and living environments.

Ergonomics Is Multidisciplinary

Ergonomics involves knowledge of sociology, psychology, anthropology, anatomy, physiology, chemistry, physics, mechanics, statistics, industrial engineering, biomechanics, and anthropometry. Principles and practices are applied to the industrial environment through activities associated with human factors, such as engineering, industrial engineering, occupational layout and design, product design, safety engineering, occupational medicine, or industrial hygiene.

Ergonomics Objective

The primary objective of ergonomics is to improve human health, safety, and performance through the application of sound people and workplace principles. The goal is to help production managers improve productivity and efficiency. Ergonomic services should not be viewed as add-on activities.

Ergonomists can best serve as part of a team including engineers, managers, medical personnel, and even line workers. The ergonomics team will systematically analyze job requirements from a worker capability and limitation perspective, analyze workplace layout and design, and recommend improvement of the production process. Ideally, these activities are per-formed proactively by integrating them into both the safety and production processes. The goal is to eliminate problems before they occur.

APPLYING ERGONOMICS: AN OVERVIEW

Effectively pinpointing ergonomic hazards depends on the triad of *recognition*, *evaluation*, and *control*. **Recognition of ergonomic hazards** usually involves the search for symptoms. Physiological stresses and muscular strains, psychological stresses, and general complaints or discomfort are typical of ergonomic problems. Excessive movement of the product on the production floor can also be an indicator. The professional asks why a product is moved before packing or why it goes from point A to point B before inspection. Anytime a product is handled, value should be added; otherwise, handling is occurring for no reason. Through review of records, preliminary observation of the workplace, and interviews with key per-sonnel, safety practitioners determine where employees are exposed to potential ergonomic hazards. If ergonomic exposures are a problem, it is necessary to initiate evaluation activities.

Evaluation implies the collection of information to help determine the extent and location of the problem. Review of written records, including OSHA 300 logs, first aid or nurses' logs, and workers' compensation forms should be considered the starting point in the evaluation process. If there are significant numbers of ergonomic injuries and illnesses, more detailed evaluation activities are performed. Indicators include back injuries, carpal tunnel syndrome, tendinitis, tenosynovitis, and muscle strains or sprains. Written records should contain the name of the individual involved in the incident, the job title, and the job location at the facility. Follow-up activities are initiated for jobs or locations where the greatest frequency of incidents or the most severe ones occur. Detailed field data collection through personnel interviews, supervisor interviews, workplace observations, and employee health surveys are typical follow-up evaluation activities to provide insight into the extent of the ergonomic problem.

The evaluation process includes job or task analysis activities. Frequently, tasks are video recorded to permit a detailed evaluation of key ergonomic risk factors. Recorded jobs are broken down into discrete behavioral steps so each step is monitored and frequency measures noted.

Factors studied during each step of the job include:

- Frequency of potentially harmful motions
- Length of time the task is performed
- Pace employees maintain during the shift
- Internal muscle forces employees exert to perform tasks
- External forces exerted on employees, such as from the weight of objects carried

With job-related data in hand, control measures are determined. The preferred intervention is the adoption of *engineering controls*, such as the design and utilization of ergonomically correct workstation and workplace layouts. If engineering controls are not feasible, administrative controls are considered. These include *job rotation* or requiring employees to perform several different jobs during a shift to reduce repetition in a task. Another option is to require "two-man lifts" when a product exceeds a certain weight. As a last resort, *personal protective equipment* (PPE) can be used. It is especially helpful when protection of a body part, such as the hands through the use of gloves, is needed. Care must be given not to mask a problem or transfer it to another body part. A splint on the wrist may alleviate carpal tunnel syndrome but, in turn, generate or aggravate a condition in the shoulder.

APPLYING ERGONOMICS: IN DETAIL

Operator-Machine Systems

Once potential ergonomic problems have been recognized and superficially evaluated for specific jobs or locations, the operator-machine system analysis is used to categorize major conditions contributing to ergonomic problems. *Operator-machine system analysis* examines factors associated with the people; equipment/machinery, including layout of the workplace; and environment. The approach directs the evaluator toward an understanding of the cause of and potential solutions for ergonomic problems by studying the interaction of the operator with the machine. Consideration is given to the quality of interaction based on expectations of machine operation, actual machine operation, and output, as well as other factors related to the environment. It provides safety professionals with a systematic procedure to study the three categories of causes resulting in ergonomic hazards in the occupational setting: people, machines, and environment.

PEOPLE VARIABLES

The people or operator variables associated with a system are composed of the human factors contributing to the ergonomic problem. These include physiological dimensions, capabilities, and limitations; psychological capabilities and limitations; and psychosocial factors. One or more of these conditions has the potential to cause injuries or illnesses in the workplace. ***Examination of mental and physical job demands*** is done to determine whether job demands exceed human capabilities. Knowledge of human physiological dimensions (***anthropometry***) and movement (***biomechanics***) is a critical component. The actions of employees while performing their tasks are compared to movement categories, based on anthropometric and biomechanical classifications. Types of movements, whether they are performed properly and efficiently, and their frequency are noted. A quantitative evaluation of motions, forces, or other contributing factors can then be performed.

Anthropometry

Anthropometry is the measurement and collection of the physical dimensions of the human body. It is used to improve the human fit in the workplace or to determine problems existing between facilities or equipment and the employees using them.

   There are two types of anthropometric dimensions useful for the study of human physiology and its effect on workplace layout and design:

1. ***Structural or static anthropometry:*** The body measurements and dimensions of subjects in fixed, standardized positions are referred to as "structural" or "static" anthropometrics. Common *structural* anthropometric measurements include stature (height), sitting height, body depth, body breadth, eye height sitting or standing, knuckle height, elbow height, elbow to fist length, and arm reach.
2. ***Functional or dynamic anthropometry:*** Functional or dynamic anthropometry refers to the body measurements and dimensions taken during physical activities. Frequently used functional measurements include

crawling height, crawling length, kneeling height, overhead reach, bent torso height, and range of movement for upper-body extremities.

Anthropometric measurements help designers determine furniture or workplace layout requirements based on typical human body sizes. Prior to the utilization of ergonomics and anthropometrics, many machines and workplaces were designed for the so-called "average" employee. Unfortunately, average statistical measurements represent less than 1 percent of the normal distribution of body measures. For example, if a standing workplace were designed for an average American male, it would not fit 99 percent of the American population.

The evolution of automobile interiors illustrates the value added by the application of ergonomics. Years ago, seats could be adjusted forward and backward to accommodate the leg length and pedal reach requirements of the driver. Height, pedal, and steering-wheel adjustments were not typically available. Cars were designed for the "average" person in the population, causing short individuals to look through the steering wheel and tall individuals to strike their heads against the roof. Today, ergonomists use anthropometric measurements to include at least 90 percent of the population. Most workplace designs attempt to achieve this goal by including people dimensions between the 5th and 95th percentiles. This can usually only be accomplished by providing adjustable devices, as they have done in automobiles with multidirectional power seats, moveable pedals, and adjustable steering wheels, seat backs, and head rests.

Percentile statistics are used extensively in anthropometrics to represent the number of people with measurements less than or equal to the dimensions of interest. If a man has an anthropometric measurement that places him in the 95th percentile, his measurement is as large as or larger than 95 percent of the population. Ninety percent of this male/female population could be accommodated for height if the workplace is designed to be adjustable for people between the 5th and 95th percentiles. As the size of the average American increases, measurements of each percentile increase accordingly. In other words, individuals within a given range continue to grow larger.

Taking anthropometric measures and adjusting the workstation may help eliminate ergonomic problems. An example is in trying to eliminate back injuries associated with lifting boxes onto a shelf. According to the recommendations associated with the NIOSH lifting equation, *knuckle height standing* at the start of a lift produces the least amount of stress upon a worker's back.

In addition, if one can minimize the destination (height of the shelf) where the box has to be placed, this will also reduce back stress. Thus, back muscle stress could be greatly reduced if the boxes could be removed from a scissors pallet at knuckle height and stored at knuckle height. This requires taking employee measurements, or when working with large numbers of workers, referring to anthropometric data for knuckle height standing. *Adjustability* can be a key solution to many ergonomic problems.

Biomechanics

Mechanics is the science of motion and force. *Biomechanics* is the human equivalent of this concept. It is the study of the mechanical operation of the human body or the science of motion and force in living organisms. In biomechanics the function of the body components is monitored and job requirements are modified to lower internal and external stresses. The musculoskeletal system provides the foundation data for the study of biomechanics. As in anthropometrics, *there are two types of biomechanical measurements:*

1. *Dynamics*: the study of moving bodies
2. *Statics*: the study of bodies remaining at rest (equilibrium) as a result of forces acting upon them

The two measures provide the ability to determine how moving body components contribute to ergonomic injuries. Factors, such as extension and force applied by the arm and leg muscles while pushing a hand truck, can be calculated using biomechanics. The load on a hand truck, friction coefficient of the walking surface, and other external forces affecting employee performance can also be determined. The measurement of primary concern is force, especially as it relates to loads and stresses on the body. *In biomechanics there are two categories of force creating motion of biological matter or movements like walking or lifting:*

1. *Load*: the external forces upon a structure or organism
2. *Stresses*: the internal forces generated in the structure as a result of loading

In human bodies movement is possible due to the application of load and stress to biological levers. The field of biomechanics teaches about joints, bones, and muscles and provides safety professionals with an understanding of how the musculoskeletal levers of the body are designed to work. This awareness can be applied to the identification and elimination of unnatural movements resulting in ergonomic problems. Evaluation activities such as monitoring frequency and duration of movement or examining postures and positions can be initiated to determine the level of ergonomic risk. Internal and external forces should also be evaluated as a part of a job or task analysis.

Twisting and other unnatural movements or postures observed during a task analysis are warning signals. They will eventually result in ergonomic injuries. Ease of work activity or biomechanical advantage is only possible when the weight is held and moved using natural posture and body position.

CLASSIFICATION OF BODY MOVEMENT, POSTURES, AND POSITIONS

The level of ergonomic risk can effectively be determined by performing ergonomic task analyses. This requires the breakdown of an activity into discrete behavioral steps. Once the steps have been sequentially developed, each can be logically broken down into the movements observed. Steps are recorded and used for calculating repetition rates and duration. Ergonomic problems may correlate with frequency and duration of biomechanical motions.

There are two different systems used to classify and evaluate body movements:

1. *Physiological classification systems* examine the way body parts move.
2. *Operational classification systems* group body movements by the particular work activity being performed.

Physiological Categories of Movement

Listings of body movements are used to classify the biomechanical movement of the individuals performing a task. The following is a cursory list of physiological movements with a brief description of each term (Kroemer, Kroemer, and Kroemer-Elbert, 1990). See figures 7-1, 7-2, 7-3, and 7-4 for clarification of the movements represented.

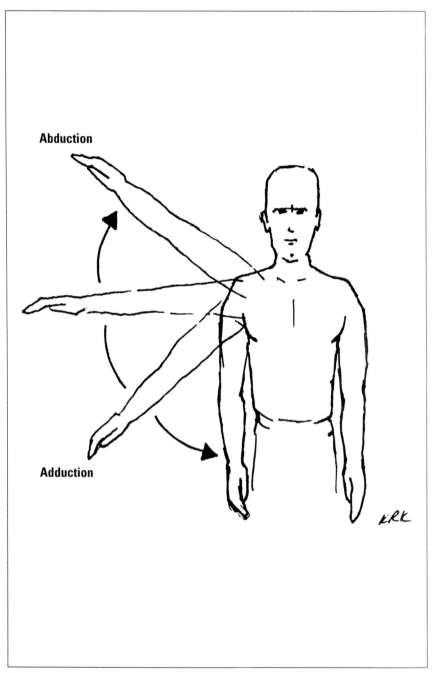

Figure 7-1.   An example of arm abduction and adduction

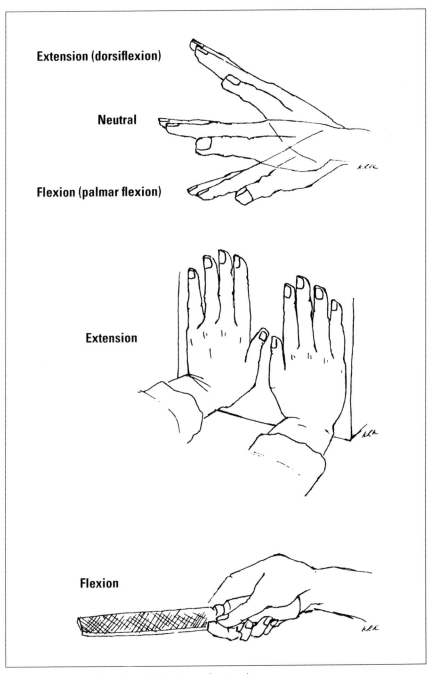

Figure 7-2. Examples of wrist flexion and extension

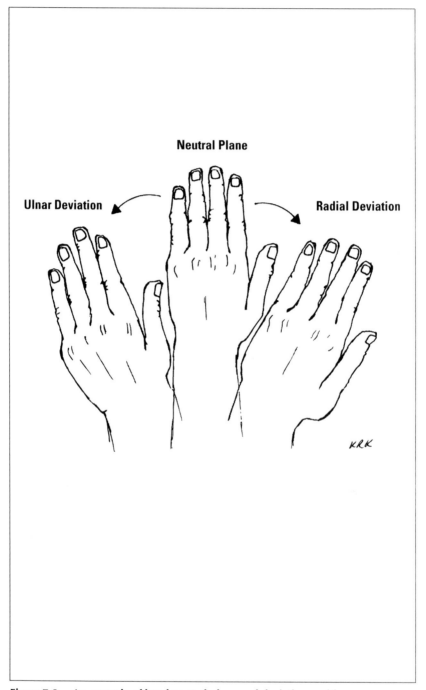

Figure 7-3. An example of hand neutral plane and deviation positions

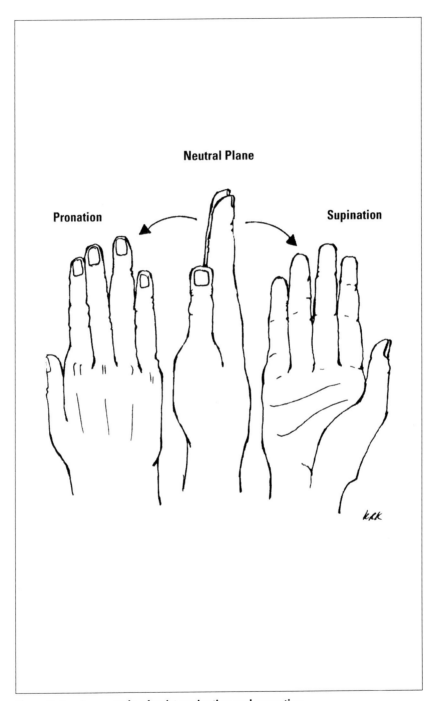

Figure 7-4. An example of wrist supination and pronation

Abduction/Adduction

Abduction: The movement of a body part away from the center plane of the body. Lifting the arm outward away from the body is an example of abduction.

Adduction: The opposite of abduction. This is the movement of the body part toward the center plane of the body. Lowering the arm toward the body is an example of adduction.

Circumduction

Circumduction: Rotary movements which circumscribe an arc. Swinging the arm in a circle is an example of circumduction.

Flexion/Extension

Flexion: The movement of a joint that decreases the angle between the bones. Bending the arm at the elbow such that the hand moves closer to the upper arm region is one example.

Extension: The opposite of flexion. The movement of a joint that increases the angle between the bones. Straightening the arm is an example of extension.

Neutral Plane/Deviation

Neutral Plane: The normal and low-stress position of segmental physiological components. Maintaining the hand, wrist, and forearm at a 180-degree angle or in a straight and linear plane or dropping the hand, wrist, forearm, elbow, and upper arm at one's side is an example of neutral planes.

Deviation: The movement or position of a body part away from the neutral plane. Bending the wrist with the hand bent toward the thumb is referred to as radial deviation. Bending the wrist with the hand bent toward the small finger is referred to as ulnar deviation.

Rotation

Rotation: A movement in which a body part turns on its longitudinal axis. The turning of the head or arm is an example of rotation.

Supination/Pronation

Supination: The turning of the forearm or wrist such that the hand rotates and the palm is facing upward.

Pronation: The opposite of supination. The turning of the forearm or wrist such that the hand rotates and the palm is facing downward.

Operational Categories of Movement

Operational classification of movement refers to the task being performed by the operator at the time of the job observation. *The following is a list of the terms used to represent the operational classification of movements:*

- *Positioning*: Moving an object and corresponding extremity from one position to another. An example of positioning would be reaching for a bolt stored in a bin to the right of an employee.
- *Continuous movement*: A single movement involving muscle control to adjust or guide a machine or other piece of equipment. An example of continuous movement would be the steering of a forklift.
- *Manipulative movement*: The handling or assembling of parts. This movement classification is usually limited to hand or finger movement. An example of manipulative movement would be the assembly of component parts.
- *Repetitive movements*: The same movements recurring repetitively. Hammering or using a screwdriver are examples of repetitive movements.
- *Sequential movements*: A series of separate movements joined in a specific order to complete a given task. Reaching for a tool with the right hand, grasping a component in the left hand, moving the two hands toward one another, and adjusting the component using the tool are examples of sequential movements.
- *Static movements*: Maintaining the position of a body member in order to hold something in place. Though movement may not be involved, the muscles are required to maintain the steady position of the object. Holding a board or plasterboard in place on the ceiling of a room while screwing it into position is an example of static loading of muscle groups or static movement.

ADDITIONAL CHARACTERISTICS OF THE PEOPLE VARIABLE

The people component of the operator-machine system also considers psychological factors affecting worker performance. ***Psychology*** is the science studying human behavior. From an ergonomic perspective psychological factors such as memory, attention, fatigue, boredom, job satisfaction, future ambiguity, expectation, and stress contribute to a wide variety of potential problems. Psychology incorporates such behavior as employee expectations for machines. Employees expect *up* to equal *on* and *down* to equal *off*. *Forward* on a lever is *faster* or *ahead*, and *backward* is *slower* or *reverse*.

The people component of the operator-machine system also incorporates ***physiology***, the branch of the biological sciences concerned with the function and processes of the human body. Physiological capabilities and limitations include the structure, strength, and movement of anatomical components. ***Physiology*** also includes the study of the human body at the cellular level. Neurological activity associated with light stimulating the retina when a warning signal flashes is an example. It can also include such complex phenomena as gross motor functioning when studying how various muscle groups of the body tense and relax to provide movement and balance while lifting and carrying an object.

Cumulative Trauma Disorders

Cumulative trauma disorders (CTDs), also referred to as ***repetitive-motion injuries***, result from excessive use of the hand, wrist, or forearm. As with over-exertion injuries, the frequency and costs of CTDs are growing to epidemic proportions. Some of the most common cumulative trauma disorders are carpal tunnel syndrome, cubital tunnel syndrome, tendinitis, and tenosynovitis.

Carpal Tunnel Syndrome

Carpal tunnel syndrome is a common wrist injury caused by compression of the median nerve in the carpal tunnel. The carpal tunnel is an opening in the wrist surrounded by the bones of the wrist and the transverse carpal ligament. The sensation in the thumb, index, and middle fingers is generated by the median nerve. When the wrist is forced to flex, extend, or deviate toward the ulnar (small finger) or radial (thumb) position, the ligament compresses the median nerve. In effect, these motions pinch this nerve as it passes through the carpal tunnel. Symptoms include tingling, pain, or numbness in the thumb and first three fingers.

The level of repetitive motion tolerated without undue risk of injury varies widely depending on age, sex, and other health factors. While no specific limits have been established to avoid CTD symptoms, factors increasing risk include repetition of motion, work-rest cycles, force, and duration of the task.

Carpal tunnel syndrome has been increasingly found among data-entry clerks, cashiers, and individuals who perform keypunch tasks due to the static, restricted posture, and high-speed finger movements.

Cubital Tunnel Syndrome

Cubital tunnel syndrome is compression of the ulnar nerve in the elbow, thought to be caused by resting the elbow on a hard surface or sharp edge. Symptoms include tingling in the ring finger and little finger.

Tendonitis

Tendonitis, the most frequently diagnosed CTD, results from inflammation of tendons due to excessive use. Common symptoms are burning sensations, pain, and swelling at various sites in the hand, wrist, or arm. Tendinitis is common for workers required to extend their arms overhead during assembly activities.

Tenosynovitis

Tenosynovitis is soft tissue trauma and injury to tendons or tendon sheaths, frequently occurring in the wrists and ankles where tendons cross tight ligaments. The tendon sheath swells, making it more difficult for the tendon to move back and forth inside. It often results from the overworking of muscles or the wrenching and/or stretching of tendons. Common forms of tenosynovitis are de Quervain's disease and trigger finger (Parker and Imbus, 1992). These disorders are commonly found in employees whose jobs include buffing/grinding and packing because of repetitive wrist motions, vibration, and prolonged flexing of shoulders. Tenosynovitis is often associated with ulnar deviation during rotational movements such as using a screwdriver.

CTD Symptoms

Symptoms of repetitive motion disorders have been described as appearing in three stages (Chatterjee, 1978).

Stage 1: Victims may experience aches and tiredness; at this time, the disorder is fully reversible and may even subside after periods of rest.

Stage 2: There is the addition of swelling and pain. The symptoms do not dissipate overnight; they usually last for several months and result in a reduction in job performance.

Stage 3: Victims may experience constant pain even while performing light duties. This condition can last for months or years.

The key to an effective ergonomics program is to identify problems before they begin. Safety professionals should look for potential CTD problems during task analysis activities by using physiological motion categories that focus attention on awkward (unnatural) postures and frequent repetitions of motions.

Psychosocial factors, the third element of the people component of the operator-machine system, refer to an individual's behavior in a group environment. The term "psychosocial" is derived from the phrase "social psychology." The ergonomist is interested in attitude formation, attitude change, leadership styles, power and influence, conflict, occupational stress, organizational structure, employee motivation, and organizational reward systems. All can influence employee performance in the social context of an organization. For example, a glass plant manufacturing bottles for soft drinks was concerned about workers not paying enough attention during the final inspection. Three employees were placed on each side of the assembly line and were assigned to watch the bottles for imperfections. Management was concerned boredom would cause them to talk among themselves and provide a distraction from their job. The management team thus had isolation booths built around each worker, so communication among them was impossible. Although this eliminated the problem of talking, it increased boredom to the point that increased efficiency became questionable.

Psychological, physiological, and psychosocial elements can occur either in isolation or in combination with one another, as they do in shift work. "Shift work" refers to work performed during nontraditional employment hours of a day. It is "any regularly taken employment outside the day working window, defined arbitrarily as the hours between 7:00 AM and 6:00 PM" (Monk and Folkard, 1992). It can refer to working on a fixed shift where individuals maintain the same schedule, like a cat-eye shift. "Cat-eye shift" is a phrase used in coal mining, where miners typically work a fixed schedule from 11:00 p.m. to 7:00 a.m. Shift work also refers to rotating shifts where employee schedules are changed on a regular basis. An example of a rotating shift is when employees work midnight to 8:00 a.m. for one week, 8:00 a.m. to 4:00 p.m. the next week, and 4:00 p.m. to midnight the third week. Nontraditional hours may lead to ergonomic problems.

Physiological hazards and strain are created when the body attempts to fight its natural circadian rhythm. *Circadian rhythms*, as a subcategory of biological rhythms or biorhythms, are the body's attempt to conform to the 24-hour day in terms of metabolism, physiology, and psychology. Susceptibility to fatigue increases when people have work schedules conflicting with these rhythms. Physiological strain is manifested by an increased number of cardiovascular and gastrointestinal problems for individuals required to perform shift work (Rutenfranz, Colquhoun, and Knauth, 1977).

Psychological factors associated with shift work include the stress and depression accompanying unusual work schedules. Individuals performing shift work are less effective in terms of cognitive abilities and attentiveness on the job. Safety records for shift workers are worse than those for day-shift employees (DeVries-Griever and Meijman, 1987), possibly due to the strains on interpersonal relationships. The psychological problem becomes a psychosocial one when shift workers do not have the opportunity to interact with spouses or friends for long periods of time, creating a void in their social support system. This loss of emotional support can result in marital difficulties and elevated frequencies of divorce. The resulting stress, in turn, manifests itself as a workplace safety or ergonomic problem.

MACHINE VARIABLES

Machines or equipment includes a variety of variables that must be evaluated. Examples are as follows:

Machines:

- the position of the displays and controls associated with the operation and monitoring of machinery

- distance of part storage bins from the worker
- distance of the conveyor systems from the worker area
- table height of the work surface

Tools:

- shape and size of handle for correct fit
- vibration translation to hands or other parts of the body
- correct use of tool

Offices:

- design and layout of computer components
- back support of a chair
- height of keyboard surface
- position of the monitor
- layout of aisles and walkways

A worker operating a bucket lift may find himself using different controls than he did with the last bucket he used. A simple layout includes a vertical stick that he pushes back and forth if he wants the bucket to move backward and forward. He also has a horizontal stick he moves up and down if he wants the bucket to move in either of those directions. He deals with a dilemma when, unfamiliar with the operation of this particular bucket, he is faced with a series of levers, all in a row, with the labels and instructions worn away over the years. Although the worker may practice in a wide-open space, the possibility exists that in the middle of the job he'll forget how each lever works. A wrong move could send the bucket into a high-voltage wire or through a window. When the accident occurs, it is often attributed to operator error, when, in fact, it is also a result of poor equipment design and an improper interface between the person and the machine.

Machines or jobs requiring large numbers of repetitions can also be problematic. ***Task requirements should be modified if any of the following factors are observed:***

- Tasks requiring over 2,000 manipulations per hour
- Manual task work cycles that are 30 seconds or less in duration
- Repetitive tasks whose duration exceeds half of the worker's shift

Problems may be exacerbated when excessive force must be applied or frequent use of stressful postures is necessary. Studies indicate that arthritis; diabetes; poor renal, thyroid, or cardiac function; hypertension; pregnancy; and fractures within the carpal tunnel increase the chances of an employee experiencing CTD symptoms (Parker and Imbus, 1992).

ENVIRONMENTAL VARIABLES

Environmental variables include temperature, lighting, noise, vibration, humidity, and air contamination—any of which can result in elevated ergonomic risks. Fatigue often accompanies high temperatures and levels of humidity, ultimately reducing the worker's mental capability to focus on a task. Health concerns such as heat stress, heat exhaustion, and heatstroke are additional dangers. Lighting or workplace illumination is another risk factor. Computer glare on a monitor in an office environment can cause eye fatigue and eyestrain problems. Insufficient illumination in a warehouse can result in slipping and tripping incidents. Excessive illumination or glare can result in operator errors while using heavy equipment outdoors.

WORKPLACE LAYOUT AND DESIGN

Once the task analysis has been performed and ergonomic problems have been observed, a determination is made as to how to eliminate the hazards. One method to achieve this goal is by examining how the work area is laid out and redesigning the workplace to eliminate the problems. Knowledge of workstations will help the safety professional in this process.

Interventions—Workstations

Ergonomists recommend seated workstations when:

- All tools, equipment, or components required to perform the tasks are easily reached from the seated position
- The primary tasks require fine manipulative hand movements or inspection activities
- Frequent overhead reaches are not required

- The force exertion requirements to perform the task are minimal
- Less than ten pounds of lifting force is required
- There is adequate leg space available (adapted from Eastman Kodak, 1983)

Workspace design specifications include a minimum of 20 inches in width, 26 inches in depth to allow for adequate leg clearance, a minimum of 4-inch clearance from the edge of the workstation, and an approximate ideal work area of 10 by 10 inches where activities are performed. Seated workstations are recommended for detailed visual tasks, for precision assembly work, or for typing and writing tasks.

Standing workstations are typically recommended when:

- The operator is required to move around a given area to perform the given task
- Extended and frequent reaches are required to perform the intended tasks
- Substantial downward forces or lifting heavy objects, typically weighing more than ten pounds, is required
- There is limited leg clearance below the workstation surface (adapted from Eastman Kodak, 1983)

Standing workstations are for precision and detailed work and heavier work activities. Because of anthropometric differences in male and female workers, the ideal working height for tasks will vary, but anthropometric charts can help determine measurements. Generating the maximum force in the upper body extremities requires employees to keep elbows close to normal elbow heights or slightly below elbow height while standing. When employees stand in a normal, relaxed position and then bend their elbows at a 90-degree angle, the measurement between the elbow height and the floor will furnish the elbow-height standing position. This measurement is the ideal height for light-work-related activities while standing. Elbow standing height is reduced 4 to 12 inches for heavier work requirements or raised four to eight inches for more detailed work activities.

The combination workstation is recommended when:

- Several tasks requiring mobility are performed
- Reaching overhead or at levels below seated positions is required—especially when forward or side position reaches must be accomplished

at a variety of levels above the work surface (adapted from Eastman Kodak, 1983)

Grandjean (1988) recommends seven guidelines for workplace layout and design:

1. Avoid any kind of bent or unnatural posture. (Bending the trunk or the head sideways is more harmful than bending forward.)
2. Avoid keeping an arm outstretched either forward or sideways. (Such postures lead to rapid fatigue and reduce precision.)
3. Work sitting down as much as possible. (Combination workstations are strongly recommended.)
4. Use arm movements in opposition to each other or symmetrically. (Moving one arm by itself sets up static loads on the trunk muscles. Symmetrical movements facilitate control.)
5. Maintain working fields (the object or table surface) at an optimal height and distance for the eyes of the operator.
6. Arrange handgrips, controls, tools, and materials around the station to facilitate the use of bent elbows close to the body.
7. Raise arms where necessary by using padded supports under the elbows, forearms, or hands.

Interventions—Manual Material Handling

Manual material handling is a common activity in most occupational environments. Workers handle raw materials, tools, finished materials, containers, and packing materials on a daily basis. Moving objects, regardless of weight, can result in arm, back, and leg strain. The costs of injuries due to material handling are enormous—from both medical treatment and workers' compensation. Losses from these injuries are compounded when the costs of days absent from work are added.

NIOSH lifting guidelines (NIOSH Publication No. 94-110, 1994) identify certain motions increasing the risk of back injuries. They include combinations of bending, twisting or turning, standing, and sudden position changes. Bending and twisting/turning are particularly troublesome. Lifting is hazardous due to the effects of leverage exerted on the spine during the movement. The force from the weight of the object is multiplied by a factor of as much as 10 or more depending on the distance of the object from the body. Twenty-four bones, or vertebrae, compose the spine. A fluid-filled disc with a fibrous outer layer is located between each pair of vertebrae to serve as a shock absorber preventing the rubbing and abrasion of the vertebrae. Disc rupture and herniation are injuries that are often cumulative in nature. A variety of factors, including age and repeated poor lifting practices, are prime culprits resulting in damaged discs (Mital, Nicholson, and Ayoub, 1993). Sometimes a back pain is attributed to a lift or sudden motion, but the underlying cause is the poor state of the back due to weakened muscles or previous injuries. It can be caused by the fluid in a disc pushing through the fibrous layer, applying pressure on the nerves in the spinal cord. Although only about 5 percent of back injuries are attributed to disc damage, it is often more serious and longer lasting than other forms of back injuries.

To avoid back injuries, employees must know proper lifting techniques. Traditionally, employees have been taught to use a bent knee/straight back technique with the object between the knees. However, this can only be performed if the object fits between the knees. Research indicates that the legs may not always be strong enough to lift heavy objects using this technique, causing awkward and stressful postures (Garg and Herrin, 1979). There is no single lifting technique that applies to all situations; the goal for safety professionals is to eliminate manual material handling whenever possible. Engineering controls like cranes and conveyors to eliminate lifting and carrying activities are ideal. If engineering controls are not feasible, administrative controls such as requiring assistance from coworkers for objects of a certain weight can be implemented.

Some additional considerations:

- Use engineering methods to automate lifting and moving
- Ensure the weight of the object is as close to the body as possible when employees lift manually
- Eliminate bending and twisting motions when lifting
- Require lifting aids, whether mechanical ones or coworkers, when exceeding NIOSH-recommended weight limits

Interventions—Video Display Terminal Workstation Design

As workers spend more time on computers, there is an increasing frequency of ergonomic injuries and illnesses. Poor workstation design is often the cause of both physiological and psychological problems (Danko, 1990). The layout and design must be geared to human capabilities and limitations in order to avoid physical discomforts, fatigue, mental stress, or injury to the worker (DeChiara and Zelnik, 1991). Computer workstations should be of great concern to safety professionals in both the layout of new facilities and improvement of existing office and other occupational settings.

Several individuals of varying body size, weight, shape, age, physical condition, attitude, aptitude, or mental ability may use workstations; therefore, adjustability is necessary. The work environment should incorporate anthropometric designs to enable all workers to function within their capabilities and perform their job tasks optimally and comfortably (Braganza, 1994). Anthropometric data supply a standard guide of measurements, both static and dynamic, to provide an optimal dimensional layout of the workstation (Woodson and Conover, 1964).

Other critical factors in the workplace design or redesign process are illumination, equipment and control display arrangements, warning signals, visual displays, control devices, traffic spaces, and storage requirements (Woodson, 1981). The nature of the task will determine the type of workstation layout appropriate for the setting: seated for data entry and telecommunication jobs; standing for assembly and machine control jobs; and combination for technical positions and computer design work. A combination workstation often meets the needs of tasks performed in both office and manufacturing environments (Grandjean, 1988). The key is having equipment and furnishings easily adjustable to ensure proper fit and conformance to anthropometric requirements. Seated workstations should have easily adjustable chairs, work surfaces, equipment, and footrests (Eastman Kodak, 1983).

A good chair is probably one of the most important workstation investments. When a chair is properly adjusted with feet firmly on the ground, the hips and knees will assume a horizontal plane position at a 90-degree angle and the pan of the seat will measure 15 to 20 inches from the floor. In the correct working position, workers' arms should dangle by their sides, and then be raised to a horizontal plane position at right angles to their bodies. When the body is assuming its neutral position, stress is relieved from the shoulders and back area, resulting in a decrease in fatigue and tension (Grandjean, 1988).

To ensure that a chair has as many ergonomic comfort and safety qualities as possible, it should have the following features:

- adjustable backrest and seat (to fit different body sizes)
- good lumbar support (lower back)
- five legs (for better stability)
- wheels (for mobility)
- reclining seat backs and adjustable seat pans (to control the pressure on the back and thighs)
- armrests and footrests (if necessary to the situation) (Barnes, 1994)

Numbness, pain, and fatigue may still occur due to reduced blood circulation if an individual has been seated for an extended amount of time. Therefore, the furniture or equipment being used should have rounded edges to prevent discomfort (Braganza, 1994). In addition, frequent breaks and stretch activities should be incorporated into the employee's daily activities. Breaks do not necessarily imply coffee breaks. Changing activities can meet the needs of an employee. Filing information while standing constitutes a break from computer-data entry activities and will also provide therapeutic benefits.

The work surface area, configuration, and height are additional ergonomic concerns safety professionals should not neglect (Braganza, 1994). Pencils, paper, and other frequently used materials should be located near the body in a range of reach from 16 to 18 inches. Arm movements should also be made in a symmetrical or opposing action while the elbows are slightly bent to prevent unnatural twisting or stretching of the body. Moving one arm by itself can cause static loads on the trunk muscles, whereas symmetrical movements allow for added control (Panero and Zelnik, 1979). Standard anthropometric data suggest that most desks are 30 inches high—too high for an ergonomically correct computer workstation. Depending on the height of the individual, the adjustability of the work surface should be optimally between 23 and 28 ½ inches high. This satisfies the seating requirements of 90 percent of seated adults (Barnes, 1994).

Thigh and knee clearance is another key consideration in designing a workstation (Barnes, 1994). A full range of movement and sufficient clearance for the thighs are necessary while one is sitting. Without sufficient clearance, an employee might twist the torso in order to get closer to the workstation. This twisting could cause injury. The minimum thigh clearance (measured from beneath the work surface to the floor) is 20 2/5 inches (Braganza, 1994). If this clearance is not available, a standing workstation might be the layout of choice.

Anthropometric measurements should also be used to determine the proper angle and height of the monitor to reduce fatigue and eyestrain. The top of the monitor should be aligned with the standard sight line, and the center of the screen should be 15 degrees below eye level (DeChiara and Zelnik, 1991). The screen should be tilted back 0 to 7 degrees to eliminate glare, and should be located between 18 and 28 inches from the eyes for ease in focusing (Grandjean, 1988). In addition, document holders should be placed at the same height and plane as the screen to decrease neck and eye fatigue (Barnes, 1994). The monitor, documents, and keyboard should be positioned to best suit the task being performed, reducing twisting and bending while one is using the equipment (Grandjean, 1988).

Purchasing ergonomically adjustable furniture will not necessarily eliminate ergonomic problems. Employees must be instructed as to how to adjust workstations to meet their specific needs. For ergonomic programs to be effective, employees must also know how to ergonomically layout and design their workstations, as well as adjust their furniture to eliminate stresses. Training is an important part of all ergonomics programs.

CONCLUSION

Ergonomics has become an increasingly important part of the safety professional's job responsibilities. With an effective implementation plan, ergonomic activities can make a significant difference in the loss-control effectiveness of the organization. Through workplace layout and task analysis activities, the safety professional may identify many ergonomic hazards. Consideration is given to the operator-machine systems, anthropometry, and biomechanics during walk-through inspections.

For some complex ergonomic hazards, organizational ergonomic factors must also be considered. Some of these hazards may be associated with factors such as shift work schedules or product flow problems. Work-rest cycles may also be an issue. Depending on the nature of the work activities, psychological factors such as fatigue and boredom can impact worker productivity and efficiency. No matter what the ergonomic problem might be, it is important for safety practitioners to use the operator-equipment system. This will ensure that all of the variables affecting ergonomic performance will be examined.

Management practices, worksite analysis, and control methods are critical in reducing ergonomic-related injuries and illnesses. Although there is no specific ergonomics standard, OSHA may look for certain practices under the general duty clause. Ergonomic tools for baggage handling, beverage delivery, computer workstations, grocery warehousing, hospitals, poultry processing, and sewing can be found on the OSHA website at www.osha.gov.

QUESTIONS

1. What are the components of the operator-machine system? Provide an example of a work activity using this system.
2. How can the use of the industrial hygiene triad be applied to ergonomic problems?
3. How can videotaping jobs be useful in the evaluation of ergonomic problems?
4. How can anthropometrics and biomechanics be used to solve ergonomic problems?
5. How does anthropometrics use statistical measurements and methods?
6. What are the costs associated with material handling and repetitive motion injuries? Why are these two types of ergonomic problems of concern to the safety professional?

REFERENCES

Barnes, K. 1994, August. Is Your Office Ergonomically Correct? *HR Focus*, *71*, 17.

Braganza, B. J. 1994, August. Ergonomics in the Office. *Professional Safety*, *39*, 22–27.

Chatterjee, D. S. 1978. Prevalence of Vibration-Induced White-Finger in Flourospar Miners. *British Journal of Industrial Medicine*, *35*, 208–18.

Danko, S. 1990, Fall. Indoor Ecology: Designing Health and Wellness into the Workplace. *Forum*, 19: 3–8.

DeChiara, P. J. and Zelnik. M., 1991. *Time Saver Standards for Interior Design and Space Planning*. New York: McGraw-Hill.

DeVries-Griever, A. H. and Meijman, P. F. 1987. The Impact of Abnormal Hours of Work on Various Modes of Information Processing: A Process Model on Human Costs of Performance. *Ergonomics*, *30*, 1287–99.

Eastman Kodak Company. 1983. *Ergonomic Design for People at Work*. Vol. 2. New York: Van Nostrand Reinhold.

Garg, A., and Herrin, G. 1979. Stoop or Squat? A Biomechanical and Metabolic Evaluation. *Transactions of American Institute of Industrial Engineers*, *11*, 293–302.

Grandjean, E. 1988. *Fitting the Task to the Man*, 4th ed. London: Taylor & Francis.

Hodkiewicz, T. S. 2000. Healthcare Hazards. *Industrial Safety and Hygiene News, 36* (8), 38.

Kroemer, K. H., Kroemer, H. J., and Kroemer-Elbert, K. E. 1990. *Engineering Physiology: Bases of Human Factors/Ergonomics*, 2nd ed. New York: Van Nostrand Reinhold.

Mital, A., Nicholson, A. S., and Ayoub, M. M. 1993. *A Guide to Manual Materials Handling*. London: Taylor & Francis.

Monk, T., and Folkard, S. 1992. *Making Shiftwork Tolerable*. London: Taylor & Francis.

National Institute of Occupational Safety and Health. 1994. APPLICATIONS MANUAL FOR THE REVISED NIOSH LIFTING EQUATION Publication No. 94-110, 1994.

Panero, J., and Zelnik, M. 1979. *Human Dimensions and Interior Space*. New York: Watson-Guptill.

Parker, K., and Imbus, H. 1992. *Cumulative Trauma Disorders*. Boca Raton, FL: Lewis Publishers.

Rutenfranz, J., Colquhoun, W. P., and Knauth, P. 1977. Biomedical and Psychological Aspects of Shiftwork: A Review. *Scandinavian Journal of Workplace Health*, *3*, 165–82.

Woodson, W. E. 1981. *Human Factors Design Handbook*. New York: McGraw-Hill.

Woodson, W. E., and Conover, D. W. 1964. *Human Engineering Guide for Equipment Designers*, 2nd ed. Berkeley: University of California Press.

BIBLIOGRAPHY

Bureau of Labor Statistics. 1993. *Occupational Injuries and Illnesses in the United States by Industry, 1991.* Washington, DC: U.S. Government Printing Office.

Laflin, K., and Aja, D. 1995. Health Care Concerns Related to Lifting: An Inside Look at Intervention Strategies. *American Journal of Occupational Therapy*, *49*, 63–72.

National Safety Council. 2001. *Injury Facts.* 2001 Edition. Ithaca, IL: Author.

Putz-Anderson, V. 1988. *Cumulative Trauma Disorders: A Manual for Musculoskeletal Diseases of the Upper Limbs*. London: Taylor & Francis.

Thompson, C. 1989. *Manual of Structural Kinesiology*, 11th ed. St. Louis, MO: Times Mirror/Mosby College Publishing.

Waters, T. R., Putz-Anderson, V., Garg, A., and Fine, L. J. 1993. *The Revised National Institute of Occupational Safety and Health (NIOSH) Lifting Equation*. Bristol, PA: Taylor & Francis.

Watt, S. 1986/1987. Nurses' Back Injury: Reducing the Strain. *Australian Nurses Journal*, *16*, 46–48.

8

Fire Prevention and Protection

CHAPTER OBJECTIVES

After completing this chapter, you will be able to

- List the components of the fire tetrahedron
- Define key fire terms
- Identify the classes of fires and extinguishers
- Explain the purpose of the NFPA and prominent safety-related regulations
- Identify placards and labels
- List the components of the fire program

CASE STUDY

The Imperial Food Products plant, located in Hamlet, North Carolina, was owned and operated by Emmet J. Roe. Its 30,000-square-foot, one-story, brick-and-cinder-block facility had no windows but had nine exits, including loading docks. The 200 employees were predominantly female. Chicken breasts were cooked, frozen, and packaged for resale to restaurant chains and frozen food companies. The facility did not have a building-wide sprinkler system at the time of the fire, but a carbon dioxide fire extinguishing system had been installed in the cooking area following a fire in the early 1980s. It was not known if the system was operational at the time of the fire (Rives and Mather, 1991, p. 1A).

The incident began when an overhead hydraulic line ruptured and sprayed flammable hydraulic fluid on the floor in the frying area. Natural gas burners, used to heat the large chicken fryers, ignited the vapors of the fluid and caused a fire, killing 25 workers and injuring an additional 49 employees (Hughes, 1991, p. 8J). As employees tried to leave the burning building, they found the fire exit doors locked. Management had locked several of them to prevent employees from stealing chickens. One door was kicked down as employees fled from the fire and suffocating smoke. A fire investigator later found a door marked "Fire Exit—Do Not Block" padlocked. Some employees could not find a way out and tried to hide in the freezer; most of them died by suffocation. It was later learned the employees had never received any type of fire evacuation training (Hughes, 1991, p. 8).

The Imperial Food Products plant opened its doors ten years prior to the fire. Over those ten years, the plant never had an inspection by the state, local, or federal inspectors. The town of Hamlet was not required to make inspections, although it was up to the municipality to enforce fire regulations. Ironically, every day the plant was in operation, there was a federal inspector on site to make sure the chicken processed was acceptable for consumption. Locking a fire door is a violation of North Carolina's fire code, a misdemeanor in this case. The state fire code requires all municipalities to have a fire inspector, even though a minimum number of inspections is not specified. Under the occupational safety regulations, a locked exit door constitutes a serious violation, and because management locked fire exits knowingly, it would have been a willful violation (Diamond, 1991, p. 7A).

OSHA's General Industry Standards contain the *Means of Egress* regulations for evacuating workplaces during emergencies. The rule requires "a continuous and unobstructed way of exit travel from any point in a building" (Occupational Safety and Health Administration [OSHA] 2003, p. 33). In 1980, OSHA added requirements for employee emergency plans and fire prevention plans to the *Means of Egress* regulations.

Employers must:

- Develop emergency plans
- Create evacuation routes
- Install alarm systems
- Create regular emergency response training programs

For the first time, victims of work-related injuries and deaths could seek prosecution for willful violations causing bodily injuries (Diamond, 1991, p. 1A).

Emmett Roe, the owner of Imperial Food Products, was sentenced to 19 years and 11 months in jail as part of a plea bargain that let his son Brad, the plant's operations manager, stay out of prison. The sentence was for illegally locking plant doors and not having a sprinkler system, leading to the death of the 25 workers.

The tragedy of the fire lies in the similarity to the infamous Triangle Shirtwaist fire that occurred in New York City in 1911. There were 146 workers killed and another 70 seriously injured when fire broke out in a fireproof building where garments were manufactured. Some exits were only twenty inches wide and the employer locked others to prevent theft. Workers were eventually able to break through a door to reach a fire escape, but it soon collapsed.

FIRE TETRAHEDRON

Four components are necessary to sustain combustion:

1. fuel
2. heat
3. oxygen
4. a chemical reaction

A fire will not burn without the presence of all four. Fuels include numerous materials, both liquid and solid, but all must be in a vapor state before they can burn. Once vaporized, they mix with oxygen to form a combustible mixture capable of burning when exposed to heat. During the burning process, a chemical reaction among the components of fuel, oxygen, and heat causes the chemical components to change form and release other gases. The following example will illustrate this process.

An owner of a boat sales and repair shop inspected a boat his assistant was repairing in the shop. During the inspection, he smelled gasoline vapors, so, before opening several doors to air out the place, he instructed his assistant not to activate a grinder. As the owner turned to enter the showroom, he heard the grinder and spun around to see about a dozen fires burning throughout the repair shop. Both men escaped with minor burns, but the boat was destroyed and the shop suffered extensive damage. The assistant may have assumed there were not enough vapors to cause a fire. He may also not have considered that the electrical grinder might spark when activated, and ignite those vapors. Had the vapors not reached the lower flammable limit, they would not have ignited.

The ***lower flammable limit***, also known as the *lower explosive limit*, is the lowest concentration of gas or vapor (percentage volume in air) that burns or explodes if an ignition source is present. The ***upper flammable limit*** or *upper explosive limit* is the highest concentration of a gas or vapor that burns or explodes if an ignition source is present. A mixture can have too little of the concentration (and be too lean) or too much of the concentration (and be too rich) to burn. In the case of the boat repair shop, the concentration was between the limits, so it ignited. Incidentally, the closer the mixture gets to the limits, the less complete the combustion. The optimal mixture for complete combustion is near the midpoint between the limits.

Gasoline is hazardous because it has a low ***flash point***, the temperature at which a substance gives off enough vapors to form an ignitable mixture, causing the substance to burn or explode. The source of ignition might be a flame, hot object, spark from a tool, or static electricity. Although it is possible for the mixture to exceed the upper flammable limit where it becomes too rich to burn, any mixture above 10 percent of its lower flammable limit

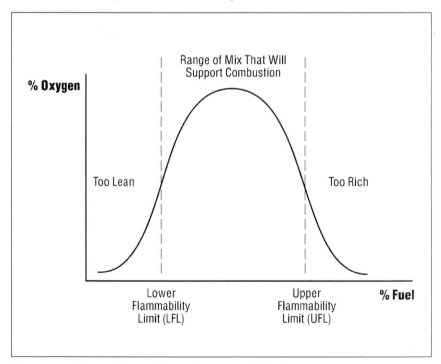

Figure 8-1. Flammability relative to the oxygen/fuel mixture

should be considered a hazard and treated accordingly. Fuel, oxygen, and an ignition source combined to cause an explosion when the assistant activated the grinder.

Vapor density becomes a key issue in determining where a vapor might be found in the atmosphere. If the vapor has a low density (below 1.0), it will float in the air; therefore, ignition of such a fire might occur anywhere in a room or at a point above the source of the vapors. If the vapor has a high density (above 1.0), it will tend to move downward. Vapors released from a gasoline spill will often float downhill. An ignition source below the gasoline may ignite it and cause the fire to spread from the point of ignition to the source of the spill. Police often evacuate a wide area on a highway surrounding a gasoline spill because a nearby car could become a source of ignition and cause the remaining contents of the tank to burn, explode, or both.

A liquid having a flash point at or above 100 degrees (Fahrenheit) is *combustible*. If the liquid ignites at a temperature below 100 degrees (Fahrenheit), it is referred to as *flammable*. The National Fire Protection Association (NFPA) subcategorizes the classifications of combustible and flammable. Knowing these subcategories can be useful in determining how to protect some properties.

CATEGORIES OF FIRES AND EXTINGUISHERS

Fires are categorized according to types of materials involved:

- *Class A fires* involve ordinary combustible materials such as paper, wood, cloth, and some rubber and plastic materials.
- *Class B fires* involve flammable or combustible liquids, flammable gases, greases and similar materials, and some rubber and plastic materials.
- *Class C fires* involve energized electrical equipment where safety requires the use of electrically nonconductive extinguishing media.
- *Class D fires* involve combustible metals such as magnesium, titanium, zirconium, sodium, lithium, and potassium.

Fire extinguishing agents are categorized by the types of fires they extinguish; that is, Class A, Class B, Class C, or Class D extinguishers are used on corresponding types of fires. Some extinguishers can be used on

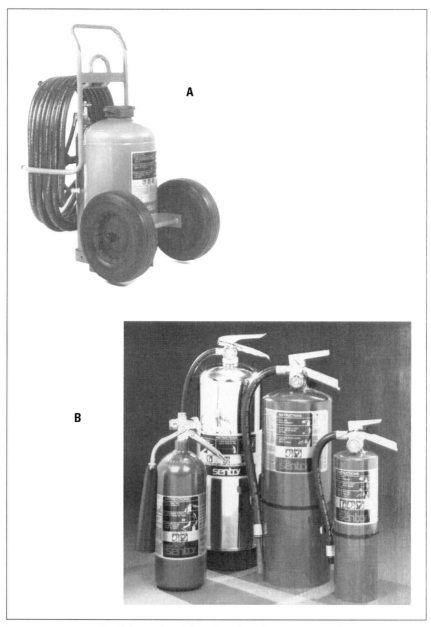

Figure 8-2. Example of fire extinguishers found in the occupational environment. (A) 150-pound wheel fire extinguisher for protection of facilities where larger fires could occur. (B) Hand-held fire extinguishers.

different classes of fires; therefore, Class A-B and Class A-B-C extinguishers are available.

Each type can be recognized as follows. An extinguisher for Class A fires may be rated as 1-A, 2-A, 3-A, 4-A, 6-A, 10-A, 20-A, 30-A, or 40-A. A 4-A extinguisher will extinguish about twice as much fire as a 2-A extinguisher. Class B extinguishers are rated similarly. Class C extinguishers are tested only for electrical conductivity; however, no extinguisher gets a Class C rating without a Class A and/or Class B rating. Class D extinguishers are tested on metal fires. The agent used in a Class D extinguisher depends on the metal for which the extinguisher was designed. The extinguisher faceplate will indicate the effectiveness of the unit on specific metals.

Fire extinguishers are distributed in the workplace so that the travel distance for Class A extinguishers is 75 feet or less. Class B and D extinguishers require travel distances of 50 feet or less. Class C travel distances are based on the appropriate pattern for Classes A or B. Extinguishers shall be mounted and located so they are readily accessible. They require a monthly visual check and an annual maintenance check in addition to any applicable hydrostatic testing.

A good example of a typical heavy-duty fire extinguisher can be found in figure 8-2. Employees who use this type of extinguisher are typically members of an onsite fire brigade.

NATIONAL FIRE PROTECTION ASSOCIATION

The National Fire Protection Association (NFPA) was founded in 1896 to design sprinkler systems and develop techniques for their installation and maintenance. Since then, the organization has evolved to become the preeminent fire prevention group in the United States. Its goal is to safeguard people and their environment from destructive fires through the use of scientific and engineering techniques and education. The NFPA's activities involve development, publication, and dissemination of codes and consensus standards (Ferry, 1990).

The NFPA's *Fire Protection Handbook* was first published in 1896 and is now in a current edition consisting of thousands of pages. It has become increasingly large because of the expansion of the body of knowledge and the complexity of industries that have evolved. The NFPA also publishes other literature, including the *Life Safety Code Handbook*, *Automatic Sprin-*

kler Systems Handbook, Recommended Practices for Responding to Hazardous Materials Incidents, and the National Electrical Code Handbook. It
publishes the Industrial Fire Hazards Handbook, which is not specifically a
code handbook, but is used as a fire-protection guide for industry. This book
focuses on fire hazards and control methods associated with major industries
and their processes.

Standards and Codes

Since the inception of the association, standards and codes have become
more than simple standardization, installation, and maintenance guides for
sprinkler systems. There are currently hundreds of codes covering a large
range of fire-related topics, with numerous NFPA committees developing
standards and codes through a democratic process. All are published in the
National Fire Codes or can be requested individually in pamphlet form.
Some of the most widely used include NFPA 70, National Electric Code;
NFPA 101, Life Safety Code; NFPA 30, Flammable and Combustible Liquids Code; NFPA 13, Automatic Sprinkler Standard; NFPA 58, Liquefied
Petroleum Gases Standard; and NFPA 99, Health Care Facilities Standard.

NFPA 70

The purpose of NFPA 70 is to provide a guide for the practical safeguarding of persons and property from hazards arising from the use of electricity.
It covers installation of electrical conductors within or on public or private
structures, the installation of conductors that connect to the supply of electricity, the installation of other outside conductors on the premises, and the
installations of optical fiber cable (NFPA, 1988).

NFPA 101

NFPA 101 was designed to establish minimum requirements for life
safety in buildings and structures. Life Safety Code addresses a wide range
of topics from fire and similar emergencies: construction, protection, and occupancy features necessary to minimize the dangers from fire, smoke, fumes,
or panic. It identifies minimum criteria for design of egress facilities permitting prompt escape of occupants from buildings, or where possible, into safe
areas within the building. The Life Safety Code also recognizes that fixed

locations occupied as buildings such as vehicles, vessels, or other mobile structures shall be treated as buildings. NFPA 101 is published using mandatory language that may be adopted in a given jurisdiction. Usually the state fire marshal determines who has authority for enforcement (NFPA, 2002).

NFPA 30

The *Flammable and Combustible Liquids Code*, NFPA 30, was developed with the intent of reducing hazards to a degree consistent with reasonable public safety. It does not seek to eliminate all hazards associated with the use of flammable and combustible liquids because the goal was not to interfere with public convenience. The code sets requirements for the safe storage and use of a great variety of flammables and liquids commonly available. It establishes requirements for the safe storage and use of liquids having unusual burning characteristics or subject to self-ignition (NFPA, 2002). An approved storage cabinet for keeping small containers of flammable and combustible liquids is shown in figure 8-3. Approved containers are shown in figure 8-4.

NFPA 13

NFPA 13 covers installation of sprinklers by providing minimum requirements for the design and installation of automatic sprinklers, including the character and adequacy of water supplies. In addition, it covers selection of sprinklers, piping, valves, and all materials and accessories and specifies which buildings require sprinkler systems (NFPA, 2002).

NFPA 58

NFPA 58 pertains to storage and handling of liquefied petroleum gases. It applies to the highway transportation of and the design, construction, installation and operation of liquefied petroleum gas (LPG) systems. It addresses issues such as location of LPG storage tanks, safety analysis, and overfilling protection devices (NFPA, 2002).

NFPA 99

The *Health Care Facilities Standard*, NFPA 99, created criteria to minimize the hazards of fire, explosion, and electricity in health care facilities. It

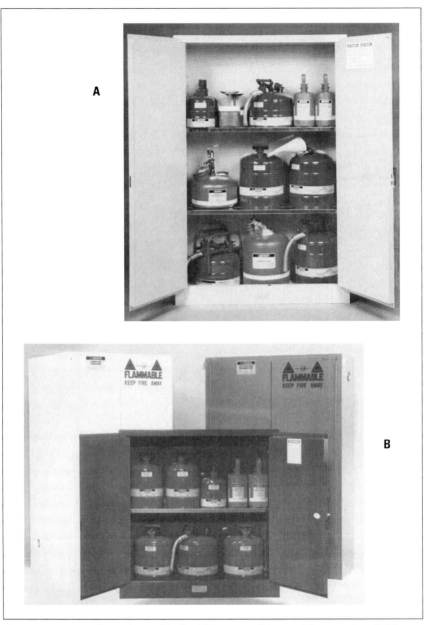

Figure 8-3. (A) and (B) Flammable liquid storage cabinets with examples of different types of safety containers

Leaktight Closure. Spring-loaded, self-closing cap prevents spillage. Cork gasket resists virtually all chemicals; won't get brittle or crack for a safe, tight seal.

Ergonomic Handle Design. Swings free for easy carrying; compound lever action makes pouring easy.

Positive Pressure Relief. Automatically vents internal pressure at 5 psig to protect against explosion.

Welded Pad. Secures handle assembly to top.

Free-flow Flame Arrester. Double mesh, laminated screen dissipates heat to prevent flashback; removes without tools for cleaning.

Double-lock Seams. Four thicknesses of metal at breast and bottom seams, flooded with solder to prevent leaks. Every can is pressure-tested for leak tightness.

Durable, Powder Paint Finish. Maximizes chemical and abrasion resistance.

Yellow Wrap-around Label. Meets OSHA requirements for cans holding liquids with flash points of 80°F and lower.

Ribbed Bottom. Provides extra strength and rigidity. Raised to prevent accidental punctures.

Heavy Gauge Terne-plate Construction Throughout.

FM Approved and UL Listed. Reliability backed by rigorous testing. Also meets OSHA and NFPA requirements.

Figure 8-4.   Example of flammable liquid safety can with illustrated safety features

specifically addresses electrical system wiring; storage and use of flammable and combustible liquids in laboratories, emergency and disaster management, oxygen storage, alarms, and design considerations (NFPA, 2002).

Educational Materials

In addition to developing codes and standards, the NFPA also produces an abundance of educational materials for all age groups and levels of competence. The aids range from fire safety materials for children to fire ground tactics for professional firefighters. These materials come in filmstrips, pamphlets, slides, movies, posters, and books. There are educational materials for training the general employee population and the plant fire brigade and for employee off-the-job fire safety education (Ashford, 1977). The NFPA website at www.nfpa.org gives insight into the offerings.

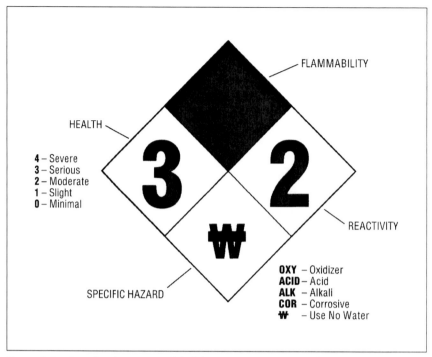

Figure 8-5. Sample of NFPA 704 label for storage containers of hazardous materials

NFPA 704

NFPA 704 is a system of marking storage containers to communicate the hazards associated with the contents of the container. It utilizes a diamond-shaped symbol broken into four quadrants at the top, bottom, left, and right of the diamond. The top portion is shaded red and represents the flammability hazard. Health hazards are indicated in the blue portion. Yellow represents the corrosivity hazard. The bottom portion is white; any special hazards will appear here. For example, a *W* with a line drawn through it means the chemical is water reactive. A radioactive, poison, or other symbol might also appear here. In the top three blocks, the extent of each hazard is represented with a numbering system of 0–4. Zero represents a minimal hazard, while four represents the greatest hazard. These are demonstrated in figure 8-5.

DOT MARKING SYSTEM

The Department of Transportation (DOT) Hazardous Materials Regulations specify the requirements for placarding and labeling of hazardous materials shipped within the United States. These labels are approximately 4 inches by 4 inches; placards are approximately 10 3/4 inches by 10 3/4 inches. Labels are attached directly to the container or on tags or cards attached to the containers. Placards are placed on both ends and each side of freight containers, cargo tanks, and portable tank containers. DOT utilizes United Nations Hazard Class numbers appearing on the bottom of the label or placard, as well as a four-digit, hazard-class identification number (see figure 8-6). This number can be cross-referenced in the DOT *Emergency Response Guidebook* to identify the material and to learn about protective measures to be taken to either avoid or respond to a spill or leak (see figure 8-7).

OSHA REGULATIONS

OSHA deals with fire protection from an employee safety standpoint, and many of the points covered in the OSHA standard are solid management practices for property safety as well. Subpart E, Means of Egress, is taken from NFPA 101, the *Life Safety Code*. The emphasis of this subpart is on protecting the employee once a fire has started. It informs the employer what to do to protect workers during the fire by addressing egress methods, automatic sprinkler systems, fire alarms, emergency action plans, and fire prevention plans.

Means of egress refers to a continuous and unobstructed way of exit travel from any point in a building or structure to a public way. *Safe exit travel consists of three separate and distinct parts:*

1. The way of exit access (such as an aisle or hallway)
2. The exit (such as a doorway)
3. The way of exit discharge (such as a sidewalk outside the building)

In addition, it should be noted that:

- Exits cannot be disguised or obstructed by mirrors, decorations, or other objects.
- They should be marked by readily visible signs.
- Any door, stairway, or other passage possibly confused with an exit should be marked "Not an Exit" or identified as "Linen Closet," "Basement," and so on, so employees can find the way out of the building.

SHIPPING DOCUMENTS (PAPERS)*

The shipping document provides vital information when responding to a hazardous materials/dangerous goods** incident. The shipping document contains information needed to identify the materials involved. Use this information to initiate protective actions for your own safety and the safety of the public. The shipping document contains the proper shipping name (see blue-bordered pages), the hazard class or division of the material(s), ID number (see yellow-bordered pages), and, where appropriate, the Packing Group. In addition, there must be information available that describes the hazards of the material which can be used in the mitigation of an incident. The information must be entered on or be with the shipping document. This requirement may be satisfied by attaching a guide from the ERG2000 to the shipping document, or by having the entire guidebook available for ready reference. Shipping documents are required for most dangerous goods in transportation. Shipping documents are kept in

- the cab of the motor vehicle,
- the possession of the train crew member,
- a holder on the bridge of a vessel, or
- an aircraft pilot's possession.

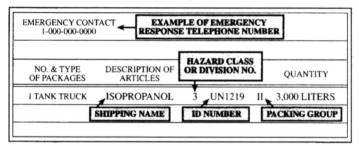

EXAMPLE OF PLACARD AND PANEL WITH ID NUMBER

The 4-digit ID Number may be shown on the diamond-shaped placard or on an adjacent orange panel displayed on the ends and sides of a cargo tank, vehicle or rail car.

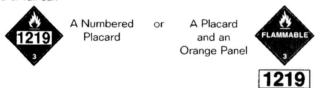

A Numbered   or   A Placard
Placard   and an
                              Orange Panel

* For the purposes of this book, the terms shipping document/shipping paper are synonymous.
** For the purposes of this book, the terms hazardous materials/dangerous goods are synonymous.

Figure 8-6.   Shipping papers and example of placard and panel with ID number

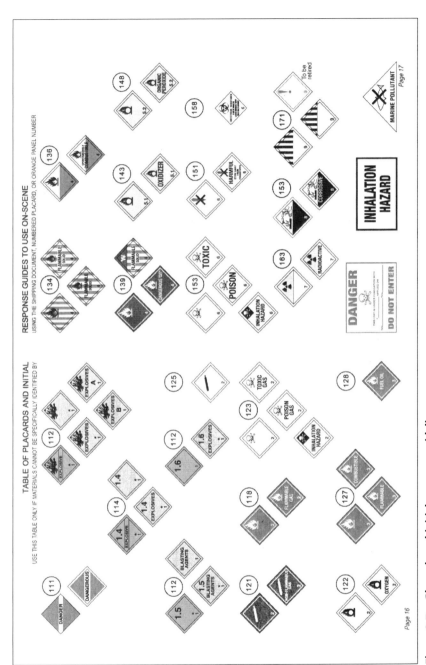

Figure 8-7. Placards and initial response guidelines

The *Life Safety Code* of the NFPA specifies the numbers and types of different occupancies and the sprinklers and alarms appropriate for each. Sprinklers and alarms are not required in all instances.

Subpart L of the OSHA regulations addresses fire protection. It contains requirements for fire brigades, all portable and fixed-fire suppression equipment, fire detection systems, and fire and employee alarm systems.

Safety practitioners have the responsibility to ensure management is aware of how to be in compliance with all applicable fire laws and how to protect workers from fire hazards. This implies instructing employees on how to prevent fires and helping ensure workers are protected if a fire begins. Typical approaches to the fire problem include ***good housekeeping***. Conditions should be monitored relative to neatness and cleanliness because fires may result from conditions such as a stack of oily or greasy rags being left in a pile or an open container. As the rags decompose, they become hot and sometimes ignite through a process known as ***spontaneous combustion***. Other materials, such as damp hay, grass, or paper products exposed to dampness or combustible liquids, can undergo the same process. The fire begins spontaneously with no apparent outside ignition source. Oily rags should be kept in approved safety cans, as shown in figure 8-8, to avoid contacts with ignition sources and incidents involving spontaneous combustion. In crowded facilities aisles and doorways are convenient drop points for tools, bicycles, and boxes because shelves and appropriate areas are remote or full. Employees may block means of egress by using these empty spaces as storage areas. These hazardous conditions must be monitored and management advised to take corrective action.

Ensuring fire alarm and suppression systems are in good working order is not only a regulatory requirement but also a requirement of insurance-company contracts. The loss-control representative from the insurance company often provides services to aid in ensuring all fire prevention and protection systems are in good working order. Both the rep and the safety professional check ***alarm systems*** and other methods of alerting workers in the event of a fire or other emergency. Alarm systems typically consist of mechanical or electronic devices such as bells or sirens located throughout the plant. Smaller businesses, when permitted by the regulations, may depend solely on someone yelling "Fire!" Suppression systems such as fire extinguishers, sprinklers, or standpipes are also inspected.

Sprinklers are the devices, often located on the ceiling, that release water when a fire occurs. Typically, the melting of a fusible link by a hot

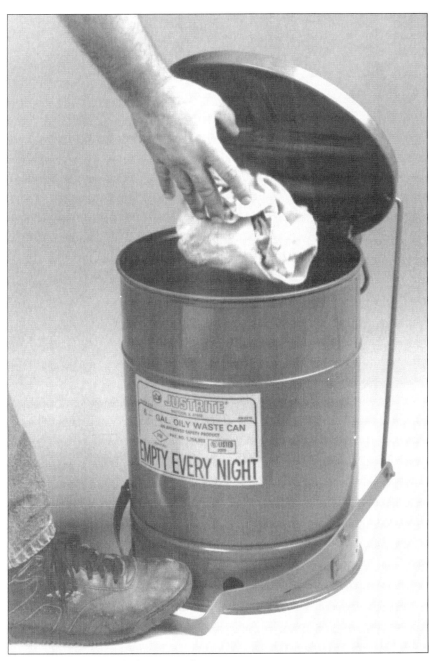

Figure 8-8.   Example of oily waste safety can

fire activates the sprinkler head. As each link is activated, water is released from it and only it. A *dry sprinkler system* is used in freezing conditions, so the water is only released into the pipes when a remote sensor opens a valve. Water will rush to all the sprinkler heads and be released through those with melted links. In a *deluge system*, all sprinkler heads are open so water will be released through all heads. Properly installed and maintained sprinkler systems are nearly 100 percent effective in protecting building occupants from fire.

Standpipes include the piping and hoses located throughout buildings, accessible during a fire. Some standpipes are *dry*, meaning no water is in the pipes until it is pumped in during the emergency. In a tall office building, for example, the hoses and pipes in the system may be for use by the fire department. When firefighters arrive at the scene, they do not want to take the time to drag hoses to the top floors. A dry standpipe system permits them to connect to a water supply at the base of the building and activate the standpipe system on the top floors. Until they connect their water supply, however, no water is available. These systems often employ large-diameter hoses at the receiving end likely to be too unwieldy for untrained persons.

Wet standpipes are typically designed for use by properly trained occupants to protect the people and contents of the building. These systems have a constant supply of water and are smaller in diameter for occupants' use. The loss-control representative of the insurance company attempts to ensure, through the insurance contract, all suppression systems are operating at all times or standby, substitute arrangements are in place during periods of shut down for maintenance and repair.

Some companies depend on *fire brigades*, consisting of employees trained to respond to fire emergencies. Incipient fire brigades may have been taught to use only fire extinguishers for response to *incipient* or small, early-stage fires. Structural fire brigades may have been prepared to a level comparable to that of members of a municipal fire department for response to large, *structural* fires. Structural fire brigades may maintain equipment, including fire trucks and turnout gear, like that found at the fire department. Unless employees have appropriate training and equipment, it is recommended to instruct them to immediately evacuate and sound an alarm when a fire occurs. Even an incipient fire can present a threat to the lives of employees when untrained personnel attempt to extinguish it. For liability and insurance purposes, many companies are reluctant to assume the risk of utilizing an in-house fire brigade as a defense for fires on their properties.

FIRE CASE HISTORY

This account was submitted by Greg Roberts, a master's degree student at East Carolina University, who had worked for a department store chain.

While safety manager, I received a call from Waldorf, Maryland, notifying me a store had been partially destroyed by fire. The details were at best sketchy, so I boarded our corporate plane immediately to survey the damage and report on the cause.

The store was of a typical design with a 54,000-square-foot interior and an additional 5,000-square-foot garden shop attached outside. It met all building codes in Maryland for fire protection and was less than one year old.

When I arrived, the store manager and the inspector in charge of the investigation met me. What we discovered was the fire began in the garden shop and spread via the cantilever roof to the inside structure. However, the fuel source was a surprise.

As many large retailers do, the buying department purchases boxcar loads of merchandise. The merchandise is then segregated and shipped in large quantities to each store. During the summer months, the store stocked and sold a great deal of peat moss, cow manure, and 10-10-10 fertilizer.

During shipping, unloading, or storage, the plastic bags containing the peat moss were damaged, allowing rain water to accumulate inside them while in our parking lot. The peat moss, usually in 40-pound bags with 40 to 50 bags per pallet, was located in the parking lot until it was needed in the garden shop area. When the restocking took place, an entire pallet (40 to 50 bags) was retrieved from the parking lot and placed under the cantilever roof for customer convenience.

The bags were protected from rain after being placed under the roofline, but they were not protected from direct sunlight during the afternoon. This allowed the sun to heat the material and, as a result, ignite the peat moss, or as described by the inspector, allowed spontaneous combustion to occur. We had never experienced an event like this, and, fortunately, no one was injured. The vast majority of the merchandise was destroyed by the sprinkler system as the valves were chained in the open position, locked, and blocked by the last shipment of merchandise that was unloaded. These violations prevented the fire department from reaching the shutoff valve and, as a result, the sprinklers operated well after the fire was extinguished.

We learned a great deal from our misfortune in Waldorf, Maryland. We were able to establish a storage policy mandating all spontaneously combustible material storage must occur away from the building in the parking lot or inside to reduce exposure to the elements. Additionally, we reinforced our policy regarding locked and blocked sprinkler valve shutoff and took disciplinary action against the store manager in Waldorf for these violations. (personal communication, March 29, 1995)

MANAGING THE FIRE PROGRAM

All in-plant and out-of-plant systems are carefully evaluated to ensure plant and emergency response personnel are notified in a timely manner. If fire disables the alarm system, backup systems should be available, and employees should know to utilize them. An effective fire plan incorporates as many of the emergency systems into the daily plant operations as possible in order to ensure they are in operational order and personnel are aware of how to use them.

Management commitment is critical to the success of the fire program and can be measured in terms of the resources and time that management makes available to the program. There should be clear lines of authority leading from manager of the organization to personnel responsible for ensuring a successful fire program.

Since fire equipment is spread throughout the plant, it may be impractical to have supervisors handle inspections in their respective sections and report to the safety department. A better approach is to work with a member of the maintenance or engineering staff to handle these responsibilities. It is generally good practice to involve as many personnel in the fire program as possible to encourage awareness and support.

WRITTEN PROGRAM

29 CFR 1910.38 requires companies to have an emergency action plan in writing covering the following elements:

- Procedures for emergency escape
- Procedures to be followed by employees who remain to operate the plant
- Procedures to account for all evacuees
- Procedures for rescue and medical personnel
- Protocols for alarm systems
- Procedures for training

A written fire prevention plan is also required and should contain the following elements:

- A list of major workplace fire hazards and their proper handling and storage procedures, potential ignition sources and their control procedures, and equipment or systems which can control a fire involving them
- The names or job titles of personnel responsible for maintenance of equipment and systems installed to prevent or control ignitions or fires
- The names or regular job titles of the personnel responsible for control of fuel-source hazards

Training is critical. Employees need awareness of the alarm systems and appropriate response to an alarm. Where the possibility exists for different types of disasters, employees need awareness of the different types of alarms. For example, a continuous bell may indicate an evacuation of the building, as with a fire. An intermittent bell may indicate employees should seek appropriate shelter in the building, as with a tornado.

Employees should also be trained in how and when to use fire extinguishers. Employees are to use extinguishers only if they have been properly trained; to do otherwise would be a violation of OSHA regulations. This is so that employees do not attempt to fight fires beyond their capabilities or the capabilities of the available equipment. If training is not mandated for employees, they have no alternative but to evacuate when a fire occurs.

CONCLUSION

Although most safety practitioners are not required to know the minute details of fire safety, it is imperative they have an awareness of the terminology surrounding fire to effectively communicate with building inspectors, fire marshals, OSHA regulators, fire equipment vendors and maintenance personnel, and insurance representatives. They must also maintain an awareness of the conditions leading to fires and work to help eliminate them throughout the facility. An effective fire program requires constant vigilance and an ongoing partnership with key representatives from industry, government, and the insurance companies. Relationships should be cultivated with as many of them as possible to maintain an effective fire prevention and protection program.

QUESTIONS

1. What is the NFPA? Contact the NFPA and request a catalog of its products. The NFPA can be reached at (617) 770-3500, or you can write to them at One Batterymarch Park, Quincy, MA 02269. Explain what products are available from the NFPA and how these products would be of help to you in industrial safety.
2. Can you find any reference to NFPA in the OSHA Standards? Where? How are you affected by the NFPA and ANSI standards? How are they the same or different from laws?
3. Why do you think NFPA evolved as it did? Are the issues tackled by the NFPA issues that the federal government should be addressing? Explain your answer.

REFERENCES

Ashford, N. 1977. *Crisis in the Workplace.* Cambridge, MA: MIT Press.

Diamond, R. 1991, September 4. Plant Never Had Safety Inspection. *Raleigh News and Observer*, 1A, 7A.

Ferry, T. 1990. *Safety and Health Management Planning.* New York: Van Nostrand Reinhold.

Hughes, J. T. 1991, September 15. Nowhere to Run. *Raleigh News and Observer*, 1J, 8J.

National Fire Protection Association. (various dates). *National Fire Codes.* Quincy, MA: National Fire Protection Association (http://www.nfpa.org).

National Safety Council. 1983. *Protecting Workers' Lives: A Safety Guide for Unions.* Washington, DC: National Safety Council.

Occupational Safety and Health Administration. 2003. *Code of Federal Regulations, Part 1910, General Industry.* Washington, DC: U.S. Government Printing Office.

Rives, J. P., and Mather, T. 1991, September 4. 49 Injured as Doors Bar Safety Routes. *Raleigh News and Observer*, 1A, 7A.

BIBLIOGRAPHY

Chissick, S. S., and Derricott, R. 1981. *Occupational Health and Safety Management.* New York: Wiley and Sons.

Follmann, J. F. 1978. *The Economics of Industrial Health: History, Theory, Practice.* New York: AMACOM.

Lacayo, R. 1991, September 16. Death on the Shop Floor. *Time*, 28–29.

Ladwig, T. W. 1991. *Industrial Fire Prevention and Protection.* New York: Van Nostrand Reinhold.

Meier, K. J. 1985. *Regulation: Politics, Bureaucracy, and Economics.* New York: St. Martins Press.

Price of Neglect. 1992, September 28. *Time*, 24.

Robertson, J. 1975. History and Philosophy of Fire Prevention. *Introduction to Fire Prevention.* Beverly Hills, CA: Glencoe Press.

9

System Safety

Celeste A. Winterberger, PhD

CHAPTER OBJECTIVES

After completing this chapter, you will be able to

- Describe the different elements that compose system safety
- Explain the development of system safety into a discipline
- Describe the importance of system safety today
- Identify the elements of the system life cycle
- Explain how a system safety program is managed
- List the different tools and techniques in the analysis of system safety

CASE STUDY

It was early summer evening on I-35W in Minneapolis, Minnesota. Commuters were heading home from work and the typical rush hour traffic around the metropolitan area was streaming across the eight-lane bridge crossing the Mississippi River. Without warning, the bridge collapsed. The span had been packed with rush-hour traffic, and dozens of vehicles plunged to the river and roads below, leaving drivers and the occupants of their cars scrambling to get out. Those less fortunate died from the fall or drowned. At final count, thirteen were known to be dead and 145 were injured.

Peter Siddons, a senior vice president at Wells Fargo Home Mortgage, was heading north over the bridge toward his home to White Bear Lake when he heard "crunching." "I saw this rolling of the bridge," he said. "It kept collapsing, down, down, down until it got to me." Siddons's car dropped with the bridge, and its nose rolled into the car in front of him and stopped. He got out of his car, jumped over the crevice between the highway lanes and crawled up the steeply tilted section of bridge to land, where he jumped to the ground. "I thought I was dead," he said. "Honestly, I honestly did. I thought it was over" (Levy, 2007).

The National Transportation Safety Board (NTSB) determined the probable cause of the collapse was the inadequate load capacity, due to a design error of the gusset plates. They failed due to a combination of factors including previous bridge modifications and concentrated construction loads on the bridge on the day of the collapse. The design error was permitted due to failure of quality control procedures and inadequate design review by federal and state transportation officials (National Transportation Safety Board, 2008, p. xiii).

The collapse of the bridge was a result of numerous failures in a series of systems including design, modification, and inspection. The NTSB noted the failures in the system, and they were a result of errors on the part of management. Until management errors are identified *and* corrected, systems will continue to fail.

Photo 9-1. Minnesota bridge collapse (copied from National Transportation Safety Board [2008])

DEFINITIONS

To understand system safety, one must know the fundamentals that go into defining the discipline. A *system* can be defined as "a group of interconnected elements united to form a single entity." Systems may include something as simple as a toaster or as complex as a chemical refinery. Perhaps the most important thing about systems is that they can sometimes be, and often are, further defined into subsystems, assemblies, subassemblies, and components (see figure 9-1). If the subsystems contain interdependent entities, they then also can be defined as systems. For example, a car brake can be defined as either a stand-alone system or a subsystem of an automobile.

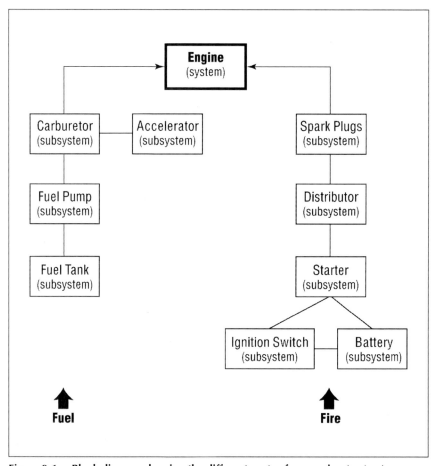

Figure 9-1. Block diagram showing the different parts of car engine (system)

Safety can be defined as making something free from the likelihood of harm. Roland and Moriarty (1990, p. 7) state that "safety in a system" is "a quality of a system that allows the system to function under predetermined conditions with an acceptable minimum of accidental loss."

A *hazard* is anything that can possibly cause danger or harm to equipment, personnel, property, or the environment. It is a circumstance that has the potential, under the right conditions, to become a loss.

Risk involves the probability that an incident will occur or the chance of occurrence *and* the resulting loss. The batting average of a baseball player can be described by a pitcher as the risk that the batter will get a hit. If someone has a batting average of .200, it means that if the pitcher throws ten pitches, it would be likely that the batter would hit two fair. Obviously, the higher the risk (including probability and cost of loss), the more important it becomes to find and mitigate the hazard. In the case of the bridge collapse, the risk was high because failure of the gusset plates would likely cause death if it occurred.

An *accident* is a dynamic occurrence caused by the activation of a hazard and consists of a number of interrelated events resulting in a loss. It can cause the injury or death of individuals, as well as property damage to equipment and hardware. A related term that is sometimes used to refer to accidents but is actually a different kind of event is an incident. An *incident* is also an unplanned event but may or may not have an adverse effect. An incident may simply be an occurrence with no losses. These are sometimes referred to as *near misses*.

System safety can be defined as "an optimum degree of safety, established within the constraints of operational effectiveness, time, and cost . . . achievable throughout all phases of the system life cycle" (Malasky, 1982, p. 17). The system-safety concept deals with the before-the-fact identification of hazards as opposed to the after-the-fact approach used for years. Consider the bridge collapse example: the accident occurred and the problem was found and resolved. Using the system-safety approach would have meant that information concerning the performance of the gusset plates was gathered prior to design and testing and was also gathered throughout implementation. There may have been a number of solutions to this problem. Even if the problem had been discovered before the fact, and no immediate solution was present, procedures might have been in place to lower the bridge load or divert traffic under appropriate load conditions—until the problem was addressed and the plates were replaced.

HISTORY OF SYSTEM SAFETY

One of the first mentions of the concept of system safety appeared in the technical paper "Engineering for Safety" presented at the Institute for Aeronautical Sciences in 1947. It stressed that safety should be designed into airplanes, and it continued by stating that safety groups should be an important part of the organization. It wasn't until the early 1960s and the development of ballistic missile systems that the concept gained a more formal acceptance. Contractors were given the responsibility for safety, replacing the practice of shared responsibility by each individual involved in the process. The first system-safety requirements were published by the Air Force in 1962. These were modified in 1963 into the Air Force specification MIL-S-38130. In 1966, the Department of Defense adopted these specifications as MIL-S-381308A. *MIL-STD-882 System Safety Program for Systems and Associated Subsystems and Equipment; Requirements for*, developed in 1982, contained the specifications for the system-safety program required by all military contractors.

NASA also implemented a system-safety program patterned after the Air Force standards. These programs were instrumental to the successful completion of many NASA projects, including the Apollo moon missions.

The private sector has begun to develop system-safety programs because of the successes of the military and NASA. Leading the way are the nuclear power, refining, and chemical industries. The adoption of system safety in those industries manufacturing consumer products has generated returns in terms of more effective products, fewer accidents, and longer product life.

IMPORTANCE OF SYSTEM SAFETY TODAY

As society becomes more technically advanced, its tools become increasingly sophisticated. In some cases, the machine has advanced further than the human capacity to control it. Jet fighters are good examples. These machines are capable of performing in G-forces that incapacitate most humans. Safety professionals need to be aware of the limits of human performance, as well as the fallibility of the individual in a mechanized system.

Product liability is also a major concern for many companies. The McDonald's coffee award is a good example. In this case, a woman placed a cup of McDonald's coffee between her legs in a car. The coffee spilled and the woman was severely burned. McDonald's was found liable for the injuries,

Table 9-1.   System life cycle

Phase	Safety Control Point	Result
Concept	Concept design review	Establish design for general evolution
Definition	Preliminary design review	Establish general design for specific development
Development	Critical design review	Approve specific design for production
Production	Final acceptance review	Approve product for release in deployment
Deployment	Audit of operation and	Control of safety operation and maintenance

even though McDonald's cups contain a warning stating the contents are hot. A systems safety approach could have prevented this event.

SYSTEM LIFE CYCLE

The system life cycle consists of six phases: concept, definition, development, production, deployment, and disposition. At the end of each phase, a safety review is conducted. A decision is then made regarding whether to continue the project or place it on hold, pending further examination.

During the *concept* phase, historical data and technical forecasts are developed as a base for a system hazard analysis. A preliminary hazard analysis (PHA) is conducted during this phase. At the gross level, a risk analysis (RA) is performed to ascertain the need for hazard control and to develop system-safety criteria. Safety management will be doing the initial work on the System Safety Program Plan (SSPP). Three basic questions must be answered by the time the concept phase is completed:

- "Have the hazards associated with the design concept been discovered and evaluated to establish hazard controls?
- Have risk analyses been initiated to establish the means for hazard control?
- Are initial safety design requirements established for the concept so that the next phase of system definition can be initiated?" (Roland and Moriarty, 1990, p. 23)

The *definition* phase is used to verify the preliminary design and product engineering. Reports presented at design review meetings typically discuss the technological risks, costs, human engineering, operational and mainte-nance suitability, and safety aspects. In addition, subsystems, assemblies, and subassemblies of the system are defined at this time. The PHA is updated and a subsystem hazard analysis (SSHA) is initiated so it can later be inte-grated into the system hazard analysis (SHA). Safety analysis techniques are used during this phase to identify safety equipment, specification of safety design requirements, initial development of safety test plans and require-ments, and prototype testing to verify the type of design selected.

Environmental impact, integrated logistics support, producible engineer-ing, and operational use studies are done during the *development* phase. The SSHA and safety design criteria are also completed during this phase. Interfaces with other engineering disciplines within the organization are fostered. Using the data collected, a go/no-go decision can be made before production begins.

The *production* phase of the system life cycle involves close monitor-ing by the safety department. In addition, the quality-control department becomes important because of its focus on inspection and testing of the new product. Training begins during this phase. Updating of the analyses started during the definition and development phases continues. Finally, all the information collected during this phase is compiled into the System Safety Engineering Report (SSER). The SSER identifies and documents the hazards of the final product or system.

When the system becomes operational, it is in the *deployment* phase. Data continues to be collected and training is conducted. If any problems occur, individuals responsible for system safety must be available to follow up and decide on possible solutions. The system safety group in the organization also reviews any design changes made on the system or product.

A sixth phase of the system life cycle, the *disposition* or *termination* phase, is the time a product or system is removed from service. In certain cases, the removal of a product may in itself create a hazardous situation. A good example is asbestos removal from a building or light transformer replacement due to PCBs. Safety professionals monitor these situations so both the worker and the public are protected.

MANAGEMENT OF SYSTEM SAFETY

Malasky (1982, p. 31) defines "system safety management" as "that element of program management which ensures the accomplishment of the system

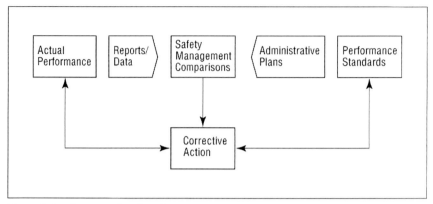

Figure 9-2.   System safety functions

safety tasks, including identification of system safety requirements; planning, organizing, and controlling those efforts which are directed toward
achieving the safety goals; coordinating with other (system) program elements; and analyzing, reviewing, and evaluating the program to ensure effective and timely realization of the system safety objectives."

Using this definition, a block diagram can be drawn to illustrate the system
safety function (see figure 9-2).

Organizational Location

The system safety organization should have a position within the company
where wide access to all areas of design, maintenance, and operation occurs. It can exist at many different levels, depending on the size of the
organization. In large companies, the system-safety role may be found at
all functional levels of the organization. In fact, MIL-STD-882B requires
that all vendors submit a system-safety plan along with their proposals for
federal contracts. Sometimes system-safety councils are used to reinforce
corporate polices throughout the different levels of the company. Smaller
companies may include the system safety work as part of each functional
department. For example, the design department should have an individual
whose responsibility is to perform product or system safety analyses during
the development phase of the system life cycle. However, it is up to each
company to decide on which levels system safety should reside. The bottom
line is to ensure that all risks have been determined and those that are unacceptable have been eliminated.

Organizational Interfaces

System safety analysis is not performed in isolation. Input is necessary from nearly all functional disciplines within the organization. During the concept and definition phases, system safety primarily interfaces with the design group. Interfaces with engineering are important during the development phase. Quality assurance, training and development, industrial safety, and manufacturing engineering are important interfaces developed during the production phase. Interfaces with maintenance and product-support disciplines are important during the deployment phase. The termination phase finds system safety interfacing with industrial safety, industrial hygiene, and product support.

Implementation Difficulties

Malasky (1982) discusses some of the difficulties sometimes encountered during implementation. Problem formulation is cited as one issue. This may be caused by the conflicting demands of the various functional departments during the design optimization process or because of inadvertent minimization of potential hazards. For example, during the design process for the Ford Pinto, a fuel-filler problem was discovered. Management decided the cost of delay would be much higher than the cost of any lawsuits resulting from the hazard and went forward with production. In the long run, the cost to alleviate the hazard turned out to be much less than the costs of the lawsuits and loss of consumer confidence.

Organizational interfaces may pose other problems. When management does not take responsibility for decision making in the process, functional and system safety groups may be at odds with each other. Management perception is another difficulty. System-safety concepts are more abstract than those in many other disciplines. Therefore, it is important for the safety professional to structure the program so its impact is clear.

Elements of a System-Safety Program Plan (SSPP)

According to MIL-STD-882B (Department of Defense, 1984), the SSPP must specify the four elements of an effective system-safety program:

- A planned approach for task accomplishment
- Qualified people to accomplish the tasks

- Authority to implement the tasks through all levels of management
- Appropriate resources for manning and funding to ensure that tasks are completed

To accomplish these objectives, the SSPP should describe:

- The safety organization
- System safety program milestones
- General system safety requirements and criteria
- Hazard analysis techniques and formats
- System safety data
- Safety verification
- Audit programs
- Training requirements
- Mishap and hazardous malfunction analysis and reporting
- System safety interfaces

For more detailed information regarding the specifics for each of these areas, the safety professional may refer to the standard.

TOOLS AND TECHNIQUES

In item four of the SSPP, a company must identify the types of techniques used in analyzing and evaluating system hazards. The following section will discuss some of the tools commonly used by the safety practitioner.

Preliminary Hazard Analysis

A preliminary hazard analysis (PHA) is the initial effort in identifying hazards which singly or in combination could cause an accident or undesired event. PHA is a system-safety analysis tool used to identify hazard sources, conditions, and potential accidents (Roland and Moriarty, 1990). At the same time, PHA establishes the initial design and procedural safety requirements to eliminate or control these identified hazardous conditions. A PHA is performed in the early stages of the conceptual cycle of system development. It can be performed by engineers, contractors, production line supervisors, or safety professionals. Management must always first look at any risk involved in the operation of the system.

After identifying hazards and their resultant adverse effects, the analyst will rate each according to the hazard classification class, which could be one of four categories:

- Class I—Catastrophic: A condition(s) that will cause equipment loss and/or death or multiple injuries to personnel
- Class II—Critical: A condition(s) that will cause severe injury to personnel and major damage to equipment, or will result in a hazard requiring immediate corrective action
- Class III—Marginal: A condition(s) that may cause minor injury to personnel and minor damage to equipment
- Class IV—Negligible: A condition(s) that will not result in injury to personnel, and will not result in any equipment damage

Roland and Moriarty (1990) show how to develop a Hazard Assessment Matrix to determine a Hazard Risk Index using frequency of occurrence and hazard category (see figure 9-3).

Frequency of Occurrence	Hazard Categories			
	I Catastrophic	II Critical	III Marginal	IV Negligible
(A) Frequent	1A	2A	3A	4A
(B) Probable	1B	2B	3B	4B
(C) Occasional	1C	2C	3C	4C
(D) Remote	1D	2D	3D	4D
(E) Improbable	1E	2E	3E	4E

The Hazard Risk Index (HRI) is then determined by reading the chart and assigning risk into one of the four levels of acceptability:

I	Unacceptable
II	Undesirable with management waiver required
III	Acceptable with management review
IV	Acceptable without review

Figure 9-3. Hazard Assessment Matrix

Subsystem Hazard Analysis

A subsystem hazard analysis is performed to identify hazards in the component systems within a larger system. For example, in the bridge collapse accident, the gusset plates could be considered a subsystem. When the plates, components of the subsystem, broke, a total system breakdown began as a cascade effect that ultimately destroyed the bridge. An analysis of all design components should be started no later than the definition phase in the system life cycle and continue until the beginning of the system production phase. Analysis techniques include fault hazard analysis (FHA) and fault tree analysis (FTA), discussed in more detail in the next section (Roland and Moriarty, 1990).

Hazard Analysis Techniques

The role of the safety professional is to anticipate, identify, and evaluate hazards; give advice on the avoidance, elimination, or control of hazards; and attain a state for which the risks are judged to be acceptable. To achieve this, the safety professional adopts a system-safety concept that includes

- an understanding of the hazards,
- an understanding of the risks,
- an identification of the hazards and risks within their system,
- an understanding of unwanted releases of energy and unwanted releases of hazardous materials being the causal factors for hazard-related incidents, and
- a knowledge of the principles and techniques used to control hazards and reduce their associated risks to an acceptable level (Manuele, 1993).

A hazard analysis is used to identify any dangers that might be present in a proposed operation, the types and degrees of accidents that might result from the hazards, and the measures that can be taken to avoid or minimize accidents or their consequences (Hammer, 1989).

System hazard analysis (SHA) is primarily performed during the definition and development phases of the system life cycle (Roland and Moriarty, 1990). However, it should be continuously implemented throughout the life cycle of a system, project, program, and activity to identify and control hazards. The purpose of performing an analysis during the early stages of the life cycle is to reduce costs. If the analysis is done after the system is in operation, the system may need to be redesigned and consequently withdrawn from service. In addition, if the system is close to the end of its life cycle, it might not be cost-effective to change it (Brauer, 2006).

A hazard analysis should contain the following information:

- Descriptive information
  —System mode
  —Subsystem mode of subsystem of hazard origin
  —Hazard description
  —Hazard effects
  —Likelihood or relative likelihood of each hazard
- Causation events of each hazard
  —Identification of events precisely as to subsystem mode, system mode, and environmental constraints
- Subsystem interface problems of special significance
  —Identification of subsystem involved
  —Identification of system and subsystem modes
- System risk evaluation
  —Severity listing of each hazard
  —Likelihood of each hazard
  (If a full quantitative evaluation is conducted, a risk evaluation should be presented for each hazard.)
- Risk summary
  —Listing risks of each hazard and for the system as a function of system modes
  —A logical evaluation of acceptability of system risks
  —Recommendations as to system risk control

This analysis can and must begin as soon as the idea for a new system or operation is conceived (Roland and Moriarty, 1990).

Technique of Operations Review (TOR)

In 1987, D. A. Weaver developed the technique of operations review (TOR) (Ferry, 1988). It was designed to uncover management oversights and omissions instead of hardware or operator problems. The four steps of a TOR analysis are state, trace, eliminate, and seek.

During the state portion of the analysis, detailed information about the hazard is gathered. If the hazard is discovered as the result of an accident, a summary of the mishap report is reviewed.

The trace portion uses a sheet displaying possible operational errors under eight categories:

- *Training* includes all errors related to inadequate preparation of the employee.
- *Responsibility* considers errors in the organizational requirements which may have contributed to the hazard.
- *Decision and direction* looks at the lack of or inadequacy in the decision-making process which causes errors in the performance of the product or employee.
- *Supervision* refers to errors due to problems with the direction of employee work.
- *Work groups* problems can be traced to the interpersonal relationships within the group.
- *Control* deals with errors related to inadequate safety precautions.
- *Personal traits* can be traced to the individual's personality and how it affects the individual's job performance.
- *Management* can be traced to poor managerial control.

Under each category is a list of numbered operational errors in that group. (A complete copy of the TOR analysis materials can be obtained from the FPE Group, 3687 Mt. Diablo Road, Lafayette, CA 94549, who are the sole distributors of this analysis instrument.) To begin the analysis, a single number or prime error under one of the eight categories is chosen. Using the Minnesota bridge collapse disaster, the prime number may be under category five, control: number 61, unsafe condition. Each subcategory number on the TOR sheet then points to other possible factors that may have contributed to the hazard or mishap. Possible factors that could have contributed in the bridge collapse include:

65. Deficient inspection, report, or maintenance (main category: control).
36. Failure to investigate and to apply the lessons of similar mishaps (main category: decision and direction).

43.　Unsafe act; failure to correct before the accident occurred (main category: supervision).

86.　Accountability; failure to develop appraisal and measurement of key goals and objectives (main category: management).

These factors are then discussed by the group. Those found to have contributed to the accident can be further broken down until all the causative factors have been identified.

When the trace portion of the analysis is finished, the "eliminate" step begins. Sometimes the trace step can identify a large number of contributing factors, a number often overwhelming to the group evaluating the hazard. Therefore, the "eliminate" step is used to discuss each contributing factor and evaluate its merit to the process.

The final step, "seek," looks for possible actions that need to be taken to correct the problem. These solutions should then be implemented.

Technique for Human Error Rate Prediction (THERP)

Swain and Rook developed the technique for human error rate prediction (THERP) in 1961 to quantify human error rates due to problems of equipment unreliability, operational procedures, and any other system characteristic that could influence human behavior (Malasky, 1982). THERP is an iterative process consisting of five steps executed in any order:

- Determine system failure modes to be evaluated.
- Identify the significant human operations required and their relationship(s) to system operation and system output.
- Estimate error rates for each human operation or group of operations that are pertinent to the evaluation.
- Determine the effect of human error on the system and its outputs.
- Modify system inputs or the character of the system itself to reduce the failure rate (Malasky, 1982, p. 257).

THERP is usually modeled using a probability tree. Each branch represents a task analysis showing the flow of task behaviors and other associations. A probability is assigned based on the event's occurrence or nonoccurrence.

Failure Mode and Effects Analysis (FMEA)

A failure mode and effects analysis (FMEA) is a sequential analysis and evaluation of the kinds of failures that could happen and their likely effects, expressed in terms of maximum potential loss. The technique is used as a predictive model and forms part of an overall risk assessment study. This analysis is described completely in the MIL-STD-1629A. The FMEA is most useful in system hazard analysis for highlighting critical components (Ridley, 1994).

In this method of analysis, the constituent major assemblies of the product to be analyzed are listed. Next, each assembly is broken into subassemblies and components. Each component is then studied to determine how it could malfunction and cause downstream effects. Effects might result on other components, and then on higher-level subassemblies, assemblies, and the entire product. Failure rates for each item are determined and listed. The calculations are used to determine how long a piece of hardware is expected to operate for a specific length of time. It is the best and principal means of determining where components and designs must be improved to increase the operational life of a product. In addition, it is used to analyze how frequently and when parts must be replaced if a failure, possibly related to safety, must be avoided (Hammer, 1989).

The FMEA is limited to determination of all causes and effects, hazardous or not. Furthermore, the FMEA does very little to analyze problems arising from operator errors or hazardous characteristics of equipment created by bad design or adverse environments. However, the FMEA is excellent for determining optimum points for improving and controlling product quality (Hammer, 1989). (An example of an FMEA for a bicycle can be found in figure 9-4.)

Fault Hazard Analysis (FHA)

A fault hazard analysis (FHA) is an inductive method for finding dangers that result from single-fault events (Roland and Moriarty, 1990). It identifies hazards during the design phase. The investigator takes a detailed look at the proposal system to determine hazard modes, causes of hazards, and the adverse effects associated with the hazards.

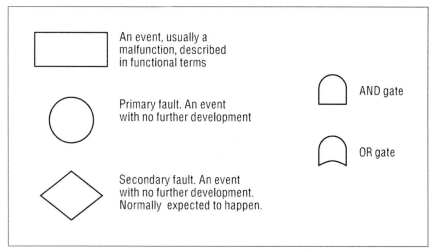

Figure 9-4. Basic FTA symbology

Fault Tree Analysis (FTA)

The most widely used analytical technique, fault tree analysis (FTA), is a symbolic logic diagram graphically depicting the cause-and-effect relationships of a system (Ferry, 1988).

The hierarchy of fault tree events can be classified as follows:

1. Head event (top event): The event at the top of the tree that is to be analyzed
2. Primary event: The main malfunction of the component
3. Secondary event: The effect that is caused on another component, device, or outside condition
4. Basic event: "An event that occurs at the element level and refers to the smallest subdivision of the analysis of the system" (Ferry, 1988, p. 153)

Using this basic hierarchy, an FTA is conducted using three steps. The first is the determination of the head event. Once the head event is set, primary and secondary events can be determined. The final step is the resolution of the relationship between the casual events and the head event using the terms AND and OR.

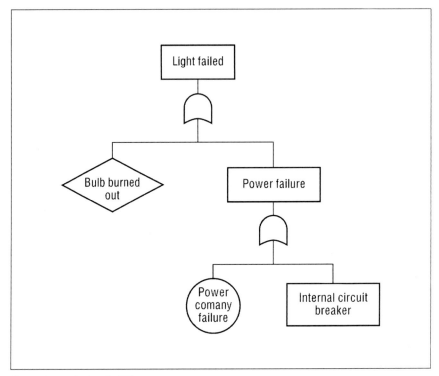

Figure 9-5.   A simple FTA

An example of a simple FTA may help explain how AND and OR logic is used to define relationships. If an analysis is done on the failure of an electric light bulb, the following fault tree can be developed.

The tree shows two OR gates. After the light fails, there are two possible alternatives. Either the bulb will burn out or there is a power failure. If there is a power failure, there are two possible outcomes: either the power has failed to come into the building or an internal circuit breaker has failed.

On the other hand, if there is an AND gate, both outcomes must occur for the event to happen. If an AND gate was placed after the power failure, both a power company failure and an internal circuit failure must happen for the event to take place.

There are limitations to this type of analysis. A specific knowledge of the design, construction, and the use of the product is necessary for successful completion of the FTA. In addition, FTA logic assumes that all events are either successes or failures. Many times such a concrete decision is not possible.

JOB SAFETY ANALYSIS (JSA)

Reasons for Conducting a JSA

Job safety analysis (JSA) is an analytical tool that can improve a company's overall performance by identifying and correcting undesirable events that could result in accidents, illnesses, injuries, and reduced quality and production. It is an employer/employee participation program in which job activities are observed; divided into individual steps; discussed; and recorded with the intent to identify, eliminate, or control undesirable events.

JSA effectively accomplishes this goal because it operates at a very basic level. It reviews each job and breaks it down into an orderly series of smaller tasks. After these tasks have been determined, the same routine of observation, discussion, and recording is repeated, this time focusing on events which could have a negative impact on each step in the task. Once potential undesirable events are recognized, the process is repeated for a third time and corrective actions are identified.

Conducting a JSA can be a valuable learning experience for both new and experienced employees. Not only does it help them understand their jobs better, but it also familiarizes them with potential hazards and involves them in developing accident prevention procedures. Workers are more likely to follow procedures if they have a voice in planning. Finally, the JSA process causes employees to think about safety and how it relates to their jobs.

Who Should Conduct JSAs?

The responsibility for the development of a JSA lies with first-line supervision. These individuals have firsthand knowledge of the process, its potential hazards, and the need for instituting corrective actions at each step. This also provides the interaction with hourly employees necessary to complete the JSA. Initially, first-line supervisors must receive training in hazard recognition and procedures necessary to perform a JSA. This training will give them the knowledge necessary to explain the JSA to employees, what it is expected to accomplish, how it is conducted, and what their part will be in the program.

A well-organized and maintained JSA program can have a very beneficial effect on accident prevention, improved production, and product quality. Responsibility for this program, as with any other program, must start at the top and be conveyed to all employees.

Procedures and Various Methods Used to Perform JSAs

A job safety analysis is a procedure used to review job methods and uncover hazards that

- may have been overlooked in the layout of the plant or building and in the design of the machinery, equipment, tools, workstations, and processes;
- may have developed after production started; or
- may have resulted from changes in work procedures or personnel.

The principal benefits of a JSA include

- giving individual training in safe, efficient procedures,
- making employee safety contracts,
- instructing the new person on the job,
- preparing for planned safety observation,
- giving prejob instruction on irregular jobs,
- reviewing job procedures after accidents occur, and
- studying jobs for work-methods improvements.

A JSA can be performed using three fundamental steps, but a careful selection of the job to be analyzed is an important preliminary step.

There are three basic methods for conducting JSAs. The *direct observation method* uses observational interviews to determine the job steps and hazards encountered. A second way to perform a JSA is using the *discussion method*. This method is typically used for jobs or tasks that are performed infrequently. It involves pulling together individuals who have done the job and letting them brainstorm regarding the steps and hazards. The third way to perform a JSA is called the *recall-and-check method*. This method is typically used when a process is ongoing and people can't get together. Everyone participating in this process writes down ideas about the steps and hazards involved in the job. Information from these individuals is compiled and a composite list is sent to each participant. Each person can then revise the list until consensus is achieved.

The following list gives the three basic approaches used to determine how to perform a specific JSA.

- By a specific machine or piece of equipment (for example, lathe)
- By a specific type of job (for example, machining)
- By a specific occupation (for example, machinist)

Selecting the Job

A job is a sequence of separate steps or activities that together accomplish a work goal. Jobs suitable for a JSA are those a line supervisor chooses; jobs are not selected at random. Those with a poor work accident history or high associated costs are analyzed first to yield the quickest possible results.

In selecting jobs to be analyzed and establishing the order of analysis, top supervision is guided by the following factors:

- **Frequency of accidents.** A job repeatedly producing accidents is a candidate for a JSA. The greater the number of accidents associated with the job, the greater its priority claim for a JSA.
- **Rate of disabling injuries.** Every job having a history of disabling injuries should have a JSA performed. Subsequent injuries prove that preventive action taken prior to their occurrence has not been successful.
- **Severity potential.** Some jobs may not have a history of accidents but have the potential for causing severe injuries. The more severe the injury, the higher the priority is for a JSA.
- **New jobs.** Changed equipment or processes obviously have no history of accidents, but the accident potential may not be understood. A JSA should be conducted for each new job. Analysis should not be delayed until an accident or near miss occurs.

After the job has been selected, the three basic steps in conducting a JSA are as follows:

- Breaking the job down into its component steps
- Identifying the hazards and potential accidents
- Developing solutions

1. *Breaking the job down into its component steps.* Before the search for hazards can be started, a job should be broken down into a sequence of steps, each describing what is to be done. There are two common errors to avoid in this process:

a. making the job breakdown too detailed so that an unnecessarily large number of steps results, and

b. making the job breakdown so general that the basic steps are not recorded.

To perform a job breakdown, use the steps in figure 9-6.

Job Name _____	JSA Number _____
Employee Name _____	Area/Supervisor _____
Employee Title _____	Last Analysis Date _____
Analysis By _____	Analysis Date _____

Job Steps	Potential Hazards	Necessary Safety Procedures	Required Safety Equipment

Figure 9-6. Job safety analysis worksheet

- Select the right worker to observe. Select an experienced, capable, and cooperative person who is willing to share ideas.
- Observe the employee performing the job.
- Completely describe each step. Each step should tell what is done, not how it is done.
- Number the job steps consecutively.
- Watch the operator perform the job a number of times until you are sure that all the steps have been noted.
- Check the list of steps with the person observed to obtain agreement on how the job is performed and the sequence of the steps.

2. *Identifying hazards and potential accidents.* The purpose of a JSA is to identify all hazards, both those produced by the environment and those connected with the job procedure. Each step must be made safer and more efficient.

   Close observation and knowledge of the particular job are required for the JSA to be effective. The job observation should be repeated until all hazards and potential accidents have been identified.

3. *Developing solutions.* The final step in a JSA is to develop a safe job procedure to prevent the occurrence of accidents. The principal types of solutions are as follows:

   - find a new way of doing the job
   - change the physical conditions that create the hazards
   - change the work procedure
   - reduce the frequency of the job

Completing the JSA

After the worksheet is complete, the data compiled is transferred to the actual JSA form. Once it is entered and verified, signature approval is obtained for the JSA from an upper-level manager. After that, the findings are discussed with employees performing the job. Any necessary safety procedures are reviewed and equipment required to perform the job is discussed with employees. In addition, a copy of the JSA is available for employees to use when they perform the job—particularly for jobs not done on a regular basis. No job is static; JSAs are critiqued on a regular basis and any necessary changes are made.

Effectively Using a JSA in Loss Prevention

The major benefits of a JSA come after its completion. Supervisors can learn more about the jobs they supervise and can use JSAs for training new employees. JSAs provide a list of steps needed to perform the job and identify the procedures and equipment needed to do it safely. Employees can use JSAs for improved safety knowledge and as reviews before performing unfamiliar jobs. Supervisors occasionally observe employees as they perform the jobs for which the analysis was developed. If any procedural deviations are observed, the supervisor alerts the employee and reviews the job operation.

JSAs should be reevaluated following any accident. If the accident is a result of an employee failing to follow JSA procedures, the fact is discussed with everyone doing the job. It can be made clear the accident would not have occurred had the JSA procedures been followed. If the JSA is revised following an accident, these revisions are brought to the employees' attention. (See figure 9-6 for a sample JSA worksheet.)

An Example

Performing JSAs is a complex process. To better understand the importance of this analysis tool, figure 9-7 shows a completed JSA for hydraulic line replacement. Please note accident categories described in the potential hazards section. These potential hazards will aid in determining what safety measures and personal protective equipment are necessary for a given job.

ACCIDENT INVESTIGATION

Accident investigation techniques are discussed in chapter 5. However, it is important to note that the accident investigation is a form of system safety analysis. During the deployment phase, accidents may occur and the findings of the investigation may result in a redesign, retrofit, and/or recall of the product or system.

JSA #: 1430	TASK NAME: HYDRAULIC LINES–CHANGING		PAGE: 1 OF 3
JOB TITLE: HYDRAULIC LINE		AREA: MAINTENANCE	
MILL:	REVISION DATE: 08/15/94	WRITTEN BY: JOE HAMMERSMITH	
INTERVIEWS WITH: ED SMITH		APPROVED BY: DAVID JONES	
1994 REVIEWED BY:	1995 REVIEWED BY:	1996 REVIEWED BY:	1997 REVIEWED BY:
1998 REVIEWED BY:	1999 REVIEWED BY:	2000 REVIEWED BY:	2001 REVIEWED BY:

TASK STEPS	POTENTIAL HAZARDS	SAFETY PROCEDURE TO FOLLOW	SAFETY EQUIPMENT
1) Put on all required safety equipment, then report to job location.	*Contact By:* Forklifts/Blind Corners	When going anywhere in the mill, always walk to one side of pathways. Look before rounding corners.	Hard hat Safety glasses Steel-toed boots Hearing Protection Gloves Lock-out tags
	Foot-Level Fall: Oil, water, or trash on floor.	Avoid wet, oily spots. Pick up or step over trash or bunks. If your shoes/boots get slippery, clean them.	
2) Lock-out the system for the hydraulic hose.	*Caught in:* You may become entangled in the machinery.	**All workers must place their Lock-out Tag on all system breakers.**	Lock-out tags
3) Bleed off system pressure if needed.	*Foot-Level Fall:* The floor is slippery from hydraulic fuel.	Walk around spilled fluid if possible. Use Oil Dri to soak up spills.	Oil Dri
4) Remove the fitting on end of hose.	*Contact By:* Hydraulic fluid will irritate and damage the eyes and skin. *Contact with:* Tools may slip and cause you to hurt your hands.	Loosen the fitting to bleed off any remaining pressure in the hose. Keep the hose away from your face and others. Wear safety glasses. Use the right size tool on the coupling. Keep tools clean from oil and grease.	Safety glasses
5) Repeat step #4 on the other hose end.	See Step #4.	See Step #4.	See Step #4

Figure 9-7. JSA for a hydraulic line replacement

TASK STEPS	POTENTIAL HAZARDS	SAFETY PROCEDURE TO FOLLOW	SAFETY EQUIPMENT
6) Cut a new hose the same size as the old hose.	*Contact By:* Your hand may get cut by the saw.	Keep your hands away from the saw blade while it is turning.	
	Contact By: Flying sparks and debris can hurt you.	Wear safety glasses. Hold the hose with one hand while cutting it.	Safety glasses
7) Press the hose end into the Coupling Press.	*Caught In:* Your hand can be crushed if the press starts.	Move hand away from the ram and hose, then start the press.	Hard hat Safety glasses Steel-toed boots Earplugs Lock-out tags
8) Lower the pressing collar onto the coupling.	*Caught In:* Your hand can be crushed if the press starts.	Make sure the press is off prior to inserting the hose and coupling.	See Step #7.
9) Repeat Step #7 and Step #8 for the other hose end.	See Step #7.	See Step #7.	Lock-out tags
10) Install the new hose onto the machinery.	*Caught In:* You may become entangled in the machinery.	**Make sure that the system is Locked-out and properly tagged.**	
	Contact By: Any sharp edges of the coupling will cut your hands.	Make sure there are no burrs on the couplings. If so, grind them off.	
	Contact With: Tools may slip and cause you to hurt your hands.	Use the right size tool on the coupling. Keep tools clean from oil and grease.	

Figure 9-7. (*continued*)

TASK STEPS	POTENTIAL HAZARDS	SAFETY PROCEDURE TO FOLLOW	SAFETY EQUIPMENT
11) Remove all Lockout Tags from the system breakers	*Caught In:* Anyone still working on the hose may become entangled in the machinery.	Make sure all workers and tools are clear of the machinery. **Do not remove another person's Lock-out Tag**	
12) Check for any hydraulic oil leaks.	*Contact By:* Hydraulic fluid will irritate and damage the eyes and skin.	Stand away from hose while system pressure increases. Wear safety glasses.	Safety glasses
	Foot-Level Fall: The floor is slippery from hydraulic fluid.	Walk around spilled fluid if possible. Use Oil Dri to soak up spills.	Oil Dr
13) Tighten any leaky couplings.	See Step #10.	See Step #10.	See Step #10
14) Return equipment to the golf cart.	*Contact By:* Dropping tools may hurt your feet.	Carry all tools with a firm grip. If necessary, make several trips to carry all tools. Wear steel-toed boots.	Steel-toed boots
	Contact With: Various hazards in the mill.	Be alert for low-hanging items and moving machinery/vehicles.	Steel-toed boots

Figure 9-7. (*continued*)

Accident Types

The following discussion of accident types assists in determining the potential hazards for the job safety analysis.

1. *Struck-by.* A person is forcefully struck by an object.
2. *Struck-against.* A person forcefully strikes an object.
3. *Contact-by.* Contact by a substance or material that is by its very nature harmful and causes injury.
4. *Contact-with.* A person comes in contact with a harmful material.
5. *Caught-on.* A person or part of the person's clothing or equipment is caught on an object that is either moving or stationary.
6. *Caught-in.* A person or part of the person is trapped, stuck, or otherwise caught in an opening or enclosure.
7. *Caught-between.* A person is crushed, pinched, or caught between either a moving object and a stationary object or between two moving objects.
8. *Foot-level-fall.* A person slips, trips, and/or falls to the surface the person is standing or walking on.
9. *Fall-to-below.* A person slips, trips, and/or falls to a level below the one the person was walking or standing on.
10. *Overexertion.* The person performs a task beyond his physical capabilities, resulting in sprain or strain injuries.
11. *Exposure.* Employee injury results from her close proximity to harmful environmental conditions in the workplace.

CONCLUSION

System safety is an important part of the product life cycle. Allowing the individual employee to find the problems associated with work is no longer acceptable. A proactive approach is necessary because of increased product-liability litigation and increased employee awareness of process-safety concerns. Therefore, the safety professional must make every effort to understand the concepts of system safety.

QUESTIONS

1. What steps make up the product life cycle? Include examples of the activities that occur during each step.
2. What is the importance of the system safety function to an organization?
3. What are the differences between system safety and industrial safety?
4. Can you develop a fault tree for a simple system? Make sure to include both AND and OR gates.
5. Using the hazard assessment matrix, what is the level of risk for the following activities?

   - Getting hit by a vehicle while walking across a busy street
   - Having a grease fire in the kitchen
   - Being involved in a plane crash

6. Can you perform a job hazard analysis of a simple system using established criteria?

REFERENCES

Department of Defense. 1984. *MIL-STD-882B: Task Section 100 Program Management and Control.* Washington, DC: US Government Printing Office.

Ferry, T. S. 1988. *Modern Accident Investigation and Analysis*, 2nd ed. New York: John Wiley & Sons.

Hammer, W. 1989. *Occupational Safety Management and Engineering.* Englewood Cliffs, NJ: Prentice Hall.

Levy, Paul, 2007. 4 Dead, 79 Injured, 20 Missing after Dozens of Vehicles Plummet into River. Minneapolis, MN: Star Tribune.com. Retrieved from http://www.startribune.com/local/11593606.html.

Malasky, S. W. 1982. *System Safety: Technology and Application*, 2nd ed. New York: Garland STPM Press.

Manuele, F. A. 1993. *On the Practice of Safety.* New York: Van Nostrand Reinhold.

National Transportation Safety Board. 2008. *Collapse of I-35W Highway Bridge Minneapolis, Minnesota August 1, 2007.* Retrieved from http://www.dot.state.mn.us/i35wbridge/ntsb/finalreport.pdf.

Ridley, J. 1994. *Safety at Work.* Oxford: Butterworth-Heinemann.

Roland, H. E., and Moriarty, B. 1990. *System Safety Engineering and Management*, 2nd ed. New York: John Wiley & Sons.

BIBLIOGRAPHY

Brauer, R. L. 2006. *Safety and Health for Engineers*. Hoboken, NJ: John Wiley & Sons.

Laing, P. M. (Ed.). 1992. *Accident Prevention Manual for Business and Industry: Engineering and Technology*. Itasca, IL: National Safety Council.

Mansdorf, S. Z. 1993. *Complete Manual of Industrial Safety*. Englewood Cliffs, NJ: Prentice Hall.

10

Managing the Safety Function

CHAPTER OBJECTIVES

After completing this chapter, you will be able to

- Explain the functions of management and how they relate to safety
- Identify the purposes of the line and staff parts of the organization
- Differentiate between audits and inspections
- Identify the role of safety in the staffing process
- Explain the OSHA guidelines for safety management

CASE STUDY

Bob Renee had worked his way up from a mid-management position to the point that he was manager of a large production facility within a major corporation renowned for its safety efforts. Bob's safety performance was considered as important as high levels of quality production. As a result, Bob developed a strong relationship with a nearby medical center and his company retained the services of an occupational physician in residency there. On a spring morning one of Bob's workers was injured on the job and transported to the medical center. That night, Bob called the occupational physician and asked her to place the injured employee in an ambulance and transport him to the plant the next day, so he could "sign in." The employee could then return to the hospital immediately, but the subsequent workday would not be considered "lost time" since the employee came to the plant. Bob was disappointed when the physician put the welfare of the patient first and refused the request. Although Bob wanted to have the appearance of having a safe facility, both he and his employer were emphasizing appearing to be safe over actually managing safely.

Safety, like many other management activities, consists of planning, organizing, controlling, staffing, directing, and communicating. All facets of management are incorporated into the safety function and safety should be incorporated into all facets of management. They must be integrated for either to be effective. Actually performing safely means more than just looking good as Bob Renee tried to do in the introductory case. This actual incident demonstrates what can happen when the emphasis is placed on appearances rather than performance. The safe welfare of the employees and the company can be jeopardized by the efforts to simply appear to be safe. Safety depends on proper performance, a function of effective management. Careful measures must be put in place to emphasize effective management performance rather than mere appearances of safety.

PLANNING

"A job well planned is a job well done." "A job well planned is a job half finished." These axioms have been repeated for decades, and they hold some truth. Planning is essential to the ongoing success of any enterprise and certainly to the components of the enterprise, including safety. A well-planned operation involves a series of deliberate steps. First, the safety practitioner must forecast the needs of the safety department for the coming year. This involves reviewing the records of successes and failures as well as all the resources used in the past. It also means anticipating the obstacles that may be encountered during the coming planning period. Most companies tend to operate on a one- to five-year cycle, with planning for budgets occurring each year. This ***forecasting of coming needs*** or predicting when they will occur is a result of looking at the past and studying the future. There are a number of texts that can guide the reader as to how to forecast.

Once forecasts are made, the practitioner must anticipate resources needed to meet those needs and make requests accordingly. A proposed safety standard coming into effect may require unusual demands, as will the planning or start-up of a new facility. ***This is part of a proactive as opposed to*** *a reactive approach to safety. The practitioner does not wait for incidents to occur, but rather attempts to anticipate problems and plans to deal with them before they result in mishaps.*

Established plans often become standards by which the practitioner can judge the performance of the safety program. They should evolve from the mission and the objectives of the organization. This can occur by carefully reviewing organizational goals with top management and writing safety objectives to complement or aid in the accomplishment of organizational goals. Once safety objectives are written and methods by which those objectives can be accomplished are laid out, budgets and timetables are formulated. A careful review of the documents is made with management, and agreement is sought. Once this occurs, the organization of resources to accomplish the objectives can begin.

Management supports safety efforts to the extent these efforts support those of the organization. Resources necessary to accomplish the task will be allocated on a basis relative to how well the accomplishment of the task helps the organization achieve its goals. Of course, personalities, personal ambitions, and politics can interfere in the process. It is critical for the efforts of the safety department to be tied to those of the organization. Safety competes with every other entity in the organization for resources. A strong business case demonstrating return on investment for funds expended helps strengthen the case for safety and shows the effectiveness of safety relative to other departments or programs.

Critical to this effort is the establishment of programs to effectively identify hazards, estimate probabilities, and minimize overall risk. The inclusion of personnel from all levels of the organization to help recognize potential problems and generate mitigating is critical for success. Specific tools, such as the job hazard analysis, checklists, employee interviews, near-miss investigations and utilization of "what if?" questions can help identify potential problems before they lead to incidents. Once potential problems and their related hazards are identified, steps must be taken to prioritize them, based upon expected severities and probabilities. Only then can a case be made to management for resources to be allocated for mitigation purposes. By attempting to hide the effects of an accident or on-the-job injury, as Bob Renee attempted to do, efforts to address the problems with appropriate resources will be stifled, since information is buried and true costs are hidden. Accurate information drives the safety process at every level.

ORGANIZING

Unlike most management activities, safety usually operates from a purely staff position. The implications of this pervade safety and affect the way it operates in the organization. *To understand the management of safety, an understanding of line and staff positions is essential.*

Line positions are charged with carrying out the major function or functions of the organization. In an army, the line fights; in a production operation, the line does the producing; in a sales organization, the line sells. First-line supervisors, plant managers, and even company presidents are considered line officers within an organization. *Staff positions*, on the other hand, are charged with supporting or helping the line positions. Typically, staff positions have no real authority over line activities; staff members only assist and advise the line officers. Any authority staff positions possess beyond that of assisting or advising is only a result of a line manager giving that authority. In some organizations, staff members are given authority over certain functions and any activities that relate to those functions. If a safety member in such an organization recognizes a problem in safety, he or she could be given authority to correct that problem. Such discretion is referred to as *functional authority*. Even in organizations where no functional authority is given, some members of line management may perceive the authority exists anyway. This could be particularly true of lower-ranking members in line positions who perceive that staff members have authority over them when, in fact, they do not.

Safety managers, safety engineers, safety professionals, and safety technicians are nearly always considered staff personnel and nearly always operate from staff positions. *A safety manager's job is to monitor safety, compare what is found against existing standards, and advise line management as to any corrective actions to be taken.* The ideal position within the organization is reporting to the chief executive officer, but typically the safety practitioner reports to personnel or human resource departments. The rationale for reporting to the top is obvious. No person within the organization is exempt from safety. If, however, safety is subordinate to personnel, human resources, or some position other than the chief executive officer, the opportunities for abuse and the potential for conflict of interest are high. Proper monitoring and appropriate advising of the enterprise cannot readily take place from the middle of the organizational hierarchy.

The full-time safety practitioner monitors safety and advises management on what changes are needed and why. To do otherwise interferes with the job of the line manager. If a conflict occurs between the safety practitioner and the production manager, the production manager's orders will take precedence. Exceptions to this rule sometimes occur when the staff person is given functional authority or authority over the safety function. He or she has authority over all matters as they relate to safety and can overrule the line manager if orders contradict those of the safety practitioner relative to safety. Of course, the safety practitioner may also have line authority over any personnel that work for and report to him or her, such as an industrial hygienist, an occupational nurse, or a safety assistant.

Generally, the safety professional is in a better position to positively influence the organization if a purely advisory role is maintained. The safety professional monitors the organization, looks for problems, and advises the line manager—essentially becoming a part of the line management team.

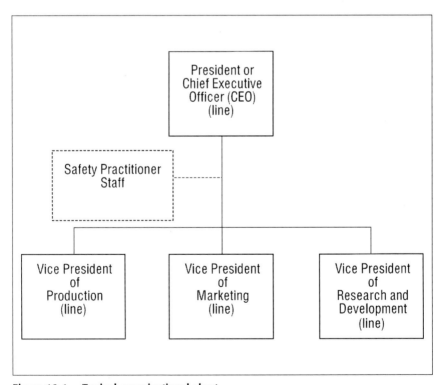

Figure 10-1. Typical organizational chart

The safety professional is simply there to help avoid and minimize safety problems that can otherwise adversely affect the line manager's resources, including personnel. By becoming part of the team, the safety professional can help a line manager perform a better job and avoid conflicts. This approach squarely places the responsibility for safety on the shoulders of line management. If an unsafe condition occurs and the line manager is informed, the manager cannot reasonably state that he or she is unaware of the condition. It is the manager's job to be aware of unsafe conditions and have them corrected.

All members of the organization must buy into safety to establish a safety culture. This only happens with support and continual input from top management. The safety professional does not lead the safety effort, but rather acts as a member of the different safety teams throughout the enterprise. Teams may be in the form of safety committees, individual crews, or work cells. It is the safety practitioner's responsibility to help those teams help themselves relative to safety. The safety professional acts as a resource or guide in helping the team establish itself. Management of safety becomes a responsibility of the line manager or leader of the work group, as well as individual members.

Carl Wagner is supervisor of an eight-member production group. The group's job is to produce a high-quality, pearl-like finish on metal furniture. Carl's responsibilities are to ensure that quality production occurs *safely.* When the plant safety manager observes one of Carl's workers not using the prescribed personal protective equipment, she addresses the problem with Carl. If Carl is not around, the safety manager discusses the problem with the worker. It's actually Carl's job to see to it the employees in his crew work safely. If Carl learns the employee is not wearing personal protective equipment because it is uncomfortable, then he discusses the issue with the safety professional. The safety professional's job is to aid Carl and help him resolve the employee's problem. Because of this relationship, Carl considers the safety manager not a threat, but rather a resource he can call on to help him and his crew to do a better job.

On Monday morning, Carl approaches Janet Reeves, the safety manager, and complains that the new respiratory equipment is too hot, and his employees do not like to wear it while working with lubricants. If Carl believes the safety equipment is simply a nuisance hindering his employees and tries to obstruct the safe-practice methods used in the plant, it is Janet's job to convince Carl the equipment is necessary to the safety of his workers and help him convince his employees. As she discusses the situation, both Janet and Carl know that top management fully supports the safety effort.

Janet discusses the problem with Carl, and they decide to take time during the next "tool box talk" to discuss it with employees. Before going into the meeting, Janet makes Carl aware that OSHA requires respiratory equipment for this operation and working without it may cause employees to develop respiratory problems. After a long discussion, Carl decides to lead the talk himself and to ask his employees if they understand the consequences of not wearing the equipment. After a little verbal sparring, the employees reluctantly agree that they recognize the equipment's importance.

Janet also suggests it is possible to shorten the time employees wear the respirators. After some discussion, all agree that two employees should perform any job demanding more than thirty minutes. After one employee wears the respirator for thirty minutes, he can switch off to permit another employee to finish. No employee will have to wear a respirator for more than thirty minutes at a time. All agree this is a workable idea and they decide to give it a try.

If Carl is unable or unwilling to comply with his safety responsibilities, Janet is in a position to discuss the matter with Carl's boss, his boss's boss, or even the chief executive officer of the company. With top management support of the safety effort, Carl's performance is measured on how well he supervises *safe* production. If he is unable or unwilling to try to produce safely, then his performance ratings will suffer. He is penalized for failing to follow procedures regarding production or any other critical activity within the company—including safety. Not only is Carl held responsible and accountable for safe production, but his direct manager and every other manager in the line production operation are also held responsible and accountable.

CONTROLLING

Controlling occurs through a number of subfunctions. It involves looking at what is happening in the organization by monitoring or comparing the results of the observations to established standards, and taking appropriate corrective actions. This occurs through inspections, audits, records reviews, interviews with employees and supervisors, and a careful watch on what is happening in the organization. The results of monitoring are compared to results from previous years, existing safety regulations or published or internally developed standards. Any deficiencies are noted and plans are made for correction. Before drastic changes occur, management approval and support are sought. Once this happens, it is up to line management to make any and all appropriate corrections. Of course, management will only make corrections when it perceives they are appropriate and beneficial. This perception is frequently based on line management's understanding of the need to make corrections and the consequences of not making corrections.

The safety professional often finds that being a persuasive salesperson is as important as knowing what to do and what to sell. The safety professional soon learns requests for change typically cost money, time, or other resources. In addition, these requests are competing with those from production, marketing, and other branches of the organization whose managers believe they can best utilize the company's resources. If money is spent on safety, it may not be available for raises for personnel, a new dental benefits plan, research on a breakthrough for a promising new drug, or a marketing effort that could reap millions. The management team will consider the safety request, weigh it against those from other parts of the organization, and respond accordingly.

If the safety professional is unable to make a convincing case for project resources, then the project won't happen. Management support for safety is a result of its perception of how well safety supports the organization. If the safety department is well aligned with the mission of the organization and the safety professional can build a strong case, usually in terms of the cost benefit to the organization, safety will be supported. However, if management does not perceive that safety supports the organization or believes that it does not provide an adequate return, it will not support it. To reiterate, the role of the safety professional is to monitor what is occurring in the organization relative to safety, compare findings to existing standards, and advise on necessary changes.

Allegedly, the major tool used by professionals to monitor the state of safety in the organization has been the audit. *Audit* is a term loosely used by professional and semiprofessional safety practitioners. It can mean anything from a cursory inspection of hand tools by shop personnel to a complete review of the safety program by the safety staff and numerous collaborating personnel. ***In reality, the audit is a tool that permits the assignment of a quantitative or numerical value to some aspect of the safety program.*** It is used to determine where that program is relative to where it ought to be. As many as a dozen or more audits might be done on an annual basis in any company. An example will illustrate this point.

Suzy Harris is the safety manager for Atlantic Widgets. She has developed nine safety programs in the following areas:

1. Hazard Recognition and Control
2. Fire Prevention and Protection
3. Industrial Hygiene Monitoring and Control
4. Ergonomics
5. Waste Prevention and Minimization
6. Security
7. Manpower Training and Development
8. Environmental Operations and Emergency Response
9. Technical Standards and Legislative Compliance

Suzy remains open to new ideas for safety programs and plans to initiate more as time permits. Each of the above programs is reviewed annually in a different month by other plant managers and members of the safety committee. Suzy always tries to bring in at least one outside person to get fresh insight into what she is doing.

In January, she audits the Fire Prevention and Protection program. During the audit, Suzy attempts to determine whether the company is in a state of readiness for any fire eventuality. She does so by evaluating the systems that are in place to ensure continuity of an effective program through interviews, observations, and a thorough records review.

Suzy and two members of her team interview Steve English, the plant manager, and ask him, "How are your supervisors specifically evaluated, for purposes of promotion or salary increases in the performance of their jobs, relative to fire prevention and protection?" She will assign a subjective rating to the following table based upon her evaluation of Steve's response: *"All plant supervisory personnel are evaluated on their performance in the area of fire prevention and protection for purposes of promotion and salary increase."*

No evaluation exists	*Evaluation occurs but there is a link to salary and promotion.*	*An effective evaluation occurs; there is a marginal impact on salary and promotion*	*An effective evaluation occurs with appropriate impact on salary and promotion*
1	2	3	4

Suzy continues with similar questions developed to measure the program. She may have developed these questions with the help of others in the plant, or she may have purchased an audit instrument from a firm specializing in safety consulting. After assigning scores to all of the questions, Suzy totals the scores and makes a determination, based on a predetermined scale, as to the strength or weakness of the fire prevention and protection program.

As a side benefit, Suzy is also in a position to make recommendations based on her findings as to what changes need to be made to strengthen the program. These can be prioritized by what the cost of not making the changes might be. The following paragraph provides an example.

Suzy learns through her audit the only protection for an infrequently used warehouse is provided by fire extinguishers. The contents of the warehouse are considered to be museum pieces and the owners are reluctant to part with them. Suzy makes a determination of the probability of a fire occurring in the warehouse in any given year. She can do this based on the past record of a large company or by estimates provided by her insurance company. The calculations indicate that a fire would likely occur only once in every 200 years with projected losses expected to run about $400,000. This amount assumes that the nearby fire department would be able to extinguish the fire with only 50 percent losses.

Next, Suzy uses the expected value technique and multiplies the probability of the loss occurring in a given year, 1/200 (or .005 times the expected loss of $400,000), to estimate the expected value of a loss at $2,000. This value approximates how much the company might be willing to invest in protecting the building from a fire in a given year. Two thousand dollars would be the break-even point in a given year. In other words, the company could only expect to save $2,000 in a given year by protecting the warehouse from fire.

How much would Suzy be willing to invest to save that money? It depends on a number of factors but primarily on how much return Suzy or the company can expect to get on its investments elsewhere. Suzy may decide that there are higher priorities for safety when she determines other findings through using this mathematical process. Even if she decides that this is a high priority, she must still convince management that it's the best investment it can make with the available funds. Whether she makes that recommendation depends on the return management normally receives on its investments. If management expects a 15 percent return in a given year, then Suzy will use the 15 percent figure as her standard of performance. She won't make a recommendation that will not result in at least a 15 percent expected return. If Suzy finds that she

Safety Audit Steps

1. Determine the area to be audited.
2. Develop or acquire an audit instrument based on the needs of the company. Be aware that no generic audit purchased or acquired from an outside firm will exactly fit the needs of the company.
3. Set up a pre-audit conference with all collaborating personnel. Discuss the purpose and scope of the audit.
4. Perform the audit, based on a predetermined evaluation system. Use interviews, observations, and records reviews.
5. Compare the audit results to the standards of performance established by the organization and the safety department.
6. Report to management based on the existing standards and variances from them. Suggest corrective actions and completion dates. Strong rationales must be prepared for any changes to which management might object. Changes must be tied to the organizational mission and objectives.
7. Follow up to ensure changes agreed on have been made.
8. Suggest corrective actions and negotiate completion dates with management.

can come up with a relatively inexpensive system, she will make the recommendation based on her comparison to the standard.

After Suzy has been at the facility for a while, plant personnel begin to understand the expectations management has for them regarding safety. Suzy finds it is necessary to perform the audit only once per year. Obviously, this is not frequently enough to monitor the overall fire program. Suzy has taught the supervisors to use their own employees to make periodic fire inspections. *Inspections* involve the monthly check of pressurized extinguishers, as well as a comparison of other safety features in the plant to given standards. *One difference between audits and inspections is that in inspections, no quantitative evaluations are made.* Suzy simply asks, "Is this done or not?" The inspection determines adherence to the operating standards. Audits are used to find out if inspections are taking place, as well as the effectiveness of those and other safety tools.

Suzy also makes periodic inspections simply to ascertain supervisors are doing their jobs and ensure compliance with the key safety standards of the organization. The supervisors realize Suzy's inspections will affect their own future in the company because she reports to management on the supervisors'

compliance with internal safety standards. Unsatisfactory findings are considered adverse performance reports for the supervisors and their superiors.
Positive reports are considered, along with other performance variables, in
promotion and salary increases. This is not the only measure of safety that
management depends on, but it is one factor in the determination of a given
supervisor's or manager's future in the company.

***One of the keys in the whole process is to take corrective action based
on the findings.*** It sounds so obvious, and yet audits and inspections are
performed regularly with recommendations reported to management, many
of which are never acted on. It isn't unusual, following a major disaster
that results in loss of lives and property, for workers to stand around saying they knew it was coming: "We tried to tell management, but no one
listened." Often the problems were identified in the regular audit or inspection procedure and recommendations were made, but no corrective actions
took place. The safety professional must successfully sell management on
the necessary course of action.

DIRECTING

The safety professional does not actively direct or lead in the organization,
unless he or she has a staff. This job belongs primarily to line management.
Ideally, the safety practitioner is given adequate resources, including a safety
budget, to help line management accomplish its objectives. A typical safety
budget would be 2 to 3 percent of sales volume.

One day, Carl Wagner approaches Suzy Harris with a safety problem.
He believes his employees would be safer if they had hard hats to wear
when they are working in "the yard." The cost of supplying the hard hats
for the whole crew is $600. Suzy believes this is a worthwhile project and
agrees to pay for the hard hats. The fact that Suzy has control of a budget
puts her in the enviable position of having managers come to her to request
not only verbal but also financial input. Safety personnel operating from
a staff position will only be successful with support, including financial
support, from line management.

STAFFING

The opportunity to hire productive, creative people is conducive to the
growth and advancement of the whole organization. The safety professional

should be aware the process contains pitfalls. The Civil Rights Act and the Americans with Disabilities Act (ADA) have requirements prohibiting discrimination against protected populations. If a job requires strength and endurance, the safety practitioner should ensure the company has a clear, written job description. Applicants are only hired contingent on their ability to meet all predetermined physical requirements. A careful review of all affirmative action regulations is necessary to ensure that company screening procedures are in compliance with the acts. Well-meaning companies have found themselves subject to lawsuits from state human rights commissions because they were unaware their hiring practices discriminated against protected populations. Specific guidelines should be obtained from the appropriate state human rights commission.

COMMUNICATIONS

The ability to communicate effectively is critical to the success of safety professionals. They must be able to speak in terms that management understands. This requires knowledge of accounting, economics, and modern production and quality theory. Strong human relations skills and related language ability are important to any successful safety effort. The safety practitioner will be working with top management and frontline workers. An effective safety professional needs to have the personality and ability to relate to both groups.

The typical safety professional spends significant time in front of people, often in training activities. Public speaking skills can be useful in these situations. Thoughts need to be well organized and the presentation should always be polished. This requires preparation and practice. Safety professionals who are not properly prepared in public speaking may join professional groups or may take courses to help them prepare. In any case, they must learn to effectively communicate in a variety of situations and with a variety of groups.

Effective writing skills are equally important. The credibility of any communication and the person doing the communicating is largely dependent upon the use of appropriate and proper grammar. The matching of subject and predicate, the use of parallel construction, integration of proper punctuation, and the application of a strong and appropriate vocabulary are particularly critical to effective written communications. Errors are easily spotted and often transmitted repeatedly. Every error tends to reduce the effectiveness of the message, so carefully proofread all written communications.

EVALUATION OF THE SYSTEM

Since the safety professional operates from a staff position, success is a result of his or her ability to enlist the support of line personnel. This is a result of integration into the organizational structure and culture and having the ability to enlist the support and cooperation of the line managers. Obviously, to accomplish this, two things are needed. *First, top management must have already endorsed safety as important to the organization, and it must have already given safety an appropriate level of support.* This occurs because management perceives that safety is a worthwhile and contributing entity within the organization and that its activities are cost-effective. Management will provide support for safety to the level and degree it perceives it is getting a reasonable return from its investment.

Support comes not only in terms of resources but also in terms of commitment to holding all members of the management team responsible and accountable for safety within their own operations. Each and every manager is responsible, not just for production, but for safe production. The manager's performance is measured relative to how well he or she performs in a manner *leading* to safe production. Within the organization, the results of this measurement are used in evaluating the manager for promotion and salary. Emphasis on final results sometimes yields cover-ups and false representations as might happen when results aren't reported or attempts are made to subvert the truth.

Second, the safety manager must be perceived as an integral part of the management team. Line managers can call on the safety manager for advice and help in making their own operations safer. The safety manager's guidance will help them to create and maintain a safe workplace for their employees. Line managers can make proposals and compete for the safety staff's time and budget to help them, but, ultimately, the responsibility and accountability for safety rest on the line manager's shoulders. Implementing the safety program is difficult because the line manager must believe the safety staff can help accomplish his or her objectives. The safety manager and staff become ex-officio members of his team. When the line manager meets with his or her own staff to discuss safety problems, a request may be made to have a member of the safety staff participate in those sessions.

A serious mistake members of some organizations make is believing they can have two separate systems for evaluating production and safety. This occurs when no measurement for safety is built into the normal evaluation system and a safety incentive program is initiated to reward safe performance.

The line manager is rewarded in terms of salary and position for having an exemplary production record. At the same time, the manager is punished by not receiving the perks associated with having no incidents or no lost-time accidents or whatever else the quota of the month is. The mixed signals do little to encourage safe production, but, instead, give the manager incentive to cover up any information he perceives will inhibit him or her from receiving the reward or bonus associated with the safety program.

OSHA GUIDELINES

On January 29, 1989, OSHA published voluntary safety and health guidelines for general industry. OSHA concluded that effective management of worker safety and health protection is a decisive factor in reducing the extent and the severity of work-related injuries and illnesses. Effective management addresses all work-related hazards, including those potential hazards that could result from a change in worksite condition or practices. It addresses hazards whether or not they are regulated by government standards.

In general, OSHA advises employers to maintain a program providing a systematic approach to recognize and protect employees from workplace hazards. *This requires the following:*

1. *Management commitment and employee involvement are complementary.* Management should value worker safety and health and commit to its visible pursuit as it would to other organizational goals. A means should be established for encouraging workers to develop and/or express their own commitment. This requires clearly stating and communicating safety and health policies and holding managers, supervisors, and employees accountable for meeting their safety responsibilities.
2. *Worksite analysis involves examining the workplace for existing and potential hazards.* Comprehensive baseline and periodic safety and health surveys should be conducted. Job hazard analysis, accident, and near-miss investigations should also be held. Workers should be able to report unsafe conditions without fear of reprisal. Trends of illness and injury should be studied over time to identify patterns and prevent problems from recurring.
3. *Once hazards or potential hazards are recognized, they should be controlled, prevented, and/or eliminated.* This requires engineering controls

where appropriate, administrative controls, or personal protective equipment where necessary. Emergency plans, complete with drills and training, should be evaluated. Medical programs should be established.

4. ***Training should address the safety and health responsibilities of all personnel.*** Managers, supervisors, and workers should understand their responsibilities and the reasons behind them. This training should be reinforced through performance feedback and enforcement of safe work practices.

In a memo dated March 12, 2012, Richard E. Fairfax, deputy assistant secretary of OSHA, advised regional administrators that certain policies could discourage reporting and be unlawful (Fairfax, 2012). These include:

1. Disciplining employees injured on the job, regardless of the circumstances of the injury. Reporting of injuries is always protected, regardless of fault.
2. Creating circumstances discouraging employee reporting of injuries or illnesses. For example, if an employee is punished for not reporting an injury quickly, then that employee might be discouraged from reporting an injury not recognized right away. That would be a violation of OSHA rules.
3. Enforcing rules more stringently against employees who are injured as a result of rules violations. In other words, whether an employer is injured or not when violating rules, the punishment should be the same; otherwise, it would be considered discrimination.
4. Creating incentive programs that discourage employees from reporting their injuries. For example, it would be a violation to reward employees for not suffering an injury or illness over a specified period of time; an employee may not report the injury or illness in order to win the award. Incentives should never dissuade reasonable workers from reporting injuries.

CONCLUSION

Because the safety practitioner is operating from a staff position, any success is a result of his ability to enlist the support of line personnel. This comes about as a result of being well integrated into the organizational structure and culture and from being able to enlist the support and cooperation of the line management. Two things are needed. First, top management must have already blessed safety as being important to the organization, and it must have already given safety an appropriate level of support. This blessing occurs because management perceives that safety is a worthwhile and contributing entity within the organization and its activities are of cost benefit. Management will provide support for safety to the level and degree it perceives a return from the same.

Support comes not only in terms of resources, but also in terms of commitment to holding all members of the management team responsible and accountable for safety within their own operations. Each and every manager is responsible, not just for production, but also for safe production. The manager's performance is measured relative to how well he performs in a manner leading to safe production, and the results of this measurement are used in evaluating the manager for promotion and salary within the organization.

Second, the safety manager must be perceived as an integral part of the management team. Line managers should be able to call on the safety manager for advice and help in making their own operations safer. The safety manager's guidance will help them create and maintain a safe workplace for their employees. Line managers can make proposals for and compete for the safety staff's time and budget to help them; however, the responsibility and accountability for safety rest on their shoulders. Implementation difficulties arise from the fact that the line manager must have reason to believe the safety staff can help him accomplish his own objectives. The safety manager and his staff become ex-officio members of the line manager's team. When the line manager meets his staff to discuss safety problems, he may request a member of the safety staff's participation in those sessions.

Lastly, the safety professional must exercise caution when advising management of programs that might appear to encourage safety, but actually discourage reporting. Information is critical to the success of any safety program and the management team should make every effort to encourage reporting in their effort to mitigate hazards and provide for a safe and healthful working environment.

QUESTIONS

1. Do you think safety practitioners should ever be given more than simple staff advisory authority? Why or why not? Do safety practitioners run the risk of assuming others' safety responsibilities when they do?

2. If there is a conflict between safety goals and production goals, which should take precedence and why? Ask the same question while playing roles of different individuals in the organization, such as different types of managers. Also assume roles of interested parties outside the organization, such as stockholders, bankers lending to the company, and government officials.

3. What steps should management take to ensure employee involvement? Should any of these steps be legal requirements?

4. How do safety practices today compare with those of several decades ago? Talk to some old-timers in industry and ask them if they are aware of any safety cover-ups that occurred in their industry. Discuss your responses.

5. What are the opinions of younger and older working adults about safety on the job, OSHA, and other safety-related issues? Do an informal survey. If a difference of opinion exists between the two groups, why do you suppose that is? As a manager, how would you approach each group to get its involvement?

BIBLIOGRAPHY

Department of Labor. *Federal Register*, 29 CRF Part 1910.
Fairfax, R. E. 2012, March 12. Employer Safety Incentive and Disincentive Policies and Practices Memorandum. Retrieved from https://www.osha.gov/as/opa/whistleblowermemo.html.

Psychology and Safety:
The Human Element in Loss Prevention

CHAPTER OBJECTIVES

After completing this chapter, you will be able to

- Define the terminology associated with the study of psychology
- Explain the concepts associated with motivation and safety
- Differentiate between behavioral and goal-directed theories of motivation
- Explain the principles associated with behavior modification and safety
- Describe the importance of establishing a positive safety culture
- Identify the pitfalls inherent in safety incentive programs
- Describe the benefits of employee empowerment and job enrichment

CASE STUDIES

An electrical maintenance supervisor with 15 years of experience ordered his assistant to de-energize a machine they were about to service. Before the assistant could "cut the power," the supervisor reached into the back of the unit, made contact with bare leads, and was electrocuted. The supervisor had been an outstanding employee with no record of accidents or related lost workdays, but while performing this task he violated several company safety policies. An electrician with his years of experience should never have made these mistakes. The results of the accident investigation revealed the supervisor was having marital problems. In addition, his spouse had indicated just

229

days before the incident that the victim had shown her the location of his life insurance policies.

*****

A small Midwestern manufacturing facility reported that 23 of their 75 employees had experienced back injuries. A job satisfaction survey indicated very low employee morale and workers who did not trust management. The survey results also indicated management was only concerned with profits and production, not employee safety.

*****

Above its main entrance a company mounted a stop light that remained green as long as there were no OSHA-recordable accidents. As it moved to break the company record of time worked with no such events, the organization provided very positive feedback and rewards anytime a major milestone was reached. When it neared a new record of 1.5 million man hours without a recordable accident, management planned a huge banquet and festivities in celebration. Many employees were shocked when the light turned red a few days before the banquet. A worker had died as a result of fatal injuries.

INTRODUCTION

It is difficult to be effective in safety without an understanding of human motivation, capabilities, and limitations because human error results in as many as 85 to 95 percent of all accidents. Traditional approaches to the study of motivation are examined in this chapter to assist safety professionals in recognizing some of the factors contributing to worker behavior. The intent is to provide useful strategies for correcting or eliminating adverse psychological safety factors from both the worker and employer perspective.

BASIC TERMINOLOGY

Psychology is the study of behavior, considering clinical, developmental, educational, experimental, industrial, social, and physiological perspectives.

Attitudes are enduring reactions toward people, places, or objects, based on beliefs and emotions. For example, if a person is taught to believe tall people are complainers, when a tall person hurts his back, the reaction is likely to be, "That tall person is exaggerating his injury; he is not really hurt. There was no problem with the job; he should have been able to lift the 180-pound load. If he really wants to work, he has to accept all of the responsibilities of the job." Notice how this attitude influences the way the injury scenario is viewed. Bias is an especially important consideration when safety professionals are concerned with hazard identification, incident-trend analysis, and accident investigation. Individual attitudes may guide attention and effort in the wrong direction.

Job satisfaction is the specific attitude and emotion individuals have about their jobs. When workers enjoy what they do, they are said to be receiving satisfaction from the job and this, in turn, may influence employee morale.

Morale is the meeting of individuals' needs and the extent to which employees recognize this meeting of needs comes from jobs. Some researchers believe that if morale is low, employee motivation will also be low.

Motivation is the inner drive, impulse, or need that creates a personal incentive toward behavior; it is an individual's tendency toward action in a given situation.

MOTIVATION

People cannot and do not motivate others. It is the individual who acts or behaves in a given situation. Environmental conditions can be established to increase the likelihood of action and performance, but it is up to the individual to respond. How an individual actually responds depends upon personal experiences forming attitudes that influence behaviors.

The goal-directed and the behavioral schools of study represent two of the most popular views of motivation. In the goal-directed school of motivation, the inner drives of individuals are examined to explain why human behavior takes place. Examples of goal-directed theories include the needs-hierarchy theory, the need-achievement theory, and the motivational hygiene theory. Examples of behavioral theories are Pavlovian theory and operant conditioning theory.

Figure 11-1.   Goal-directed model of motivation based upon the research

Goal-Directed School

Maslow's Needs-Hierarchy Theory

The goal-directed study of motivation states that people have various types of needs. Depending on the specific theory of interest, these needs can be physiological (the need for food, clothing, or shelter) or psychological (the need for love, recognition, and affiliation with others). Needs help individuals to establish goals guiding behavior to gain rewards and satisfy needs (see figure 11-1: the model first developed by Deci that pictorially describes the relationship between needs and behavior).

Dr. Abraham Maslow's **needs-hierarchy theory** is based on the goal-directed model of motivation (see figure 11-2 for a pictorial representation of Maslow's model of motivation). According to Maslow's theory of motivation, people have basic needs—hunger, thirst, and desire for warmth—that must be satisfied before more advanced needs become motivating influences. It is only after these basic physiological needs are satisfied that individuals seek safety and security, the need to be protected and preserved in terms of health and well-being. Living in a location free from turmoil and conflict is an example.

Once the safety and security needs are satisfied, individuals seek to establish close friendships and loving relationships with others. With this social support system in place, they strive to satisfy needs associated with self-esteem and esteem by others. Esteem refers to the individual's sense of personal worth. Self-respect, dignity, independence, and confidence are some of the attributes individuals are trying to achieve at this level.

Maslow theorized that the individual, having satisfied all the lower levels of needs, now has the potential for **self-actualization**, a concern for the well-being of others or the state of society. Examples of self-actualized individuals include Eleanor Roosevelt and Mother Teresa.

There is little evidence to justify the accuracy of this theory in explaining human motivation; Maslow's work is based on post-hoc studies of selected individuals. Researchers have found, however, that certain needs must be

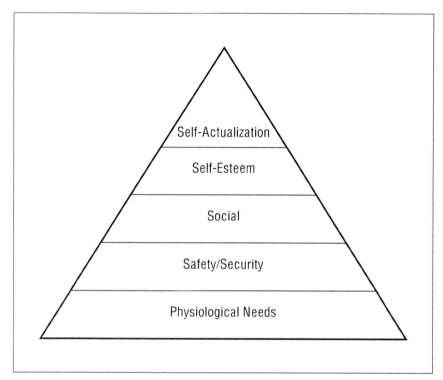

Figure 11-2.   Maslow's needs-hierarchy model of motivation

satisfied before individuals can or will pursue other needs. Any of Maslow's needs could serve as motivators for individuals, depending on their psychological state.

People or employees are motivated by unsatisfied needs only. Once needs are met in a given category and the future appears to hold the same, individuals will move to a higher level of need. In the United States, most employees in an industrial or construction setting have their basic needs met through wages and benefits. Unmet needs of employees are likely to be of a psychological rather than of a physiological nature. Motivation toward safety should likely focus on something other than meeting physical needs.

McClelland's Need-Achievement Theory

A goal-directed theory that has acquired significant attention due to tangible research supporting its premise is McClelland's ***need-achievement theory***.

According to McClelland, there are three essential motives that drive human behavior:

1. achievement
2. affiliation
3. power

Achievement is the accomplishment of tasks and activities. Individuals having a high need for achievement are very comfortable working alone. They tend to be very creative and can develop unique solutions to the challenges presented them.

Those with a need for *affiliation* seek the company of others and find satisfaction in close relationships. They may gain satisfaction from knowing or meeting famous people. The enjoyment of being a part of a certain work crew and talking with coworkers may satisfy this need. Pride associated with the wearing of team or company clothing is an example of outward display of affiliation.

The need for *power* is probably the most misinterpreted of McClelland's three needs. Power, in this context, does not necessarily mean insidious and tyrannical control over people but rather the ability to positively influence others. Managers often have high power and achievement profiles. They have a need to get tasks accomplished through the use of and influence over subordinates in the organization. According to this theory, many teachers and trainers have a high need for power. They gain satisfaction from students' successes and accomplishments.

While some individuals may have a predisposition toward one of the three needs, it is more likely they will possess two or all three of them. Frequently, we attempt to influence the behavior of others by using rewards that seem important to us. It is possible to influence the behavior of others only when you know what is important to them. By considering all three of McClelland's motives and analyzing individuals to determine what needs each has, it becomes theoretically possible to structure rewards in such a way that individuals' needs are met through higher safety performance.

Herzberg's Motivation Hygiene Theory

Frederick Herzberg, in his ***motivation hygiene theory***, states that employee motivation depends on characteristics of the job referred to as ***intrinsic job*** *factors.* When Herzberg performed his study, he asked workers what about their job motivated them. The responses included achievement, activity, authority, creativity, importance, independence, interest, knowledge, personal growth and development, promotion opportunity, recognition, responsibility, service to others, utilization, and variety. He learned that some of the factors not directly related to the job, such as a clean work environment, had no effect on motivation. These external factors can be addressed, and although the workplace may be clean or hygienic—having nothing to make the worker dissatisfied—it does nothing to satisfy or motivate the worker. A few of the less obvious intrinsic job factors or motivators deserve exploration.

Activity refers to a steady and physically acceptable level of performance. Physical and mental workloads are balanced and reasonable. When workloads become either excessive or minimal, occupational stress, resulting from fatigue or boredom, can become a safety problem.

Authority refers to the amount of power inherent in a job to influence or control the work activities of others. Here, again, occupational stress is often overlooked. When individuals have extensive responsibilities but little control over the accomplishment of tasks, stress may result. Job satisfaction may be closely tied to the appropriate balance of responsibility (accountability for performance, decisions, and outcomes) and authority.

Creativity reflects the opportunity for individuals to apply inventiveness, resourcefulness, and personal talents to the work situation. Many individuals enjoy the process of developing new and innovative solutions to problems in the workplace. Jobs that tap into an individual's creativity often seem more enjoyable because they tend to be less routine.

Importance refers to the perceived value of the work performed. When employees believe they make a difference in the organization, job satisfaction increases. When coupled with *interest*, the chance to perform activities compatible with the individual's personal preferences, importance makes employee motivation tend to soar. Employers may also affect employee motivation by offering opportunities for promotion, recognition, paths for advancement, and appreciation for strong performance. Intrinsic job factors are only motivational if the individual considers them as such. Not everyone desires advancement or promotion, so they may not be motivating to an individual who desires to remain a member of a crew or team.

The more intrinsic job factors there are, the greater the job satisfaction. *Job satisfaction* describes the positive feeling workers have about their jobs. It implies employees are meeting their needs through their work. Several studies suggest a negative correlation between job satisfaction and accidents—that is, the greater the job satisfaction of employees, the fewer the accidents. If the intrinsic job factors are removed, the worker is no longer motivated to perform. Factors not intrinsic to the job, external factors, are more related to the environment and are not considered motivators. A dirty workplace would be external. A clean environment might be pleasing, but it does not contribute to motivation. By making a dirty workplace clean, we make it less undesirable, but we do not make the worker more motivated. We can remove all of the undesirable external factors and will still not motivate the worker; all we have is a clean or hygienic work environment.

From a safety application perspective, truly effective motivators are relatively inexpensive since they have to do with the job itself. Enhancing the job, by making it more interesting or by empowering the worker to make more decisions, provides a more efficient work environment and tends to be motivational. Simply enhancing the environment of the job by providing more pleasant work surroundings may do nothing to actually provide motivation for the employee. The effective manager will attempt to determine employee needs and provide an environment that helps the employee to meet those needs by performing the job in a safe and effective manner.

Behavioral School

The behavioral school examines environmental factors influencing human performance. Two of the most widely known behavioral theories are ***Pavlovian conditioning, also known as the classical conditioning theory, and operant conditioning, or the Skinnerian theory***.

Pavlov

Pavlovian conditioning explains behavior as reflexive in nature. The traditional example is of dogs salivating when a stimulus, a bell, is paired with the presentation of food. After repeated pairings, Ivan Pavlov found the conditioned or learned stimulus produces the same response as the unconditioned or natural stimulus. An example of this type of motivation is the fear and anxiety people feel when they enter a location where they have been injured. The location becomes a conditioned stimulus or cue producing the natural emotional response.

People involved in serious automobile accidents often experience anxiety when they attempt to drive again. One industrial example is the near-fatal shock of an electrician at a high-voltage substation. When he returned to work after a six-month absence, the employee could not pass through the gates of the power substation without experiencing debilitating anxiety. The power substation became his conditioned stimulus associated with severe and intense pain.

Skinner

B. F. Skinner proposed ***operant conditioning***, a process in which the frequency of occurrence of behavior is modified by the consequences of the behavior; that is, the consequences of people's actions will determine future actions. The external consequences, referred to as the environment or environmental stimuli, may increase, decrease, or have little effect on behavior. When the likelihood of a positive response increases or a negative response decreases, following the presentation of an event or stimulus, the process is referred to as ***positive reinforcement***. Praise, recognition, and financial rewards are just a few possible examples of positive reinforcers. By definition, a stimulus is not a positive reinforcer unless it increases the likelihood of the behavior reoccurring in the future.

When the likelihood of a positive response decreases or a negative response increases following the presentation of an event or stimulus, the process is referred to as ***punishment***. Figure 11-3 shows the relationship between the presentation and removal of environmental events and their effect on behavior. When workers associate safety with pleasant results or positive reinforcement, their safe behavior is likely to continue. When safe behavior is associated with unpleasantness, it is likely to discontinue. If a worker equates PPE (personal protective equipment) with discomfort, he or she is less likely to wear it. If, however, PPE is associated with being able to spend more quality time with family, being in line for a raise, or positive feedback from someone respected, the employee is likely to utilize it. Supervisor praise for employee compliance with a company safety hard hat policy is an environmental stimulus considered positive reinforcement if it increases the likelihood of the employee wearing the hard hat in the future. If the supervisor is disliked and has no authority to discipline the employee, praise from the supervisor may be seen as negative reinforcement and it could even decrease the likelihood of employee use.

Behavior	Stimulus	
	Presentation	**Removal**
⬆	Positive	Negative
⬇	Punishment	Extinction

Figure 11-3. Operant conditioning model of environmental processes and their effect on behavior

If employees are suspended every time they are observed performing unsafe behaviors and over time the unsafe behaviors decrease in frequency, this process would be considered punishment. If an employee only wears safety gear when the supervisor is in his work area, this behavior would be considered *avoidance*, being maintained by negative reinforcement.

Extinction is the removal of a positive stimulus, reducing the likelihood of a behavior recurring in the future. Incentive programs are often used to improve safe performance. Frequently, when incentive programs are eliminated, safe performance will decline. Behaviorists could explain the reduction in desired safe performance a result of the process of extinction. Before providing substantial safety awards, prizes for safety competitions, or other expensive incentives, one must consider how long the company is prepared to maintain them. How much is the company willing to spend? What happens when the incentives are discontinued? Once rewards are given, employees view them as *entitlements*. Entitlements may cease to serve as reinforcers of behavior, but remove the entitlements and the company can face significant anger and disappointment from its workers.

THE RATIONAL EMPLOYEE: APPLYING MOTIVATION THEORIES

Employees are rational; they always do the rational thing—in their own minds. They have specific reasons for behaving the way they do. The challenge is to determine the factors supporting undesirable performance and develop strategies to modify behavior.

There are several possible factors leading to undesirable safety performance, including employee background, peer influence, and company environment. Although no two employees will necessarily respond in a like manner to the same stimulus, motivational theories attempt to apply what has been learned about a few to the whole population. What works for one won't necessarily work for another, because each person has different needs; therefore, it is important to understand each employee and his or her needs.

Suppose a new employee must work near a vat of acid. The employee has been trained in recognizing hazards in the workplace but still does not wear personal protective equipment (PPE). Is it irrational not to wear eye protection near acid? Most people would probably say "Yes!" This undesirable behavior seems to be irrational, but is it? If coworkers heckle the new employee for talking about safety and wearing PPE, is the worker being irrational for not wearing it? The new employee may have a strong need for affiliation and may choose to follow the norms established. This scenario can also be examined from the perspective of the operant conditioning theory. Coworker verbal abuse might be considered as a punishing stimulus. If the new employee is punished every time he is observed wearing eye protection, he is likely to choose not to wear it.

The safety professional must also look at the organizational environment maintaining poor performance. In the above safety scenario, the new employee will compare the importance of the social reinforcer of going along with what the crew wants with the likelihood of punishment from the employer. The selection of the social reinforcer is especially likely if inconsistent enforcement prevails or if the discipline is only a slap on the wrist. Resulting employee behavior of choosing not to wear eye protection is far more rational to the employee than working safely.

ORGANIZATIONAL ENVIRONMENT AND THE SAFETY CULTURE

From a behavioral perspective, organizational actions are just as important as organizational directives. In a Fortune 500 company, for example, senior management decided it was good for employee relations to emphasize safety. Safety slogans were visibly posted throughout company facilities and weekly *tool box safety meetings* were mandated for company work crews.

Safety messages were included in all employee payroll envelopes. The company spent freely to promote safety awareness.

During programmed maintenance shutdowns at this same company, however, employees were instructed to get equipment running again quickly. Employees were informed that approximately $1 million was lost every day this particular plant was not operating. The underlying message to the employees was to get production moving no matter what the cost. Several mechanics and utility workers were seriously injured during the maintenance shutdowns because employees took shortcuts and failed to use the correct tools and procedures to perform many of the tasks. This example is not unique. An organization's actions and the types of statistics it measures clearly communicate what is important. Several authors have pointed out that when safety performance is measured in terms of injury statistics, there is a tendency for employees not to report incidents. A company may look good on paper at the same time that *fatcats* (fatalities and catastrophic accidents) are lurking in the corner. A key part of any safety program is the upward flow of information regarding incidents and minor accidents. Programs that encourage employees not to report incidents or accidents hinder the flow of information upward. Hazards and their causes may remain hidden until a tragic accident that can no longer be disguised occurs. It isn't surprising that a program that goes for months or years without a recordable incident may suddenly experience a fatality or other catastrophic accident.

Another example would be responding to the fears associated with OSHA inspections and citations or to pressure from an insurance company to fix hazards or reduce costs. Short-term, Band-Aid approaches will not build a safety culture supported by employees. Workers recognize when budgets are provided for quick fixes and behave accordingly. Company motives are more transparent to employees than some managers believe. When a manager looks the other way to increase the flow of production or make a certain part of the organization look good on paper, it rarely goes unnoticed by the workforce.

The term **safety culture** is frequently used, but what does it really mean? When considered from a safety management perspective, safety culture should be viewed as a nurturing and cultivating environment that values low risk in actions of the organization. In his best-seller, *The 7 Habits of Highly Effective People*, Stephen Covey discusses the relationship among paradigm, values, principles, and success. As Covey points out, a paradigm is a model or frame of reference. It is used to understand what is happening at any given point in time, as well as where we want to go. Organizations need a culture or paradigm to help them interpret events. They also need a culture with clearly specified goals of what the organization wishes to achieve. *An organizational paradigm and a clear conceptualization of values, principles, and goals are the foundations of a culture that will promote the right actions to achieve excellence and success.*

An example of the influence of organizational paradigms, values, and safety cultures on management decision making was observed when a textile manufacturer was constructing a new facility. Upper-level management, including the safety and operations managers, agreed that every contractor would have to comply with all company health and safety rules during the project. In all of the written contracts, this was clearly noted. If the textile manufacturer observed a violation, the employee of the subcontractor would be temporarily removed from the job site. A repeat violation meant termination of the subcontractor's contract.

Six months into the two-year project, one of the major subcontractors responsible for multistory steel-beam erection had employees working over 30 feet in the air without fall protection. After the second observation of this violation, the plant safety manager ordered the workers down, informed the contractor of the repeat violation, and initiated the process to terminate the contract. The contractor was removed from the project and replaced by a firm that accomplished all the goals initially established by the textile manufacturer. The plant manager in this scenario did not compromise. As a result, the project was completed on time with only one OSHA-recordable injury.

The other extreme occurred when a large plant, operating in a corporate environment renowned for safety, had a worker who suffered heat stress on the job but failed to show signs of the stress until after his shift ended. Although the worker succumbed to the stress in the company parking lot and was sent to the hospital in an ambulance, the company refused to recognize the event as work-related. They also failed to take steps to correct the problems associated with the heat stress. Although the employee died on the way to the hospital, the facility continued to expose workers to the same hazardous conditions. The manager's incentive for not recording a fatality was so strong, he was willing to put his company at risk by not reporting to OSHA within the mandatory 8-hour period. In some organizations similar motivation exists at every level. Both examples demonstrate the effect of value systems and the safety cultures on personal safety behaviors

Values influence behavior. They impact the way people work, the way people treat others and the way people react to figures of authority. This paradigm also holds true for organizational cultures. Values are not part of the paradigm or guidelines that magically appear overnight. For most people, their parents, brothers and sisters, sports heroes, teachers, friends, and religious leaders are just a few of the many individuals who influence their values. For organizations, it is primarily managers' values that mold the organizational values and establish, whether consciously or unconsciously, the safety culture and how it defines success.

It is critical for the safety professional to work within the organization and its management to establish the safety paradigm and safety culture. The safety professional must cultivate the knowledge of those in influential positions, taking advantage of every opportunity to educate management on the benefits of a safety culture. Gathering necessary data, knowing the regulations, and presenting positive examples of what other organizations are doing are part of the job of safety. Sometimes actions must be taken because they are the right things to do, but this will be possible only if the organization has clearly established safety values and principles in place. The safety professional's goal must be to assist in the development of those values and principles.

INCENTIVES VERSUS INHERENT REINFORCEMENT

The premise behind the use of incentives is that employees require added encouragement to work safely. Many incentive programs use statistical measures such as recordable accidents on the OSHA 300 log or days without a lost-time accident to determine the winners. Awards have taken a variety of forms. They can be anything from baseball caps, T-shirts, or jackets to automobiles and trucks.

It is an unfortunate fact that incentive programs frequently create more problems than they solve. Research repeatedly demonstrates incentive programs improve safety numbers while they are in place, but once the incentive program ends, performance returns to previous levels at best; in some instances, safety performance is much worse. Employees may view incentive programs as entitlements and come to expect them as bonuses. According to the operant conditioning theory of motivation, *extinction* occurs when a reinforcer is removed and behavior terminates. Employees will sometimes sandbag reports of accidents and injuries until the end of the incentive program, at which time the incident is reported as if it has just happened.

Incentive programs create unhealthy competition when individuals, crews, departments, or plant managements undermine the activities of their competitors to gain the awards. At one company, the department with the cleanest work area during the previous month was rewarded with a steak dinner. Trash dumping in competing departments caused the incentive program to backfire.

Incentive programs based on statistics may produce the unwanted side effect of inaccurate recordkeeping. At another company, employees had worked over eleven months without a lost-time accident, presumably because of a new safety award program initiated earlier that year. Word circulated around the facility that the company CEO planned to fly to the facility and present the award for going one year without a recordable accident. Approximately two weeks before the big day an employee slipped while using a prybar. The prybar struck him in the face, shattering his safety glasses. While the employee avoided a serious eye injury, he received a facial laceration requiring stitches. Under normal circumstances he would have been sent home after a visit to the emergency room, but supervisors and coworkers pressured the employee to return to work. The incentive program motivated the employee to return to work; however, the message sent was that it is OK to have an accident if a lost-time workday is avoided.

Positive reinforcement by management may not be positive to the participants of the incentive program. Coal miners at one facility had the opportunity to be in a drawing for a new pickup truck if they met certain safety criteria. When asked why the incentive program failed, interviewed miners reported they received workers' compensation benefits matching approximately 70 percent of their normal salary. In addition, their homes, cars, boats, and other loans were automatically paid for by the loan insurance that went into effect when they could not work as a result of an injury. A *positive reinforcer* increases the likelihood of behavior. Winning a pickup truck was not a reinforcer; sustaining an injury and collecting benefits was a reinforcer.

Incentive programs cannot be used in place of effective safety programs and positive safety cultures. Safety incentive programs are not magic; they may have a negative effect on safety efforts. OSHA now prohibits employers from discriminating against employees because they report an injury or illness (U.S. Department of Labor, Occupational Safety and Health Administration [OSHA], 2012, March 12). Reporting a work-related injury or illness is a core employee right. Employers must know of and correct dangerous conditions that put the workforce at risk.

EMPLOYEE EMPOWERMENT AND JOB ENRICHMENT

With the current emphasis on safety culture and Total Quality Management (TQM), progressive companies are recognizing the value of employee-driven safety programs. Terms like "employee empowerment" and "job enrichment" reflect the importance of intrinsic job factors for the promotion of safety.

Employees of today are generally not receptive to authoritarian styles of management. As discussed in the motivational hygiene section of this chapter, intrinsic job factors that are inherently reinforcing are more likely to motivate employees. Incentive programs are extrinsic reinforcers "artificially" incorporated into the organization to increase the likelihood of desired performance. Once the artificial reinforcer is removed, natural behavioral activities return to the work environment.

Intrinsic reinforcements are those positive aspects of behavior and environment that are self-perpetuating. Watching television is an example. It is entertaining, relaxing, enjoyable, and has intrinsic reinforcement. Intrinsic reinforcers can exist in all activities. In an organization, job enrichment and empowerment refer to the incorporation of intrinsic job factors into the work. *Job enrichment* involves adding more tasks to the job to make it more interesting and enjoyable. For example, an employee who formerly called upon someone else to maintain his equipment is taught how to maintain it and given time to do so when the equipment needs maintenance. Taking time off from routine tasks to perform this needed function may make the job more interesting and more enjoyable to the employee, thereby enriching it. Some individuals desire jobs with responsibility and authority. *Empowerment*, or delegating the role of decision making to an employee by giving him more autonomy over the decisions governing his workday, may make him feel his job is more interesting and important. Both of these are examples of permitting the employee to participate in the organization by giving him more autonomy over his job and future. The employee participates in the success of his job and that of the organization. ***By sharing responsibility for success, including success in safety, everyone becomes a player with a vested interest.***

CONCLUSION

This chapter included a brief discussion of the theories of motivation and examined the importance of the organizational environment and the safety culture, along with the strengths and weaknesses of incentive programs. Job enrichment and employee empowerment were discussed as methods to increase employee buy-in to the safety process. References at the end of this chapter can assist the safety professional in learning more about psychology and safety.

QUESTIONS

1. How does job satisfaction influence an individual's safety? What role does worker attitude play in accident prevention?
2. Why is knowledge of worker motivation important to a safety professional?
3. How are goal-directed theories different from the behavioral theories of motivation?
4. How can the motivational hygiene theory of motivation be used to improve a safety program?
5. What is meant by the term *safety culture*? How do you establish a positive safety culture?
6. How can positive reinforcement and punishment be used in a behavior modification–oriented safety program? Give examples.

REFERENCES

Covey, R. S. 1989. *The 7 Habits of Highly Effective People*. New York: Simon & Schuster.
U.S. Department of Labor, Occupational Safety and Health Administration [OSHA]. 2012, March 12. Memorandum for Regional Administrators, Whistleblower Program Managers. Retrieved from https://www.osha.gov/as/opa/whistleblowermemo.html.

BIBLIOGRAPHY

Catania, A. C. 1968. *Contemporary Research in Operant Behavior*. Glenview, IL: Scott, Foresman.
Daniels, A. C. 1989. *Performance Management: Improving Quality Productivity through Positive Reinforcement*. Tucker, GA: Performance Management Publications.

Everly, G. S., and Feldman, R. H. L. 1985. Occupational Health Promotion. *Health Behavior in the Workplace.* New York: John Wiley & Sons.

Geller, S. 1994, September. Ten Principles for Achieving a Total Safety Culture. *Professional Safety, 39,* 19–24.

Henderson, C. J., and Cernohous, C. 1994, January. Ergonomics: A Business Approach. *Professional Safety, 39,* 27–31.

Kamp, J. 1994, May. Worker Psychology: Safety Management's Next Frontier. *Professional Safety, 39,* 32–38.

Katz, D., and Kahn, R. L. 1978. *The Social Psychology of Organizations.* New York: John Wiley & Sons.

Kohn, A. 1993. *Punished by Rewards.* Boston: Houghton Mifflin.

Krause, T. R. 1995, February. Driving Continuous Improvement in Safety. *Occupational Hazards, 57,* 47–49.

Krause, T. R. 1995. *Employee-Driven Systems for Safe Behavior: Integrating Behavioral and Statistical Methodologies.* New York: Van Nostrand Reinhold.

Krause, T. R., Hidley, J. H., and Hodson, S. J. 1990. *The Behavior-Based Safety Process.* New York: Van Nostrand Reinhold.

Lenckus, D. 1994, October. Safety Awareness Alone Not Enough to Sow the Seeds for Fewer Injuries. *Business Insurance, 42* (3), 15–17.

Liebert, R. M., and Neale, J. M. 1977. *Psychology.* New York: John Wiley & Sons.

Makin, P. J., and Sutherland, V. J. 1994, May. Reducing Accidents Using a Behavioural Approach. *Leadership & Organization Development Journal, 15,* 5–10.

Minter, S. G. 1994, January. A Safe Approach to Incentives. *Occupational Hazards, 57,* 171–72.

Nirenberg, J. S. 1984. *How to Sell Your Ideas.* New York: McGraw-Hill.

Preston, R., and Topf, M. 1994, March. Safety Discipline: A Constructive Approach. *Occupational Hazards, 56,* 51–54.

Reynolds, G. S. 1968. *A Primer of Operant Conditioning.* Glenview, Scott Foresman.

Saari, J. 1994, May. When Does Behavior Modification Prevent Accidents? *Leadership and Organization Development Journal, 15,* 11–15.

Stein, D. G., and Rosen, J. J. (1974). *Motivation and Emotion.* New York: Macmillan.

Weinstock, M. P. 1994, March. Rewarding Safety. *Occupational Hazards, 56,* 73–76.

12

Improving Safety Performance
with Behavior-Based Safety

Earl Blair, EdD, CSP

CHAPTER OBJECTIVES

After completing this chapter, you will be able to

- Define related behavioral safety terminology
- Explain how to implement behavioral safety processes
- Describe techniques and benefits of safety coaching
- Identify best measures to track behavioral safety performance
- Discuss success factors for improving safety performance
- Explain how to use behavioral approaches for continuous improvement

INTRODUCTION

Behavior-based safety is commonly called behavioral safety. Behavioral safety is not an alternative to traditional safety programs; it is one component of a comprehensive effort.

When organizations demonstrate a thorough understanding and proper application of behavioral safety, safety performance is typically enhanced. Behavioral safety is not meant to replace other safety initiatives, but rather to enhance and complement them.

There are a wide variety of theories and techniques under the name of behavior-based safety. Many of these approaches are not based on sound research or experience. However, this is not intended to imply there is only one correct way to implement and proceed with behavioral safety programs. Once basic behavioral principles and techniques are understood, flexibility is recommended. The behavioral process should be customized for each organization and site. This chapter presents basic principles, proven strategies, and practical implementation steps that have worked well for many organizations.

A common distinction is to view behavioral safety as a process rather than a program. This is not to belittle or diminish the importance of programs. *Rather, a process approach provides certain desirable benefits:*

1. It encourages everyone to recognize the daily importance of safety.
2. It views safety performance as a long-term process—a process that can be continuously improved.

McSween observes that an organization may initiate a series of programs under the umbrella of an ongoing safety process (1995, p. 34). Krause emphasizes that continuous improvement is a factor that distinguishes a genuine process from a program (1997, p. 73). Ultimately, the purpose of behavioral safety is not simply to improve individual behaviors, but to positively impact the overall safety culture and related safety systems. As these goals are accomplished, safety performance will continually improve.

MISCONCEPTIONS ABOUT BEHAVIORAL SAFETY

There are a number of misconceptions about behavioral safety that can impede progress. *The following is a brief list of common misconceptions about behavioral safety, followed by the author's comments:*

- *Behavioral safety fails to address system causes of injuries.* Many organizations can demonstrate that behavioral safety actually provides an excellent tool for addressing system causes of injuries. An organization is not using behavioral safety to its potential if system improvements do not result from the behavioral effort. We will discuss this further later in the chapter.
- *Behavioral safety is used to blame employees.* Behavioral experts agree that if behavioral safety is used to blame employees it will not work.

Blaming employees is the opposite of a proper approach to behavioral safety. Proper applications of behavioral safety generally include a "blame-free" or "no-fault" understanding as part of the process itself.

- *Behavioral interventions are the least effective intervention.* Behavioral interventions are not the first intervention priority. Elimination of hazards, substitution of less hazardous materials, and engineering solutions are all higher-priority interventions than behavioral effort. However, these higher-level solutions are not always possible or fully effective (nor "foolproof"). Once the appropriate engineering safety interventions have been implemented, behavioral initiatives are often the most effective supplemental interventions that are capable of dramatically improving safety performance. As an example, think of engineering solutions for automobiles, combined with behavioral interventions targeting sober driving. The engineering solutions and behavioral solutions work in concert; neither engineering nor behavioral efforts work effectively alone, but rather both kinds of interventions supplement and complement the other.
- *Behavioral safety allows management to abdicate responsibility for safety.* Management's behavior is without a doubt the most important behavior in behavioral safety. Unfortunately, there are some behavioral implementations that specifically target hourly employees' behavior and don't address the behavior of management and professional staff. This kind of implementation results in a number of negative consequences, including lack of effort and apathy on the part of members of management, reduced morale and interest in safety on the part of hourly workers, and, most likely, poor safety performance. Everyone's behavior is important to safety performance; appropriate safety-related behaviors should be defined, measured, and reinforced for managers and all levels of employees.
- *Behavioral safety is a magic bullet.* There is no magic bullet. Improving safety performance requires a thorough understanding of related areas of psychology, leadership, and culture. Additionally, there must be a strong commitment to safety, effective implementation, and a willingness to work both smart and hard.

For additional review of behavioral safety misconceptions, see Blair's "Behavior-Based Safety: Myths, Magic & Reality" (1999, p. 25).

BASIC DEFINITIONS AND TERMINOLOGY

At-risk Behavior: Actions increasing the potential consequence of injury or illness. The term was coined by Deborah Pivaronas as a way of describing critical, safety-related behaviors without the associated negative connotation (Krause, 1997, p. 10). The term is more neutral and less evaluative than *unsafe behavior.*

Attitudes: Enduring reactions toward people, places, or objects based on beliefs and emotional feelings. Attitudes include thoughts, feelings, and predispositions, and are difficult to change.

Behavior-based Safety: The use of applied behavior analysis methods to achieve continuous improvement in safety performance (Krause, 1997, p. 3). See additional definitions in this chapter.

Culture: Involves the things people do and say in an organization, and why they say or do those things. Organizational cultures develop over time and consist of traditions, beliefs, values, and the way things are done in the organization. The culture consists of the internal atmosphere or climate and commonly includes subcultures and countercultures.

Intervention: An action taken in an effort to correct or improve a practice or process. In safety, an intervention usually involves an attempt to eliminate or reduce a hazard or risk.

Motivation: An inner drive, impulse, or need that creates a personal incentive toward behavior. An individual's tendency toward action in a given situation.

Process: A number of steps/actions taken in order to accomplish a goal (McSween, 1995, p. 45).

Psychology: The study of human behavior. The *New Merriam-Webster Dictionary* defines psychology as "the science of mind and behavior" and "the mental and behavioral characteristics of an individual or group" (p. 587).

Values: Deep-seated beliefs that influence behavior. Core values are commitments individuals hold without compromise.

PRINCIPLES AND STRATEGIES OF BEHAVIORAL SAFETY

Behavioral approaches are based on years of research in the field of applied behavior analysis. Geller (2001, p. 21) notes, "Behaviorism has effectively solved environmental, safety, and health problems in organizations and communities; first, define the problem in terms of relevant observable behavior, then design and implement an intervention process to decrease behaviors causing the problem and/or increase behaviors that can alleviate the problem. The behavior-based approach is reflected in research and scholarship."

Applied behavior analysis methods are useful to achieve continuous improvement in safety. Krause (1997, p. 3) explains, "These methods include identifying and operationally defining critical safety-related behaviors, observing to gather data on the frequency of those behaviors, providing feedback, and using the gathered data for continuous improvement."

Research in behavioral safety was conducted in the United States in the 1970s and 1980s with documented (and often dramatic) reductions in injuries (Komaki, Barwick, and Scott, 1978; Sulzer-Azeroff and de Santamaria, 1980). Despite these successful studies, behavioral safety did not catch on and become common in the United States until the 1990s, when it became largely mainstream practice in many organizations.

COMMON PROBLEMS WITH SAFETY EFFORTS

McSween notes four common problems that seriously hinder safety efforts:

1. ***Severe Consequences for Reporting Injuries:*** There cannot be any pressure to hide injuries and incidents in successful safety efforts. It is important for organizations to learn from their incidents and prevent additional or more serious injuries. Reasons for underreporting include (1) the organization responding to reporting with punishment and (2) peer pressure when incentive programs are based on injury outcomes.

2. ***Safety Awards Not Related to Behavior:*** Safety award programs typically do not reinforce safe behavior on the job. As noted above, awards and incentives tend to influence employees not to report injuries. Many organizations base their awards on outcome measures (injury statistics) rather than process measures—people's compliance with safety procedures on the job (McSween, 1995, p. 15).

3. ***Dependence on Management or Staff for Planning and Decision Making:*** Many companies expect managers to enforce safety rules and procedures. Consequently, employees tend to rely on managers to ensure safe behavior rather than look out for each other. Employees are

not adequately challenged to improve safety, and many of them tend to look to someone else for primary responsibility for their safety.

4. ***Reliance on Punishment to Reduce Risky Behavior:*** Many safety programs are based primarily on punishment. Punishment is reactive and focuses on unsafe acts rather than emphasizing safe behavior. Mc-Sween notes there are numerous additional problems with punishment that are not readily apparent. For example:

- Punishment is effective only in the presence of the punisher
- Punishment tends to teach the wrong lesson (i.e., don't get caught)
- Punishment damages relationships and curtails involvement
- Punishment is difficult to maintain (1995, p. 17)

Dominic Cooper adds two important points:

- "The effectiveness of punishment depends on its consistency. It only works if given immediately and every time an unsafe behavior occurs."
- "Punishment serves more to suppress existing behavior than to encourage new behavior patterns. This . . . can cause anxiety and resentment that surface negatively in other areas of work activity" (1998, p. 271).

The point about punishment is not that it is never appropriate, but that it must be balanced. Punishment should be considered a last resort, not a first response.

Unfortunately, overcoming these four common problems does not ensure effective safety efforts. Solving these problems is a step in the right direction and helps establish a solid foundation to build on.

A PROCESS FOCUSING ON IMPROVING SAFETY BEHAVIOR

A brief definition of behavioral safety was provided within the "definitions" section. More detailed definitions describe activities involved in the behavioral safety process.

According to Petersen,

Behavior-based safety is, in fact, one of the primary answers to future improvement, but only if and when it does the following:

- Defines the behaviors needed at each level of the organization from bottom to top
- Ensures that each person clearly understands the required behaviors
- Measures whether the behaviors are, in fact, there
- Rewards (reinforces) the behaviors on a regular basis

Behavior-based safety systems that are real management systems do not turn safety over to any one level of the organization. It is great to get worker involvement—in fact, it is crucial. But to eliminate and isolate management from the process is dangerous to both sides. Management has the legal accountability for safety. (2001, p. 38)

Krause's definitions state,

With so many different initiatives being referred to as "behavior-based safety" it is important to discuss criteria for the definition of the activity . . . the most effective way to approach BBS is as an integrated, interdisciplinary activity, drawing not only from applied behavior analysis, but from quality management, organization development, and safety and risk management. The four key activities are 1) identify critical behaviors; 2) gather data on those behaviors; 3) provide ongoing, two-way feedback, and 4) remove barriers to safe behavior. Although this list may seem familiar, the fourth activity is the least understood and the most important for long-term success. Companies must understand the significance of using data gathered during observations to continuously improve the facility and the work process. (2002, p. 27)

The long-term goal of the behavioral safety process is to establish a strong and lasting safety culture. Erickson states, "It is the organization itself, through its culture, that determines the level of safety performance that will be achieved. There is overwhelming scientific evidence supporting this statement" (2001, p. 142). She continues by advising safety professionals that no matter how advanced their knowledge and skills are for building a strong safety program, they cannot, by themselves, effectively provide a safe and healthful working environment.

Manuele confirms the notion that behaviors are determined by the culture. He states, "An organization's culture consists of the values, beliefs, legends, rituals, mission, goals and performance measures, and the sense of responsibility to the employees, customers and community, all of which are translated into a system of expected behavior" (1997, p. 18). He further believes that "a company must have established—as part of its culture—

that employees are expected to, and have responsibility to, report risky behavioral situations to management. Furthermore, management must have established through its actions, that such references will receive proper consideration. If this is not the case, the behavioral problem is principally one of culture, for which action at the first-line level of employment will have limited success" (1998, p. 33).

BEHAVIOR SAMPLING FOR PROACTIVE MEASURES

Obviously, it is unrealistic to think that every behavior or even a majority of behaviors can be observed. Organizations simply do not have the resources to observe all behaviors. The strategy then is to observe a sample of behaviors. The behaviors that are observed are the ones that have been identified as critical to safety performance.

One of the main purposes of behavioral observation is regular sampling of the safety process (Krause, 1997, p. 195). Selectively sampling safety-related behaviors is a powerful tool. First, conduct of critical behaviors is a predictor of future safety performance. In this respect, behavior sampling is a leading indicator, rather than a lagging indicator such as injury rates. Trends can be analyzed and an organization can know precisely what needs to be addressed to be proactive. Second, behavior sampling focuses primarily on the safety process, not outcomes. By specifically focusing on key process activities, the outcomes can be improved.

EMPLOYEE-DRIVEN PROCESSES AND PARTIAL EMPOWERMENT

Another strategy used in most behavioral safety processes is the employee-driven or bottom-up approach. This does not mean that management abdicates responsibility and fails to be actively and visibly committed to and involved in the safety effort. Rather, employees (including management) drive the behavioral safety process, with support and resources provided by management. This is contrary to many traditional approaches to safety that are driven from the top down, or from outside influences such as OSHA. Safety scholars believe in the employee-driven process strongly enough to write books about it: Krause wrote *Employee-Driven Systems for Safe Behavior*, Petersen recently wrote a book titled *Authentic Involvement*, and

Geller wrote *The Participation Factor: How to Increase Involvement in Occupational Safety.*

Krause presents the following formula in *Employee-Driven Systems for Safe Behavior*: "Employee Involvement + Scientific Method = Continuous Improvement." Furthermore, he states that the proper use of methods to manage safety hinges on two factors:

1. scientific measurement and management of all levels of employee workplace behaviors, and
2. the involvement of all employees in this ongoing feedback and problem-solving process.

Petersen discusses the concept of limited empowerment versus total empowerment. Limited empowerment suggests we live within some ground rules. We all play with the same rules that make the game fair and safe. Most football rules are, in fact, safety rules, and most teams find them easy to follow. Limited empowerment means management establishes some ground rules for everyone to follow (Petersen, 2001, p. 63). Employees are empowered within this framework to participate in decision making for improving safety performance.

IMPLEMENTING BEHAVIORAL APPROACHES

How to Implement Behavioral Safety—Common Steps

Cooper recommends a nine-step implementation process (Cooper, 1998). *The following is an abbreviated adaptation of the recommended steps, plus the important step of solving system problems and pursuing continuous improvement:*

1. Seek/gain workforce buy-in to the behavioral process prior to implementation. Conduct briefings and educational awareness sessions so everyone is informed and knows what behavioral safety entails as well as its potential benefits.
2. Select a project team or steering team to implement and run the system.
3. Identify the critical safety behaviors that should be addressed. This may be accomplished similar to the way priorities are chosen for job

hazard analysis: review the past three years' injury records to determine behaviors and system factors contributing to injuries.

4. Develop specific behavioral checklists that cover the critical behaviors identified.
5. Train personnel from each workgroup how to conduct behavioral observations and how to provide feedback to others.
6. Establish a baseline of safe behavior by monitoring behavior for four to six weeks. Determine the average safe behavior levels during this baseline period.
7. Ask each workgroup to set a safety improvement target, using their baseline average as the starting point for comparison.
8. Monitor progress on a daily basis and provide detailed feedback to each workgroup on a weekly basis.
9. Review performance trends to identify barriers to improvement (Cooper, 1998).
10. Concentrate on continuous improvement of the system and behavior at all levels by solving any problems in the management system that encourage risky behavior or create/allow hazards to exist.

Organizations may be unsuccessful in implementing behavioral safety if they omit recommended steps. For example, if an organization fails to gain personnel buy-in (step #1), there will likely be a high level of resistance and possibly widespread efforts to defeat the process. If organizations omit step #10, they will not realize the full benefit that behavioral safety approaches can provide. Applying all these steps increases the probability of a successful behavioral effort as well as the sustainability of the process.

How to Conduct a Safety Assessment

Assessments are commonly conducted prior to the implementation of behavioral safety processes. Assessments may be conducted internally, or one of numerous safety consultants who perform safety assessments may be hired for this purpose. The purpose of safety assessments is to determine an organization's current level of safety performance and provide recommendations for improvement.

McSween (1995, 64–82) notes that making an assessment is somewhat analogous to completing a puzzle and recommends the following steps for conducting an assessment:

Step 1. Review the organization's safety data, including injury statistics and actual accident reports. Determine if the organization is above or below industry average. Identify groups of workers with the highest and lowest incidence rates.

Step 2. Conduct interviews with people from a diagonal slice of the organization. The interviewer should get very precise answers to such questions as how often area safety meetings are held, whether leadership regularly talks to employees about safety, and whether safety data is graphed and reviewed. McSween notes, "Constructing a well-designed survey requires experience or training in survey construction" (1995, p. 74).

Step 3. Observe safety meetings, safety audits, and safety practices in work areas. The observer should ask himself or herself if meetings are well run and if employees participate in meetings, and watch for safe and unsafe work practices.

Step 4. Analyze information and develop an improvement plan. McSween (1995, p. 79) recommends that ideal areas for pilot testing recommendations are those areas that are representative of the organization and have high incidence rates and management personnel willing to support the behavioral process.

Step 5. Make a final report and presentation. This is essentially a sales presentation, as one of the main purposes of assessment is to create management support. Stress the realistic benefits of systematic implementation prior to discussing the details of your recommendations.

For additional details about this five-step assessment process (such as recommended survey questions and an outline for your assessment report), refer to McSween's book *The Values-Based Safety Process: Improving Your Safety Culture with a Behavioral Approach* (1995).

An alternative assessment approach for determining organizational readiness is suggested by Cooper. This approach is based on a ***Praxis Six Cell Analysis Model*** that allows companies to conduct their readiness assessment at minimal cost.

The Praxis Six Cell Analysis Model is based on a performance equation:

$$\textbf{Motivation} \times \textbf{Ability} = \textbf{Performance}$$

The Praxis Model asks six questions at three organizational levels:

To Individuals:

1. Am I happy to behave safely?
2. Do I know how to behave safely?

To Workgroups:

1. How will others respond if I behave safely?
2. Will others provide the help, authority, information, and resources I need to behave safely?

To the Workgroups from an Organizational Perspective:

1. What rewards will we get for behaving safely?
2. Do our structures, systems, and environment facilitate or block us from behaving safely?

Truthful and positive responses to all six questions suggest readiness for a behavioral safety approach. If one or more of the questions is answered negatively, it is in the organization's interest to address the issues involved prior to implementation of a behavioral approach. For additional details, including how to use the readiness survey, visit www.behavioral-safety.com.

Developing an Inventory of Critical Safety-Related Behaviors

Behavioral safety involves measuring safety-related behaviors. Krause notes that the behavioral inventory is how we measure behavioral safety performance. *Benefits of using a critical safety behavior inventory for continuous safety improvement include:*

- Problem solving driven by data accumulated is accurate instead of superstitious.
- Performance feedback based on these data is true, rather than false feedback that impairs people's judgment about the causes of accidents.

- Safety training and incident analysis based on the behavioral inventory are useful for new employees and veterans alike (adapted from Krause, 1995, pp. 112–13).

What Are Critical Behaviors and Why Develop an Inventory of Critical Behaviors?

Critical behaviors are those actions that contribute to good safety performance or, conversely, that lead to injuries. Krause observes, "In most accidents at-risk behavior is the final common pathway. Each facility, characterized by particular production processes, products, and workforce, has a characteristic cluster of these final common pathways that are responsible for a highly significant percentage of its safety incidents" (1995, p. 113).

The challenge is to (1) identify the specific safety-related behaviors for a particular site, (2) establish an inventory of operational definitions for these behaviors, and (3) prepare a checklist based on these critical behaviors for observers to use. ***Developing this critical behavior inventory serves the following purposes:***

- Directs the steering team where to focus its efforts for maximum impact
- Writes operational definitions that provide detailed descriptions of the safe way to perform critical behaviors
- Establishes the criteria for behavioral observations and checklists
- Provides an upstream measurement for safety performance

How Can We Identify Critical Behaviors?

Krause (1997, p. 180) suggests four ways to identify critical behaviors at a site:

1. Make a behavioral analysis of incidence reports
2. Interview workers
3. Observe workers while they work
4. Review work rules, job safety analyses, and procedure manuals

These four approaches are rather self-explanatory, but the reader interested in learning more detail is referred to *The Behavior-Based Safety Process:*

Managing Involvement for an Injury-free Culture (Krause, 1997). Ultimately each site should develop their own inventory of critical behaviors, and the list is usually not that long—perhaps 15 to 25 behaviors that are genuinely crucial to safety performance. Earnest advises that the list be kept to five or ten critical behaviors (1994, p. 3).

It is worth noting again that developing job-specific behaviors is very similar to performing a job safety analysis (JSA) or job hazard analysis (JHA). JHAs and JSAs, however, are not normally written as operational definitions of behavior. It is suggested that organizations upgrade their JSAs to behavioral standards (Krause, 1997, p. 185).

Steps of the Observation Process

There is no one best way to perform observations—techniques and methods depend on the organization and the existing safety culture. It is recommended each site tailor the observation process to its particular needs. If a site benchmarks to another site, the suggested strategy is to use the benchmark sites' approach as a guide but develop specific procedures to accommodate the process locally.

The following steps are provided as a broad guide for observations:

1. *Select specific behaviors to observe as derived from the Critical Behavior Inventory.* The strategy for this step is to pinpoint safe practices (McSween, 1995, p. 106). These behaviors should be very clear and specific so that everyone trained in the observation process can agree on the safe way to perform a task. If the organization has developed a critical behavior inventory, and established operational definitions as part of the inventory, then critical safe practices have already been identified.

2. *Develop behavioral checklists for particular jobs and departments.* When checklists are developed, they are not cast in stone. First draft checklists, and then revise and improve them as necessary to aid the observation process. In general, checklists should be developed for each area. Keep the list of checklists as small as practical. A good practice is to keep the checklists short and manageable—ensure checklists focus on the critical behaviors. Don't forget to include management and leadership levels in the mix for critical behavior checklists.

3. *Develop specific procedures for the observation process.* Determine who will observe and who will be observed, when observations will

take place, and how long observations should take and other details of the process. In general, it is recommended that behavioral observations be conducted at least once per shift (Earnest, 1994, p. 4). Various organizations follow different schedules depending on the size of the workforce and the level of risks existing at the site. A problem with infrequent sampling is that it is not as effective. Some organizations conduct weekly, daily, or various other schedules of observations.

4. ***Determine procedures for data processing and feedback.*** Who will collect the data? Who will analyze the data? How will the data be shared with workgroups and the site as a whole? How often will the data be shared? What will be the criteria for the data to be tracked and shared?

Steps for Continuous Improvement

Establish a system to continuously improve the observation process. ***Items to consider for inclusion in this system:***

- Perform trial runs and fine-tune the observation checklist and process (McSween, 1995, p. 134).
- Conduct a management review of the process to encourage management input.
- Analyze data to identify areas for follow-up. Examples include trend analysis and Pareto charts to determine priorities.
- Follow-up on targeted items. This is an opportunity for problem solving so that the same problems do not occur over and over.

Organizations may use continuous improvement tools, such as Geller's ***DO IT*** process. ***This tool presents behavioral safety as a continuous four-step process as follows:***

- **D** Define the critical target behavior(s) to increase or decrease.
- **O** Observe the target behaviors during a pre-intervention baseline period to set behavior-change goals.
- **I** Intervene to change target behavior(s) in the desired directions.
- **T** Test the impact of the intervention by continuing to observe and record the target behavior(s) during the intervention program (Geller, 2001, p. 131).

In addition to or in place of DO IT, organizations are encouraged to use their existing continuous improvement systems, such as Deming's **Plan, Do, Study, Act—PDSA**, to improve their behavioral safety process (Deming, 1986). This helps to integrate the behavioral safety process into the management system and culture. Two important, related points to keep in mind:

- Behavioral safety is about integrating behavioral management into the management of safety in your company.
- Behavioral safety is a process to improve behaviors at all levels, not just a sequence of activities, meetings, and observations (Pounds, 1997, p. 7).

SAFETY COACHING

Geller defines coaching as "essentially a process of one-on-one observation and feedback" (2001, p. 239). One-on-one coaching is a high-level intervention and involves imparting both direction and motivation. Behavioral safety coaching today commonly involves peer-to-peer coaching, although supervisor-to-worker coaching is not uncommon.

Sports teams need good coaches to perform at their best. Similarly, workgroups and organizations need good coaches to meet the objective of improving behavior and reducing injuries.

In sports, athletic coaches normally have a higher status than their players. In the workplace, safety coaches are more similar to **player-coaches**: they are at the same level as the coworkers whom they observe and provide feedback to. An employee does not become an expert or superior just because he or she functions as a safety coach. It is normally a peer-to-peer process.

The long-term effectiveness of behavioral observations is contingent on appropriate feedback or coaching following an observation. Feedback is a powerful way to influence future behaviors.

Steps in the Coaching Process

Sarkus reveals "Seven C's" for coaching in his book *The Safety Coach.* The book is divided into three sections illustrating (1) the power of reinforcement, (2) the power of relationships, and (3) the power of information.

The Power of Reinforcement

Positive feedback tends to reinforce safe behavior. This feedback should be confirming, or socially rewarding, between coaches and coworkers. Sarkus notes that ***confirming feedback should be immediate, frequent, and favorable to be most effective*** (2001, p. 32). Sarkus recommends consistently confirming in favor of correcting to help coworkers become more committed to safety and less fixated on mere compliance.

Correcting feedback may be needed when employees are working at risk. Sarkus suggests starting by asking questions to determine if there are obstacles to working safely. Correcting feedback must be delivered in a constructive manner, or it may be perceived as punishment. The safety coach should skillfully open the door for safer performance by asking questions, encouraging two-way conversation, and redirecting behavior as appropriate.

There are usually many more opportunities for confirming feedback than for correcting feedback. In either case the feedback should be very specific. Vague feedback, such as "Good job" or "Be more careful," has no place in safety coaching.

The Power of Relationships

Good coaches build positive relationships. Sarkus discusses four Cs that help safety coaches develop their relationships:

- ***Caring***: Caring leaders develop a sense of community where people support each other. Caring is a foundation that provides meaning to the actions safety coaches take to prevent injuries.
- ***Collaborating:*** Collaborating means involving workers in meaningful ways to improve safety. Survival in today's competitive society, much less great safety performance, requires collaborating employees.
- ***Coaching:*** Coaching is about modeling the way and setting a great example. Great coaches set high standards, and they have a strong desire to bring out the best in others.
- ***Conciliating:*** To conciliate means to resolve disputes or conflicts so that relationships can be repaired. Some tips Sarkus gives for conciliating include focusing on the issue, listening up, and sticking to the facts (2001, p. 90).

The Power of Information

The final "C" Sarkus discusses is *clarifying*. Safety coaches and leaders influence behavior by clarifying safety values. Values in particular, and feedback in general, may be imparted through one-to-one communication or in group settings.

How to Provide Meaningful Feedback

Safety coaches observe and then give feedback. Feedback must be meaningful and meet certain criteria to be effective. The following are characteristics of meaningful feedback, whether the feedback is confirming, constructive, or a combination of the two:

- *Feedback should be specific.* Vague feedback is essentially useless, and potentially counterproductive or harmful.
- *Feedback should be immediate or quick.* In general, the sooner feedback is provided, the more effective it is. Sports coaches have a good opportunity to help players improve by providing immediate feedback during practice and games.
- *Individual feedback should be given privately.* There may be occasional exceptions to this, but the outcome or intent of feedback, even if inadvertent, should not embarrass others.
- *Coaches should listen actively.* Allow your coworker to finish what he or she says without interrupting. Maintain eye contact and ask open-ended questions to clarify the meaning of statements. As Covey (1989) advises, "Seek first to understand, then to be understood."

Common Performance Metrics for Behavioral Safety

Most organizations measure safety performance based on injuries and injury rates. Measuring behaviors adds another dimension—upstream or leading metrics—to the existing trailing metrics of injuries. Probably the most common behavioral measure is *percent safe*, a measurement of the critical behavior inventory calculating the percentage of safe behaviors from the total behaviors observed (or sampled). Although the percent safe measure is unlikely to be absolutely accurate, it's usually close enough to give a good idea of actual behaviors.

Safe behavior results should be tracked, graphed, and posted regularly for everyone to see. ***There are at least two benefits expected from posting safe behavior percentages:***

1. Posting provides everyone with knowledge of results. No one knows what the percent of safe behavior is in various areas without behavioral sampling, tracking, and posting. This awareness of risky behavior and trends is a beginning step toward improvement.
2. Tracking and posting safe behavior scores can be motivational. If behavior scores are tracked on a regular basis, everyone can see if performance is improving, remaining steady, or declining. People recognize they have an opportunity to contribute to the safety process.

It should be noted that percent safe should not be overly emphasized, as there can be a temptation to falsify the numbers.

Other behavioral measures include, but are not limited to:

- Recording the number of behavioral observations performed over a period of time. Organizations may choose to establish goals for the number of behavioral observations to conduct per week or per month. Too much emphasis should not be placed on the number of observations conducted, as the quality may suffer in order to produce a large quantity of observations.
- Another positive measure is the percentage of employees participating in the behavioral process over a period of time. Each site can establish its own criteria for participation.
- Other behavior/activity measures include participation in safety meetings, submission of safety suggestions that are implemented, number of near-hit reports over a period of time, number of job safety analyses performed or updated, and number of safety corrections made from work orders or similar avenues.

As noted earlier, these measures add a proactive dimension to traditional safety measures, which tend to be based on injuries, failures, and mistakes. These behavioral measures are all positive and measure what an organization and individuals achieve and accomplish, rather than their failures.

Potential Barriers to Successful Implementation of Behavioral Safety

Although the concepts and principles of behavioral safety are simple and straightforward, implementation issues can be difficult to handle and complex to understand. It only takes one persistent problem to undermine a behavioral safety effort. *The following list gives common examples of problems that may be encountered:*

- Failing to adequately plan and train prior to implementation. Employees must know what to observe and how to effectively intervene (Petersen, 2001, p. 53).
- Failing to provide planned, ongoing feedback to measure the effectiveness of the behavioral approach.
- Treating behavioral safety as a separate program rather than integrating it into existing management systems.
- Overemphasizing results (injury measurements). In numbers-oriented companies, when the results are not achieved the tendency is to find someone to blame (McSween, 1998, p. 47).
- Looking at behavior only without looking at the cause of the behavior. There is a void in the worker-focused model of behavioral safety when evidence of root-causal analysis of behavior is missing. Manuele stresses that "analysis be performed to identify root-causal factors of at-risk behavior. It should also be made clear that, for systemic causal factors, engineering and work method revisions must be the first considerations" (Manuele, 1998, p. 37).
- Overemphasizing the process (clarifying requirements, standardizing procedures, and establishing measures for key steps). Many of the problems in organizations are not in a particular process but in relationships between employees and managers. In many cases the cause of problems is managerial practices that destroy interpersonal relationships, creating distrust and discouraging employees from bringing safety matters to the attention of management. Most current problem-solving tools are useful for addressing process issues, but they are often inadequate for addressing behavioral issues (McSween, 1998, p. 48).
- Failing to get workforce buy-in. This may come about because the management team, without consultation, has imposed the system on the workforce (Cooper, 1998).

- Failing to target behaviors leading to injuries with observation checklists, and not defining critical behaviors with sufficient precision.
- Failing to enlist workforce participation in developing behavioral checklists, and the targeting behaviors not acceptable to the workforce.
- Using observation checklists to focus on unsafe conditions instead of risky behaviors. Behavioral safety should not be used as an audit of unsafe conditions, although sometimes behaviors lead to unsafe conditions.
- Trying to convey an optimistic picture of safety in their area in the use of percent safe scores that do not reflect the reality of the workplace.
- Punishing for failure to behave safely as indicated on the behavioral checklist. Punishment will undo everything the behavioral safety system tries to accomplish.
- Inadequately dealing with problems in safety improvement target-setting meetings. Common problems include insufficient preparation, insufficient time to discuss the issues, inconvenient meeting times (everyone cannot attend), and one or two individuals hijacking the sessions to air their grievances concerning what management has traditionally done (or not done) about safety.
- Performing observations at the same time every day. Observation sampling should be performed at random times throughout the week so it is unpredictable.
- Failing to institute a standardized procedure for people to hand in their completed observation checklists. This may result in the data for feedback becoming lost or mislaid.
- Having no computerized means to calculate and analyze the observation scores. It is cumbersome to process the data by hand and the quality of feedback suffers.
- Failing to conduct regular weekly feedback sessions. People are busy and stretched for time. The resulting perception is that management does not value the behavioral safety process, creating lack of buy-in.
- Failure to obtain ongoing management support. Managers do not see themselves as part of the problem and conclude they do not have anything to offer (Cooper, 1998).

The last twelve bullets above are adapted from an article by Sulzer-Azeroff and Lischeid (1999) and may be viewed in more detail on Professor Cooper's website at www.behavioral-safety.com. Behavioral safety efforts must anticipate and overcome these barriers to succeed. The next section gives advice on specific success factors.

Success Factors for Behavioral Safety

For a successful behavioral safety effort, numerous factors must work together in harmony. Ultimately, the key to safety success lies with management. *The following are essential features of the behavioral safety process:*

- Management must be visibly committed to the process. Examples of support include allowing people time to conduct observations, encouraging everyone to behave safely, setting a safe example personally, facilitating target-setting and feedback sessions, providing needed resources for the effort, and being willing and able to handle any barriers that arise. Krause notes, "The ability of leadership to demonstrate its willingness to address issues is probably the single most important (success) factor" (Krause, 1998, p. 41).
- There must be a significant level of workforce participation in and understanding of the behavioral safety process. Workers must be partially empowered in meaningful ways. They must clearly understand specifics of the behavioral process, such as how observation data will be used, what their own role in the process is, and how strongly management is committed to the process. Workers must possess a strong technical knowledge of how behavioral safety works. "Profound knowledge drives success, not buzzwords" (Hansen, 2001, p. 114).
- Selection, training, and guidance of the Implementation Team are predictors of success (Krause, 1998, p. 41).
- Data must be collected and used for decision making and for continuous improvement. The collection, analysis, and publication of data are integral components of the process. All employees need to be aware of the progress or lack of progress in safe behaviors.
- The process must be well planned in advance. It is better to make sound decisions at the outset rather than to be blindsided by unexpected crises once the process has begun. Behavioral safety initiatives are major undertakings requiring thorough planning and understanding prior to their beginning.

- Training and communication must be adequate for all levels to teach the necessary skills to identify critical behaviors, conduct observations, provide feedback, and perform problem-solving activities.
- All levels of personnel must be involved in the process.
- The behavioral process must be designed to meet the specific needs and peculiar circumstances of the organization. "One size does not fit all . . ." (Hansen, 2001).
- The basic premise and key objective of a behavioral safety initiative must be clearly established. "It's about process improvement . . . not finding and fixing careless employees" (Hansen, 2001, p. 114).
- A high level of trust is ordinarily a prerequisite to successful implementation.
- Leadership must address the safety issues (hazards) existing in the environment and risks that occur in working situations. Leadership must deal with management system inadequacies.
- Safety management systems must be aligned with behavioral safety principles. When there is a lack of alignment the likelihood of success is reduced.
- There must be an emphasis on long-term continuous improvement of behaviors and performance rather than short-term gains.
- There must be systematic procedures for structured problem solving and resolution/follow-up for process improvement.
- Safety champions must be carefully selected and groomed. Individuals with leadership skills, even if not in positions of authority, should be considered for roles as safety champions. These individuals are enthusiastic about safety, well respected by their peers, and considered team players, and they carry considerable influence by their words and actions. Safety champions are needed at every level of an organization.
- The superior approach is to focus on positive achievements rather than lack of failures. Emphasis should be placed on safe rather than unsafe and achievements should be celebrated. Failures should be viewed as temporary obstacles or challenges that will be overcome during the long-term process of continuous improvement.
- Recognition for safe behavior and safety-related accomplishments should be integrated into the daily work culture. Safe behaviors may cease if individuals are not recognized and encouraged to continue working safely. Peer recognition and feedback are powerful tools to improve safety performance.

- Patience and persistence are required. Organizations must allow time for trust to evolve and allow the process to work. "Behavior change in an organization is a long-term process that needs continuing investment over time" (Hansen, 2001, p. 115).
- Finally, as noted under "Barriers to Successful Implementation," there must be a systematic analysis of *why* behaviors occur. This requires looking for root causes in the system and in the culture. This is a leadership issue—leadership is responsible for establishing the safety culture. Pounds explains how this is ultimately a leadership issue: "A culture is, at its core, the sum total of those behaviors that leadership rewards and punishes . . . consequently, the behaviors and results that a front-line supervisor rewards or punishes (values) reflect the behaviors, results and values demonstrated by the top of the organizational hierarchy. For the behaviors, changing a culture, therefore, means systematically determining the behaviors that are wanted by an organization, then systematically rewarding those behaviors when they occur" (1997, 13).

CONCLUSION

Experience and research verify the potency of behavioral safety. According to McSween, research findings demonstrate the value of behavioral safety:

> The only empirical approach to improving safety that has proven to be effective is a behavioral safety process. Behavioral safety is the only approach that has routinely produced significant reductions in incidents in well-designed research studies. The approach involves employees using a systematically developed checklist as the basis for feedback on critical safety practices observed in work areas. (1998, p. 49)

QUESTIONS

1. Why should people view safety as a process rather than a program?
2. Describe some ways to get employees to develop ownership in workplace safety.
3. What is involved in effective safety coaching?
4. Discuss the use of specific behavioral metrics in safety. Why should behavioral measures be considered in addition to traditional measures?

5. What barriers do you think are the greatest impediment to behavioral safety success? How would you propose to overcome these barriers?
6. List and explain five critical success factors for behavioral safety.

REFERENCES

Blair, E. H. 1999. Behavior-based Safety: Myths, Magic & Reality. *Professional Safety*, 44(8): 28–32.

Cooper, M. D., 1998. *Improving Safety Culture: A Practical Guide*. Chichester, UK: John Wiley & Sons.

Covey, S. R. 1989. *The Seven Habits of Highly Effective People*. New York: Simon & Schuster.

Covey, S. R. 1990. *Principle-Centered Leadership*. New York: Simon & Schuster.

Deming, W. E. 1986. *Out of the Crisis*. Cambridge, MA: Massachusetts Institute of Technology, Center for Advanced Engineering Study.

Earnest, R. E. 1994. What Counts in Safety? *Insights into Management*. National Safety Management Society.

Erickson, J. 2001. *Proceedings of the ASSE Safety Management Symposium*. Orlando, FL: American Society of Safety Engineers.

Geller, E. S. 2001. *The Psychology of Safety Handbook*. Boca Raton, FL: Lewis Publishers.

Hansen, L. H., 2001. Behavioral Safety—Does It Work? "Yes and K(No)w" . . . The Difference. *Behavioral Safety Symposium: The Next Step*. Des Plaines, IL: American Society of Safety Engineers.

Komaki, J., Barwick, K. D., and Scott, L. R. 1978. A Behavioral Approach to Occupational Safety: Pinpointing and Reinforcing Safe Performance in a Food Manufacturing Plant. *Journal of Applied Psychology*, 63(4): 434–45.

Krause, T. R. 1995. *Employee-Driven Systems for Safe Behavior: Integrating Behavioral and Statistical Methodologies*. New York: Van Nostrand Reinhold.

Krause, T. R. 1997. *The Behavior-Based Safety Process: Managing Involvement for an Injury-free Culture*. New York: Van Nostrand Reinhold.

Krause, T. R. 1998. The Challenge of Behavior-based Safety. In *Proceedings of Light Up Safety in the New Millennium: A Behavioral Safety Symposium*, 26–42. Des Plaines, IL: American Society of Safety Engineers.

Krause, T. R. 2002. Cross-functional Improvement: Behavior-based Safety as a Tool for Organizational Success. *Professional Safety*, 47 (8): 27–33.

Manuele, F. A. 1997. *On the Practice of Safety*, 2nd ed. New York: John Wiley & Sons.

Manuele, F. A. 1998. Perspectives on Behavioral Safety: Observations of ASSE's Behavioral Safety Symposium. *Professional Safety*, 43 (8): 32–37.

McSween, T. 1995. *The Values-Based Safety Process*. New York: Van Nostrand Reinhold.

McSween, T. 1998. Culture: A Behavioral Perspective. In *Proceedings of Light Up Safety in the New Millennium: A Behavioral Safety Symposium*, 43–49. Des Plaines, IL: American Society of Safety Engineers.

Petersen, D. 2001. *Authentic Involvement.* National Safety Council: NSC Press.

Pounds, J. 1997. Behavioral Safety and Future Trends for Change Management. *Performance Management Magazine*, 16 (1).

Pounds, J. 1996. High-Risk Safety: The Six Biggest Mistakes in Implementing Behavior-based Safety. *Performance Management Magazine*, 15 (4).

Sarkus, D. J. 2001. *The Safety Coach: Unleash the 7 C's for World Class Safety Performance.* Donora, PA: Championship Publishing.

Sulzer-Azaroff, B., and de Santamaria, C. M. 1980. Industrial Safety Hazard Reduction through Performance Feedback. *Journal of Applied Behaviour Analysis*, 13 (2): 287–95.

Sulzer-Azaroff, B., and Fellner, D. 1984. Searching for Performance Targets in the Behavioral Analysis of Occupational Health and Safety: An Assessment Strategy. *Journal of Organizational Behaviour Management*, 6 (2): 53–65.

Sulzer-Azeroff, B., and Lischeid, W. E. 1999. Assessing the Quality of Behavioral Safety Initiatives. *Professional Safety*, 44 (4): 31–36.

13

Workplace Violence

J. Brett Carruthers, CSP

CHAPTER OBJECTIVES

After completing this chapter, you will be able to

- Describe the extent of the workplace violence problem in the United States
- Compare the frequency of workplace violence fatalities to other causes of death on the job
- Identify the high-risk work environments experiencing workplace violence
- List several occupations where workplace violence statistics indicate a problem of concern for the safety professional
- List some of the factors that contribute to workplace violence
- Explain the importance of establishing a workplace violence prevention program
- Describe some of the strategies companies can use to prevent or minimize the effects of workplace violence

CASE STUDY

Jane Doe had walked to work on a January evening because she could not afford to own a car. Minimal skills and education left her vulnerable in a number of ways. Jane worked as a clerk in a combination gas station–convenience store. At approximately 1:30 a.m., an unknown assailant walked into the store, ostensibly to make a purchase, but in reality to rob the store. Maybe Jane knew the assailant, or possibly he was afraid that she would recognize him later. In either case, before he left, Jane was dead, the victim of multiple stab wounds to the chest and neck. A few hundred dollars were missing from the cash drawer. Jane's body was discovered when the next customer came in for a purchase; her assailant has yet to be found.

INTRODUCTION

Robbery and criminal acts are often motives for homicide at work, accounting for a large number of the deaths. Many homicide victims work in retail establishments such as grocery stores, restaurants, bars, and small gas stations, making employees of traditional nighttime retailers particularly vulnerable.

Managers and small business owners who have done little or nothing in the past to protect their property and, more importantly, their employees from robbery find themselves being attacked on all fronts. Not only are they more vulnerable to losses from criminal activity, but they also find themselves increasingly subject to litigation and Occupational Safety and Health Administration (OSHA) scrutiny.

WORKPLACE EPIDEMIC OF VIOLENCE

Dennis Johnson, clinical psychologist and president of Behavior Analysts & Consultants, a management firm that closely tracks workplace violence, stated, "Workplace violence is the new poison of corporate America. It is not just a reflection of a violent society, but of that violent society interacting with workplace dynamics that are significantly changed from 10 or 15 years ago" (Dunkel, 1994, p. 40). Since that time, the numbers have increased, especially when terrorist actions are considered forms of workplace violence.

Background

Homicide remains a leading cause of job-related deaths, with many violent crimes committed while the victim is working or on duty. The motive behind many of these homicides is disputes among coworkers, customers, or domestic partners. The vastness of the workplace violence is epidemic:

- There are now nearly two million violent victimizations per year in the workplace.
- Occupational homicide has been a leading cause of death by injury in the workplace for women.
- One in four U.S. workers will be attacked, threatened, or harassed during their work career.

VICTIMIZATION OF THE AMERICAN WORKFORCE

Profile of Victims

The victims of workplace violence can be anybody, and no one is immune. *Examples of certain activities that tend to put workers at risk are:*

- Working with the public
- Handling money, valuables, or prescription drugs (e.g., cashiers, pharmacists, pharmacist technicians)
- Carrying out inspection or enforcement duties (e.g., government employees)
- Providing service, care, advice, or education (e.g., health care staff, teachers)
- Working with unstable or volatile persons (e.g., social services or criminal justice system employees)
- Working in premises where alcohol is served (e.g., food and beverage staff)
- Working alone, in small numbers (e.g., store clerks, real estate agents), or in isolated or low-traffic areas (e.g., washrooms, storage areas, utility rooms)
- Working in community-based settings (e.g., nurses, social workers, and other home health care providers)
- Having a mobile workplace (e.g., taxicab, delivery personnel)
- Working during periods of intense organizational change (e.g., strikes, downsizing) (www.ccohs.ca/oshanswers/psychosocial/violence.html)

Cost to Business

Workplace violence has a staggering impact on the emotional and fiscal health of an organization. Direct and indirect costs include direct legal and medical expenses, out-of-court settlements, death benefits, employee assistance programs, security-related services and products, business disruption, negative public and media relations, diverted resources, emotional scarring, extended litigation, fear, loss of personnel, name/product tainting, facility repairs, turnover, and wasted knee-jerk reaction expenses.

One incident of workplace violence can decimate a smaller business and ravage a larger one. Victims of violence have to undergo not only the pain of the attack but also the resulting personal losses, including lost workdays and mental anguish. The number one source of security liability claims is rape/sexual assault of women. The average settlement award for these claims is well into the hundreds of thousands of dollars, whereas the average verdict award is in the millions (Workplace Violence, 2002).

OSHA PERSPECTIVE

OSHA has developed several different guidelines and recommendations to reduce worker exposure to violence, but it has not yet initiated rulemaking. The various guidelines are available at OSHA's website, www.osha.gov. The guidelines developed are being enforced via the Occupational Safety and Health Act's general duty clause.

High-Risk Workplaces

All workplaces are vulnerable to violence: family-owned businesses, government, major corporations, manufacturers, the military, nonprofit, private, and public organizations, and retail, service, and other small businesses.

Historically, the occupation with the highest rate of occupational homicide has been taxicab drivers/chauffeurs. Other historically high-risk occupations include law enforcement officers, hotel clerks, and gas station workers. Retail trades have some of the highest number of occupational homicides. Parking lots and garages on work premises account for a large percentage of violent crime locations.

NIGHTTIME RETAILING

Due to the problems of violence, OSHA has paid particular attention to retailing in recent years and has developed guidelines accordingly. The reasons are obvious and are based on the dangers faced by employees, particularly of nighttime retail operations. Employees like Jane Doe can and must be protected. Here's how it can be done.

As in any other area of safety, a written plan outlining policies designed to deal with the anticipated problem is written. The plan includes procedures for cash management, handling customers, and generally minimizing the likelihood of robbery. More importantly, employees are trained, educated, and drilled on the steps to take should a robbery occur.

Planning includes acquiring and installing certain basic equipment. *Many nighttime retailers, particularly convenience stores and gas stations, benefit from the following:*

- Installing security camera(s) capable of recording robberies or criminal activities. This acts not only as a deterrent but also as a means of preventing a perpetrator from returning to the establishment or other operations if apprehended. As a complement to the camera(s), height markers displaying measurements from the floor are placed adjacent to the entrance. The camera and the clerk can use these to help identify assailants.
- Installing a drop safe or other cash management device to limit the amount of accessible cash. Conspicuous notices stating the cash register contains small amounts of money, such as $50 or less, are placed near the entrances and cash registers. Employees are carefully instructed in cash management procedures—particularly to keep cash on hand low (OSHA National News Release, USDL 98-179, April 28, 1998).
- When feasible, installing a silent or personal alarm system to notify police or private security forces when a crime is in progress (OSHA National News Release, USDL 98-179, April 28, 1998). With the advent of wireless devices, wireless personal alarms are now available that can be worn by personnel and activated by depressing the device button. Prompt notification and response by law enforcement may save a life.
- Ensuring the store and parking lot are well illuminated during all operating hours. Limit the number of access doors that are open. Install mirrors to observe areas not directly observable from the checkout area.

Avoid permitting employees to exit into poorly lit, unmonitored areas or isolated spots such as garbage areas and outdoor freezers. Employees might be assaulted leaving the premises and the assailant may use the door as a means of entrance (OSHA National News Release, USDL 98-179, April 28, 1998).

- Having at least two employees working at night to help protect each other. Many times a clerk is the victim of two would-be shoppers. One will distract the clerk while the other pulls a weapon, attacks, and disables the victim. If a second employee is not an economically feasible alternative, then protective enclosures for the lone attendant are a must (OSHA National News Release, USDL 98-179, April 28, 1998).

Shields can convert a nighttime retail establishment into a veritable fortress in short order. They can stop the penetration of clubs, knives, and small-caliber firearms. Enclosures should meet Underwriters Laboratory (UL) or American Society for Testing and Materials (ASTM) standards. All monies and the worker stay behind an open counter during the day and early evening. After traffic thins out and the store becomes more vulnerable, a bullet-resistant, composite-glass shield is dropped into place from overhead. With plate steel behind the counter and a clear shield between the employee and would-be assailant above the counter, a high level of protection is afforded against robbery and violence. In some cases, the unit is placed at the door of the store, with a transparent turnstile permitting the employee to pass goods to and receive money from customers. If the owner believes this will inhibit the sale of merchandise from the store, then the enclosure can be placed around the employee inside the store.

The Gas Mart where Jane was killed had advertising signs plastered all over the windows. Automobile and pedestrian traffic were unable to see into the store or to observe any crime in progress, permitting the murderer to commit his crime in relative privacy. His only concern was that a random shopper might enter the store before he left. Most convenience stores have large expanses of windows, and if they do not, they should. These windows should be kept open to the street with signage removed. A would-be thief is more likely to avoid a well-lit, windowed store because a crime in progress can be more easily spotted by a passerby or police patrol in the area.

Jane was wearing street clothes when the crime in her store took place. All store personnel should wear distinctive smocks or uniforms. Many stores engage in this practice for image purposes, but it is also a sound safety practice.

In the event that the police are summoned to a crime in progress, it will be easier for them to distinguish the perpetrator from the victim(s).

In spite of the above precautions, if a robbery does occur, store personnel should be instructed to fully cooperate with the robber. The silent alarm should be activated only if it can be done so discreetly. Employees are instructed to speak only in direct response to the perpetrator's questions and not to volunteer additional information. They should be as observant as possible and attempt to avoid confrontation (Butterworth, 1993).

Once the robber leaves the store, employees should immediately lock the doors and call the police. They should ask any witnesses to stay until police arrive and to not discuss the incident until that time. Everyone should avoid touching any surfaces the perpetrator may have touched (Butterworth, 1993).

Anticipating and preparing for events such as robberies should be an important part of any employer's plans—especially for nighttime retailers. The costs of formalizing and executing these plans are minimal compared to the potential saving of the lives and well-being of employees. Relatively inexpensive steps to protect the establishment's business investment and employees against robbery and associated violence can save in terms of criminal losses and litigation.

HEALTH CARE AND SOCIAL SERVICE

OSHA has also paid particular attention to health care and social service workers due to the high number of violent acts directed toward these groups in terms of both homicides and non-fatal attacks. These workers face an increased risk of work-related assaults stemming from several factors to include:

- The prevalence of handguns and other weapons among patients, their families or friends
- The increasing use of hospitals by police and the criminal justice system
- The increasing number of acute and chronic mentally ill patients being released from hospitals without follow-up care (these patients have the right to refuse medicine and can no longer be hospitalized involuntarily unless they pose an immediate threat to themselves or others)
- The availability of drugs or money at hospitals, clinics and pharmacies, making them likely robbery targets

- Factors such as the unrestricted movement of the public clinics and hospitals and long waits that lead to client frustration; increasing presence of gang members, drug or alcohol abusers, trauma patients, or distraught family members; low staffing levels; isolated work with clients during examinations or treatment; and solo work, often in remote locations with no backup or way to get assistance, such as communication devices or alarm systems (this is particularly true in high-crime settings)
- Lack of staff training in recognizing and managing escalating hostile and assaultive behavior
- Poorly lit parking areas

At a minimum, workplace violence prevention programs should:

- Create and disseminate a clear policy of zero tolerance for workplace violence, verbal and nonverbal threats, and related actions. Ensure that managers, supervisors, coworkers, clients, patients, and visitors know about this policy.
- Ensure that no employee who reports or experiences workplace violence faces reprisals.
- Encourage employees to promptly report incidents and suggest ways to reduce or eliminate risks.
- Require records of incidents to assess risk and measure progress.
- Outline a comprehensive plan for maintaining security in the workplace. This includes establishing a liaison with law enforcement representatives and others who can help identify ways to prevent and mitigate workplace violence.
- Assign responsibility and authority for the program to individuals or teams with appropriate training and skills. Ensure that adequate resources are available for this effort and that the team or responsible individuals develop expertise on workplace violence prevention in health care and social services.
- Affirm management commitment to a worker-supportive environment that places as much importance on employee safety and health as on serving the patient or client.
- Set up a company briefing as part of the initial effort to address issues such as preserving safety, supporting affected employees and facilitating recovery (OSHA, 2004).

Proper Security Measures

Security is an issue that has been especially problematic in the health care industry, but the principles of prevention apply to other industries as well. Sound security measures begin with the most basic and yet the most difficult measure—controlling access to the facility. Each entrance to a building should have some type of access control device. These devices can be as simple as keyed entrances or as complex as bar code or magnetic card readers or closed circuit cameras with intercoms and buzzers. Diligent access control will be the most daunting element an organization must overcome when implementing improved security measures.

Another access measure is controlling the visitor entrance to the facility by having specific procedures for handling visitors. Visitors are required to sign in at specified point(s) and, in sensitive areas, must be accompanied by an authorized escort at all times. They are never to be given the freedom to roam at will within a facility.

One final means of controlling access to a facility is to ensure that devices installed to control facility access are not compromised. Compromising security includes purposefully leaving doors open or revealing access codes, duplicating keys, and leaving fence gates unlocked. *Access control is only as good as management's commitment to facility security management.* Security is not a matter of convenience!

Another basic security measure is to issue color photo identification cards. The following information should be on the identification badge:

- Employee's name and signature
- Employee's identification badge serial number
- The signature of the person authorized to authenticate and issue identification badges

The key to a successful security program is to customize the measures necessary for your organization's needs. Finally, whether a company takes a low-profile approach (access control, identification badges) to security or a highly visible approach (closed circuit television cameras, private security officers), a facility's security management program should be given the same management attention as safety, operations, sales, and marketing (Vincent Piazza, personal interview, May 20, 2013).

PROBLEM EMPLOYEES: FORMULA FOR FAILURE

Violence in the workplace sometimes involves employees only—one or more against another. Jim Gary was a large man, not easily intimidated. As foreman of a relatively rough bunch of workers, Jim made sometimes unpopular decisions and took criticism personally. After a particularly loud and confrontational disagreement, he wasn't surprised to see one of his crew waiting for him after work with a jack handle in one hand, slapping it against the other. The crew member stood just outside the company gate, and when he caught a glimpse of Jim, he stood in front of the open gate, waiting for Jim to leave company property. Jim's approach did not follow company policy. He hopped in his pick-up truck and slowly drove toward the employee. When he was within 25 feet, he gunned the engine and tried to run over his aggressor. Although he missed, he had made his point, and wasn't threatened again. This behavior—on either party's part—could have easily escalated into a situation in which one or both were injured or killed, and the company may have become the final victim through financial liability and adverse publicity.

The Ingredients List

Fear and violence among workers can be ascribed to various factors:

- Loss of a job or position
- Interpersonal conflicts on the job
- Technological innovations
- Prevalence of violence in society
- Diversity and change in the workplace
- Shifts of responsibility at home
- Job loss or demotion

Historically, researchers have attempted to profile perpetrators of workplace violence. They have considered such characteristics as having a fascination with guns. This could apply to hunters, military officers, or members of police departments. Many of the other characteristics are even less useful. Knowing someone is from a broken home is no reason not to hire that individual or to use him or her in a sensitive position. The same could be said about aggressiveness. On one hand, a person with a history of violent behavior is more likely to exhibit the same characteristics in the future. However, otherwise docile employees can be pushed to the breaking point under high-pressure conditions. Descriptions by neighbors of murderers from their neighborhoods bear this out: "He was a likeable individual." "He pretty much kept to himself and didn't bother anyone."

Economic demands put considerable pressure on otherwise stable employees. Displacement or wage loss through company relocation, downsizing, or market changes creates extremely adverse conditions for workers. Domestic disturbances and problems are being carried into the workplace at an alarming rate. With schedules and demands on American workers pushing them to the breaking point, some will and do suffer breakdowns. For the typical wage earner with a mortgage, car payments, and other debts, even small changes in the monthly paycheck can create considerable frustration and devastation.

Other Ingredients

A highly stressed workplace is the most susceptible to violence. *The following factors add considerably to the problem:*

- Management does not talk with or delegate control to employees.
- Employees are micromanaged by doting managers.
- Employee work is fast-paced and performed in poor environmental conditions.
- Overtime is frequent and mandatory.
- Employee benefits have recently been cut (Joyner and McDade, 1994, A8).

If workplace downsizing is coupled with a combination of any or all of the above, an extremely stressful situation has been created. As stress from the workplace and family builds, the displaced worker, who may already be on the edge, becomes more desperate and isolated. This combination can become lethal—making the time ripe for acts of violence to occur.

RECOGNIZING THE POTENTIAL AGGRESSOR

Aggressors can be anyone: clients (current and former), competitors, criminals, current and former employees, current and former relationship partners, customers, drug addicts, gangs, or terrorists. Whoever the aggressor may be, there are usually red flags present.

Disgruntled Employee Red Flags

Employees exhibiting marked changes in their demeanor and performance—for example, those who are normally quiet suddenly becoming loud, boisterous, or disgruntled—may signal the beginning of a volatile situation. During the satellite video conference on workplace violence presented by the George Washington National Satellite Network (1994), FBI Special Agent Eugene A. Rugala tendered the following list of red flags of human behavior that should be closely monitored:

- History of exhibiting violent behavior
- Obsessing with weapons, and collecting and compulsively reading gun magazines
- Carrying a concealed weapon
- Making direct or veiled threats
- Using intimidation or instilling fear in others
- Maintaining obsessive involvement with job
- Acting as a loner
- Exhibiting unwanted romantic interest in coworker
- Sexually harassing opposite gender coworkers
- Exhibiting paranoid behavior
- Exhibiting an lack of willingness to accept criticism
- Carrying a grudge
- Exhibiting recent family, financial, and/or personal problems
- Undue interest in publicized violent events
- Testing limits of acceptable behavior
- Suffering from stress in the workplace such as layoffs, reduction in forces, and labor disputes
- Showing extreme changes in behavior or stated beliefs (Rugala, 1994)

Special Agent Rugala emphasized one key point: "An employee can manifest one or all of these traits and never act out violently! This is where the importance of having solid employee/supervisor relationship cannot be underscored!"

Employee Disenchantment

Employee disenchantment can be a major source of stress, and over time, stress can build to the point of the person "snapping," according to Special Agent Rugala. The following list highlights causes of employee disenchantment:

- Confusion
- Lack of trust
- Not being listened to
- No time to solve problems
- Office politics
- Someone solving problems for you
- Not knowing whether you are succeeding
- Indiscriminate application of rules
- Boss takes credit for others' ideas and work
- Believing you cannot make a difference
- Meaningless job (Rugala, 1994)

The response to stress may be an act of violence, often exhibited as revenge (B. Walton, personal communication, 1994).

Revenge

Revenge comes in a variety of forms. Carefully plotted revenge can destroy a company. The following are different types of revenge:

- Product tampering
- Rumor mongering
- Theft of property
- Theft of secrets
- Stalking
- Kidnapping executives
- Threats
- Harassment

The following are examples of revenge:

- The Tylenol tampering incident in the 1970s had a profound effect on Johnson & Johnson, as well as every other manufacturer of consumer medical products, and on how these products were sealed and packaged.
- Encyclopedia Britannica experienced a serious product-tampering incident when a discharged employee changed all references of Allah to Jesus. Thousands of copies were printed before the change was discovered.
- A terminated employee on his way out of the company's headquarters building lobby gouged several priceless Van Gogh paintings, destroying them.
- A discharged telephone company employee tapped the telephones of several prominent citizens in the community and then attempted to extort money from them to keep the information he acquired quiet.
- An employee who was fired from a fast-food restaurant changed the slogan on the cash register receipt to a racial epithet, insulting numerous customers.
- An information technology director, knowing he was about to be terminated, inputted the system "key," wiping out all company servers and the data on them. When this "key" was inputted and acknowledged, all data was wiped clean with no means of recovery.

These are only a few examples of thousands of acts of revenge that take place in the workplace annually. A malicious employee wanting to pay back the company because he was discharged can cause significant damage to an organization's image and reputation.

Violence

Unfortunately, some individuals resort to violent acts, including:

- Intimidation
- Assault
- Rape
- Robbery
- Property destruction

Proactive intervention is required on the part of the employer.

DEFUSING A TIME BOMB: THE VIOLENT EMPLOYEE

While there is no cure for workplace violence, ***there are proactive, preventive steps that can be implemented to provide a realistic approach to defusing workplace violence before it occurs:***

1. Establish a clear non-harassment policy
2. Perform pre-employment screening
3. Establish a drug testing program
4. Conduct employee and management training in stress management and communications
5. Plan for crisis management
6. Establish proper security measures
7. Foster a working liaison with local law enforcement
8. Conduct security-related drills (Vincent Piazza, personal interview, May 20, 2013)

Nonharassment Policy

Senior management must make it crystal clear to its employees that harassment and threats will not be tolerated. Garry Mathiason, a San Francisco lawyer who specializes in workplace liability law, states, "If you have a policy or plan in place, then what is tolerated and what is not become part of the culture" (Dunkel, 1994, p. 70). This is especially important with verbal threats. Company policy must be firm that verbal threats will not be tolerated. Mathiason goes on to say, "If you doubt me, [ask yourself] when was the last time you made a joke going through an airport metal detector?" (Dunkel, 1994, p. 70).

Pre-employment Screening

The significance of this tool cannot be understated. How involved your organization gets with pre-employment screening will be a function of the nature of your company's business, its culture, and the sensitivity of the position being filled.

Before any pre-employment screening is done, consult with your organization's legal counsel on the proposed pre-employment screening to be conducted. This is vitally important and can save your organization significant expense should a pre-employment screening tool be considered to discrimi-nate in your state (Susan McClaren, personal interview, April 18, 2013).

Target Stores in Oakland, California, asked prospective security guards to take the Rodgers Condensed CPIMMPI written prescreen (a psychologi-cal screening tool). An attorney for the plaintiffs, in a class-action lawsuit, hinted that a number of the questions on the long-used exam were "ex-tremely invasive on matters of sexuality, religion, bodily functions, and the like" (Albrecht and Mantell, 1994, p. 50). Under the settlement agreement, Target Stores paid $1.3 million to 2,500 awardees and agreed to ban the test in its 113 California stores.

What pre-employment screening tools should be used? The following are a minimum:

1. Work history verification
2. Military history verification
3. Credit history
4. Driving record check
5. Criminal history check

Before using any of the above screening tools, check with legal counsel because all the above may not be legal in every state. Red flags are unex-plained gaps in employment of greater than 30 days; scant details of disci-plinary matters in military service; dire financial circumstances (high debt, home foreclosure, etc.) in the credit history; and numerous accidents or traffic violations on the driving record. Obviously, one must evaluate these from a big-picture perspective; one negative report may not be too bad, but taken as part of the whole it may be significant (Vincent Piazza, personal interview, May 20, 2013).

Drug Testing

As with the pre-employment screening, company legal counsel and person-nel experienced with Employee Assistance Programs and workplace drug testing must be consulted before a drug-testing program is started (Bruce Wilkinson, personal communication, February 8, 2013).

Drug testing is conducted in three ways: pre-placement, random, and for-cause or reasonable suspicion. The total testing procedures are typically handled by an outside firm to avoid casting any suspicion on the process, including the selection of employees for random testing. Employment offers are conditional upon negative results from candidates' drug tests.

Employee and Management Training

Employee and management training are critical to successfully handling a potentially explosive situation. Areas include stress management, conflict resolution, negotiating, and interpersonal communications. The training must be tailored to the organization's philosophy and culture.

Training should be presented at two levels. First, it should be presented at the employee level to establish a baseline—a common foundation. Second, it must be presented at supervisory and managerial levels to establish a consistent manner in which various situations will be handled. Additionally, the training should include role playing to provide participants the opportunity to experience a variety of situations in a controlled learning atmosphere (Vincent Piazza, personal interview, May 20, 2013).

Crisis Management Planning

Planning for an incident involving workplace violence should be no different than planning for a fire or chemical spill. As a result of the terrorist events on September 11, 2001, in this post-9/11 world, preplanning for an incident of workplace violence is mission critical. Having a plan of action in place that is easily executable will provide structure to an otherwise chaotic situation.

What should be in a crisis management plan for handling a workplace violence incident? In their book *Ticking Bombs: Defusing Workplace Violence in the Workplace*, Steve Albrecht and Michael Mantell provide the following list:

- Telephone numbers of local law enforcement and emergency services
- Notification of key company personnel (this list must include the company's legal counsel)
- Procedure for protecting the scene of a workplace violence incident for investigators
- Procedure for checking the integrity of your company's data and computer systems
- Arrangements and retainer contracts for on-scene employee counseling immediately following a significant event (24-hour coverage)

- Training of senior site management in dealing with the media
- Designating an official company spokesperson and media relations procedure
- Procedure for providing grief- and trauma-recovery time for victims and related witnesses, bystanders, and employees
- Procedure for company sanctions and/or punishment for the instigators in lesser cases
- Arrangements with company Employee Assistance Program and psychological counseling programs to help employees cope with post-incident stress management
- Procedure for handling family notification(s) of the victim(s) and the company's aid package
- Procedure for handling cleanup and scene restoration (1994, pp. 221–42)

While the steps above deal specifically with handling an incident of workplace violence, another important procedure that must be developed is an effective termination procedure.

The termination procedure should include details on how to handle the termination process and ***must include at a minimum:***

- Handling the individual's personal effects
- Preventing facility and information system access
- Collecting company property (credit cards, keys, access cards, identification badges, proprietary information, etc.)
- Handling outstanding expense reports
- Handling severance, vacation pay, and other continued benefits
- Handling difficult employee terminations (Vincent Piazza, personal interview, May 20, 2013)

The above steps are not all-inclusive and may not apply in all situations. They must be tailored to each company's structure and uniqueness.

Liaison with Local Law Enforcement

During a crisis situation, it is helpful to know the key personnel in the local law enforcement organization and their commitment to assisting your organization. This is not the time to be exchanging initial introductions (Arthur Kelly, personal communication, February 22, 2013).

This plan does not guarantee workplace violence will not occur; however, it will better prepare the organization should an event occur.

PLANNING FOR SECURITY EMERGENCIES

There are different responses and protocols employed when an emergency occurs. Some, such as a response to a building fire or weather emergency, are easy to initiate. Others, such as armed intruders, active shooters, or emergencies in the community, are more difficult to assess. Sadly, more shooting and violent incidents (such as the Boston Marathon bombings) appear to be occurring throughout the country. A proactive approach requires emergency planners to develop plans to respond to three security emergency scenarios: lockdown, lockout, and shelter-in-place.

Lockdowns

There are different reasons to initiate a lockdown: armed intruder, active shooter, civil expression, domestic violence, or an irrational visitor. To do it right, procedures need to be developed and implemented, training conducted, and plans tested. A lockdown is initiated when there is a crisis or physical threat inside the building and movement by occupants may place them in jeopardy. Lockdowns can help contain the threat, keep individuals away from where the threat may be present, and protect individuals by getting them away from places where the threat may be approaching. Depending on the situation, lockdowns may last a short time or could go on for hours.

Law enforcement must be notified immediately. An ongoing relationship with all first responders will ensure that all parties are familiar with each entity's plans, roles, and responsibilities, and that this information is integrated into the municipality's plans.

Each facility and emergency is different. The following recommendations should be used as a starting point or reference to evaluate existing plans.

During a lockdown the following actions must be taken immediately:

- Immediate notification of building employees and visitors that a lockdown is in place. Specified personnel must secure building perimeter doors.

- Individuals inside the building should remain in the room or office where they are located and lock the door.
- If in a corridor, elevator, or stairwell, proceed to the nearest office to take shelter and ensure that the door is locked.
- All individuals should move down onto the floor and remain away from windows and doors.
- All individuals should make as little noise as possible. Individuals should set cell phones and pagers on "vibrate" or "silence" mode.
- Secure all windows and close blinds and/or curtains. Turn lights off.
- Do not open the door for anyone. When the building has been cleared, your door will be unlocked by law enforcement personnel and you will be given instructions regarding the next action to take.

Lockouts

A lockout is initiated when there is a physical threat outside the building and evacuation may place building occupants in jeopardy. Lockouts are typically implemented when a crime occurs near a building. They prevent the physical threat from gaining entry. As with lockdowns, lockouts may be quick or protracted. During a lockout the following actions must be taken immediately:

- Immediately notify building employees and visitors that a lockout is in place. Specified personnel must secure building perimeter doors.
- Individuals inside the building will not be able to leave the building.
- Secure all windows and close blinds and/or curtains. Turn lights off.
- Keep away from windows.
- Be prepared to move into lockdown.

Shelter-in-Place

Shelter-in-place is initiated in the event of an external event, such as a release of hazardous chemicals or a large, yet-to-be controlled fire, when it is not possible or advisable to evacuate a building. Shelter-in-place is also used where internal control is needed because of a medical emergency in the building. Shelter-in-place events are usually shorter in duration.

During a shelter-in-place event, the following actions must be taken immediately:

- Immediately notify building employees and visitors that a shelter-in-place is under way. Specific instructions will follow if building systems need to be turned off.
- In a medical emergency, individuals and visitors inside the building should remain in the office or building area where they are located until the all-clear signal is given.
- In an environmental event or large fire in a multi-story building, all individuals and visitors should move upward to a higher floor, since most chemical agents are heavier than air.
- Close all windows and perimeter doors.
- Turn off heating, air conditioning, and ventilation systems.
- Have a designated individual check all building openings to ensure that none have been overlooked.
- A designated individual should monitor radio or television stations for further updates and have occupants remain in the shelter-in-place until authorities indicate it is safe to leave.

Developing the Plan

Emergency procedures must consider the specific construction characteristics of the facility. When the plan is developed, the following need to be considered:

- **What are the structurally safest areas of the building?** Develop diagrams that identify these areas and, as necessary, identify the appropriate routes to proceed to these areas.
- **How good is the intercom system?** Does it reach all interior areas? Will it reach people who are outside the building?
- **Is the facility shared with other organizations?** If so, what are their emergency plans? As the controlling interest in the building, you will need to assist these organizations with their plans and include them in drills.
- **Developing site-specific plans along with other stakeholders in your buildings**. These individuals are key assets and will have the best knowledge of the facility. This knowledge will be of use in making these plans more effective and ensure that key responsibilities are assigned to appropriate building personnel.

- **Developing a communication plan.** Communication with non-affected facilities/buildings is important so they may prepare to take action as appropriate.

Communicating the Plan

Once the plan is developed and implemented, it must be communicated to building employees. They must know and understand the differences between the three key security emergency scenarios: lockdown, lockout, and shelter-in-place. Employee training sessions need to be held to communicate these plans and answer employee questions that may arise.

Practicing the Plan

Finally, conduct lockdown, lockout, and shelter-in-place drills to ensure that employees understand these procedures, as they are very different from evacuation drills. If drills indicate that further training is needed, conduct refresher training sessions to review procedures. Poor performance during a drill usually indicates poor performance and reaction during an actual emergency. These are not times for complacency.

Use the following link to access an excellent video of an active shooter drill, which includes different first responders. It is an excellent example of inter-agency response and cooperation: www.cffjac.org/go/jac/media -center/video-gallery/tcm-active-shooter-scenario/.

CONCLUSION

This chapter reviewed the occupational epidemic sweeping the United States commonly referred to as "workplace violence." It is a problem many safety professionals address in an attempt to reduce the risk of serious injury to thousands of working men and women across America. The workplace aggressor can be anyone from a disgruntled employee to a psychologically disturbed competitor. Workplace violence can be as traumatic as a seriously deranged man shooting an ex-boss and coworkers, as insidious as product tampering, or as horrendous as a terrorist attack. As with any other loss in the occupational environment, the prevention of workplace violence requires a comprehensive approach that includes everything from pre-employment screening to crisis management planning. No workplace is immune. With workplace homicide being the number one cause of occupational death for women and the third leading cause of death for all workers, safety professionals' responsibilities include workplace security.

QUESTIONS

1. What is workplace violence? How far-reaching is this problem in the United States?
2. Why is knowledge of workplace violence important to a safety professional?
3. What are some of the factors that contribute to workplace violence?
4. List and describe some of the red flags that could indicate a disgruntled employee might become violent.
5. What are some of the acts of violence associated with workplace violence?
6. List and briefly describe the components of a comprehensive workplace violence program.

REFERENCES

Albrecht, S., and Mantell, M. 1994. *Ticking Bombs: Defusing Violence in the Workplace.* Burr Ridge, IL: Irwin Professional Publishing.

Bureau of Justice Statistics. *Violence and Theft in the Workplace.* U.S. Department of Justice: Office of Justice Programs. Available at http://www.bjs.gov/index.cfm?ty=pbse&sid=56.

Butterworth, R. A. 1993. *Convenience Business Security Act: Robbery Deterrence and Safety Training Guidelines.* Tallahassee, FL: Office of the Attorney General.

Chavez, L. J. 1999. *Workplace Violence Defined.* Available: http://members.aol.com/endwpv/definition.html (accessed November 13, 1999).

Dunkel, T. 1994, August. Danger Zone: Your Office. *Working Woman, 19,* 39–71.

Florida Department of Legal Affairs, Division of Victim Services. 1993. *Additional Statement to the Secretary of State Rule* (Chapter 2A-5). Tallahassee, FL: Legal Document Reproduction.

Joyner, T., and McDade, S. 1994, March 31. Workplace Violence Blamed on Stress. *Niagara Gazette.* p. A8.

OSHA. 2004. *Guidelines for Preventing Workplace Violence for Health Care & Social Service Workers.* Available at http://www.osha.gov/Publications/osha3148.pdf (accessed July 29, 2009).

OSHA National News Release, April 28, 1998. USDL 98-179.

OSHA National News Release, May 9, 2000. USDL.

Rugala, E. A. 1994. *Recognizing the Potential Aggressor*. Address presented at the George Washington National Satellite Network Video Conference: Washington, DC.

Toscano, G., and Windau, J. 1994, October. The Changing Character of Fatal Work Injuries. *Monthly Labor Review*, *117*, 17–18.

Warchol, G. 1998, July. Workplace Violence, 1992–1996. *National Crime Victimization Survey*. (Report No. NCJ-168634). Washington, DC: Bureau of Justice Statistics, U.S. Department of Justice. Available at http://www.bjs.gov/content/pub/pdf/wv96.pdf.

Workplace Violence: http://www.workplace-violence-hq.com.

BIBLIOGRAPHY

Bachman, R. 1994, July. Violence and Theft in the Workplace. *Crime Data Brief*, 1–4.

Bell, C. A. 1991, June. Female Homicides in US workplaces. *American Journal of Public Health*, 729–32.

CAL/OSHA. 1995. CAL/OSHA Guidelines for Workplace Security. California Department of Industrial Relations, Division of Occupational Safety and Health. Revised March 30, 1995.

CAL/OSHA. 1995. Model Injury and Illness Prevention Program for Workplace Security. California Department of Industrial Relations, Division of Occupational Safety and Health. March 30, 1995.

Carder, B. 1994, February. Quality Theory and the Measurement of Safety Systems. *Professional Safety*, *39*, 23–28.

Catania, A. C. 1968. *Contemporary Research in Operant Behavior*. Glenview, IL: Scott, Foresman and Company.

Daniels, A. C. 1989. *Performance Management: Improving Quality Productivity through Positive Reinforcement*. Tucker, Georgia: Performance Management Publications.

Firearm Injury in the U.S. National Center for Injury Prevention and Control, CDC. http://cdc.gov/injury/wisqars/index.html.

Gunman Kills 2, Wounds 2 at Ford Union Meeting. 1994, August 14. *Buffalo News*, A7.

Homicide in US Workplaces. 1992, September. U.S. Department of Health and Human Services, 1–7.

"How Prevalent is Gun Violence in America?" National Institute of Justice (26 October 2010). U.S. Department of Justice. http://www.nij.gov/topics/crime/gun-violence/welcome.htm.

Johnson, D. L., Kiehlbauch, J. B., and Kinney, J. A. 1994, February. Break the Cycle of Violence. *Security Management*, *38*, 24–28.

Johnson, D. L., and Kinney, K. A. 1993. *Breaking Point*. Chicago, IL: National Safe Workplace Institute.

Jorma, S. 1994, May. When Does Behavior Modification Prevent Accidents? *Leadership & Organization Development Journal*, 15.

Lincoln, B. 2006. *Holy Terrors—Thinking about Religion after September 11*, 2nd ed. Chicago: University of Chicago Press.

Meyer, C. 2012, June. How Awareness Stops Negligence, Presents Workplace Violence in Healthcare. *Security Magazine.*

Nater, F. P. 2013, February. 10 Ways to Prevent Workplace Violence Escalation. *Security Magazine.*

NIOSH Urges Immediate Action to Prevent Workplace Homicide. 1993, October 25. *NIOSH Update.*

NIOSH Current Intelligence Bulletin 57. 1996. Violence in the Workplace: Risk Factors and Prevention Strategies.

Northwestern National Life Employee Benefits Division. 1993. *Fear and Violence in the Workplace.* Minneapolis, MN.

Preventing Homicide in the Workplace. 1993, September. *NIOSH Alert.* U.S. Department of Health and Human Services.

Simonowitz, J. A. 1993. *Guidelines for Security and Safety of Health Care and Community Service Workers.* California Department of Industrial Relations. Division of Occupational Safety and Health, Medical Unit.

6,271 Job-Related Deaths Counted in 1993; Vehicle Crashes, Homicides Lead All Causes. 1994, July 27. *Niagara Gazette*, A6.

Sprouse, M. 1992. *Sabotaging the American Workplace.* San Francisco, CA: Pressure Drop Press.

Survey Reveals the Cost of Workplace Violence to Employers. 1995, May 15. *OSHA General Industry News.* Santa Monica, CA: Merritt Publishing.

Sygnatur, E., and Toscano, G. 2000. Work-Related Homicides: The Facts. Compensation and Working Conditions (available at the Bureau of Labor Statistics website, http://www.bls.org).

Tarasuk, V., and Eakin, J. 1994, May/June. Adding Insult to Injury. *OH&S Canada, 10*, 44–47.

U.S. Department of Justice. 2010, October. How Prevalent Is Gun Violence in America? National Institute of Justice. Available at http://www.nij.gov/topics/crime/gun-violence/welcome.htm.

Terrorism Preparedness

CHAPTER OBJECTIVES

After completing this chapter, you will be able to

- Understand the compelling reasons for preparing for terrorist attack in business
- Identify some of the key historical events causing the United States and its citizens to prepare for terrorism
- List critical areas of concern that need to be addressed in terrorism preparation
- Understand the role of the safety and health professional in company terrorism planning activities

CASE STUDY

Steve Smith is sitting at his desk on Monday morning when the call comes through. A bomb has been placed in the pesticide plant where he works. A quick mental review of chemicals stored on-site brings to mind vessels of materials that have IDLHs (Immediately Dangerous to Life and Health) as low as 1 part per million. Steve closes his eyes for five seconds before dialing 911. His panicked voice has more questions for the dispatcher than the dispatcher has answers. Steve is quickly transferred to a police official with little knowledge of pesticides or their potential for damage to the environment and the surrounding community. Steve is searching for answers as to what to do next, but he is having trouble finding those answers.

HISTORIC INFORMATION

February 26, 1993: An explosion roars through the underground parking garage at the World Trade Center in New York City. Smoke rises and begins filling corridors and some escape routes. Six people are killed and over a thousand are injured, primarily due to the effects of smoke inhalation. The device—made primarily of urea nitrate, a fertilizer-based explosive—also included three large metal cylinders of compressed hydrogen gas. All of this was placed inside a rental truck and detonated by fuse. The explosion punched a hole approximately 150 feet in diameter and 50 feet deep. This was the first major attack in the United States to bring the reality of international terrorism home to all United States citizens.

April 19, 1995: An explosion rips through the Murrah Federal Building in Oklahoma City, Oklahoma. It completely tears off the north wall and leaves 168 dead, including 19 children. The terrorists had purchased ammonium nitrate from a farm cooperative and nitromethane racing fuel at a racetrack for materials to construct the bomb. It was rigged in a rental truck and parked in front of the building. A detonation-delay mechanism allowed the bomber to escape prior to the explosion. At the time, this was the most lethal terrorist attack to ever occur in the United States.

September 11, 2001: Four commercial airplanes are hijacked on the East Coast. American Airlines Flight 11 out of Boston crashes into the North Tower of the World Trade Center with 92 people on board. United Airlines Flight 172, also out of Boston, crashes into the South Tower of the World Trade Center with 65 passengers on board. American Airlines Flight 77 out of Washington Dulles crashes into the Pentagon with 64 on board. Flight 93 out of Newark, New Jersey, crashes into a rural western Pennsylvania field with 44 people on board. An estimated additional 2,666 people on the ground die in the World Trade Center and another 125 are dead in the Pentagon. Over 3,000 people die in one day as a result of these terrorist attacks (U.S. Department of State, 2002).

OVERVIEW

Nearly every American resident is touched, appalled, and angered by the tragedies mentioned above. The common thread is terrorism. The constant theme is increased violence, destruction of property, and loss of life. Each attack has taken terrorism to a new level in terms of audacity, mayhem, cost, and human suffering. The problem of terrorism is a difficult one. The terrorist lives among us. He socializes, works, and moves around with the same freedoms as nearly every American citizen. Terrorists have access to as much information and know as much about our vulnerabilities as nearly any one of us—more than most because they have been studying our weaknesses. They pose as tourists, students, employees, and concerned Americans. Prior to their attack the September 11 terrorists were living and moving among us and we, as Americans, were accepting of them and their behaviors.

Terrorists also come in domestic forms, as evidenced by Tim McVeigh, conspirator in the Oklahoma City bombing. Disgruntled employees, estranged spouses, extortionists, and others can terrorize our employees and our business enterprise. A determined terrorist can wreak havoc on the most organized and otherwise safe business or government enterprise. Two-thirds as many Americans were lost in one day on September 11 as were killed in the workplace in a whole year. Most of those Americans were at work. The attacks in Oklahoma City, New York, and Washington were similar in that each was aimed at an American institution. Our government and the capitalist system were attacked. Targets of terrorist attacks are frequently symbolic of a philosophy, system, or group the terrorist opposes. The more noted the symbol, the more satisfying the target may appear to the terrorist. Depending on the motive, future terrorist attacks may continue to target major United States symbols. These might include centers of the United States philosophy, including large buildings, tourist attractions, and centers of activity. Of course, military bases, factories, and even schools and hospitals—any institutions supporting the infrastructure of the United States—face the risk of terrorist attack. Schools have suffered in recent years, with students turning on their employers, their teachers, and their peers with acts of mayhem and violence. From sniper attacks to confrontational shootings, enterprises are faced with the problem of protecting their employees as they work and move about performing their business and that of the enterprise.

RESPONSIBILITIES

Although no written anti/counter-terrorism program is required by law, it could easily be included in any comprehensive safety or loss control program. At minimum it should be attached to the emergency preparedness program. Many companies address bomb threats or even worker unrest, as in strikes or walkouts. Companies are also expanding their programs to incorporate terrorism and acts of subversion within the company. New architecture, building plans, and company activities are viewed in light of potential adverse effects of terrorist and similar activities. Chemical, biological, radiological, nuclear, explosive (CBRNE)—all may be considered within the context of a terrorist or other attack. Following a rash of shooting events in public places, many organizations are considering events that little more than a decade would have been given no thought.

Planning

The approach to countering terrorism is similar to that of other safety problems. First is consideration of vulnerabilities and probabilities of attack. Similar industries in the United States and abroad are considered along with problems they may face. Industry associations may be able to help. Sometimes companies attempt to suppress adverse publicity, but a thorough search may turn up information not readily available. Determine whether your company or other similar companies have faced terrorist problems in the past. History may repeat itself.

In the initial stages of evaluation of your facility, carefully consider each of the following points to assess the facility for target potential.

Activities

What activities take place that might be attractive to a terrorist group? On 9/11, it was airlines transporting large numbers of peoples in airplanes that were to be used as missiles that attracted terrorists. If the goal was to hurt large numbers of innocent people, cripple the transportation system for days, make travelers reluctant to travel, slow down the trading of commodities, and strike a blow at our financial network, the terrorists were successful. Numerous organizations, engaged in various activities, were adversely affected. The ripple effect through the economy and the American public was likely unforeseen by even the terrorists. As a safety professional, consider the possibilities and how destruction of your personnel and/or your property could adversely affect not only the future of your own organization but that of the nation. The results of your contemplation will give you insight into whether or not your organization could be a potential target.

Production

If a company manufactures or maintains inventories of any products that could conceivably be turned into weapons or agents of destruction, it may be attractive as a target. From the activities in Oklahoma City and New York, it is obvious that seemingly innocuous items intended for useful activities can be turned against the company or others for terrorist purposes. Chemicals in large quantities—explosives, fertilizers, pesticides, herbicides, fuels, energy sources, poisons—or mechanical devices are just a few of the items requiring careful guarding. Follow processes from beginning to end and identify easy or high-risk targets. Work toward protecting them first. Manufacturers of food products, drugs, cosmetics, or other items of mass consumption are potential targets. Carefully assess vulnerability to determine how the product could be contaminated or altered to pose a threat to a large portion of the population. If a company hosts events or maintains facilities with large gatherings of customers or even employees, it could be considered a target. Even the name of the company could be considered a drawing card. It was more than a coincidence that United and American Airlines were targeted for attacks on September 11.

Terrorist networks like Al Qaeda seek spectacular attacks that cause mass casualties and do severe damage to the United States economy. They attempt to produce maximum psychological trauma. Prime targets may be the air transportation industry, significant landmarks, or the energy industry, especially including nuclear power plants. Of course, one of the approaches of terrorism is to strike where it isn't expected. Special precautions should be taken by the otherwise unsuspecting, such as hospitals, universities, and schools, and other areas where vulnerabilities exist and people's guards may be down. Financial institutions or those transmitting large amounts of financial data can consider themselves targets. Terrorists may operate from locations outside of the United States, and with the expertise required to hack into any computer network utilize their skills to raise funds and adversely affect institutions and individuals.

Bottlenecks

Look at the areas of operation in that might act as bottlenecks in the event of attack. Once vulnerabilities have been considered, the next step is to determine where problems are most likely to occur. During the events of 9/11, occupants of the World Trade Center had only the stairs to evacuate. Given the limited amount of time and the access denied to those on the top floor, many were unable to get out. Had the chemical leak at Bhopal, India, been a result of a terrorist attack, the results would have been no different. Multiple escape routes and consideration of how and where to flee in the event of an attack can save lives. During the planning process, carefully consider how evacuation will take place, whether the primary means of evacuation could be cut off, and whether alternate means of getting out are feasible. Once this has been determined, ensure that employees are trained on evacuation signals, routes, and gathering spots. Drilling and practice of evacuation procedures helps ensure that employees will maximize the probability of escape during a terrorist attack.

Location of Your Property

Obviously, companies in populated areas with high traffic are less likely to notice unusual activity near the facility than those located in rural areas with one road in and one road out of the plant. Terrorists planning to affect large numbers of people may well prefer to be in a more populated area for maximum effect. New properties should be carefully examined from a security standpoint, with special consideration given to the difficulty of protecting real estate. Although much of the decision may be governed by marketing or logistics, there are a few points that should be evaluated.

Are you in a high-profile area likely to be targeted? City centers, airport, and tourist destinations have been targeted in the past. Is your location easily accessible by vehicular or even pedestrian traffic? Can you keep intruders at a distance from your facilities via fences or other barriers? Remote locations are less likely to have police patrol and they can be easier for terrorists to spot from the air. On the other hand, they can be less expensive to protect from ground intruders.

Fences and Entrances

Protecting the perimeter is relatively easy and inexpensive. Once the perpetrator is inside the facility, the capacity to inflict harm increases dramatically. Consider the likelihood that a perpetrator would want to come into the facility by crashing through or climbing a fence and plan accordingly. Fences can be made as secure as you desire. Detectors are available to virtually eliminate the possibility of a perpetrator entering without security forces being made aware. Another likely scenario is one of the perpetrator entering as an employee. Careful screening of applicants, including background and criminal record clearances, may be useful, but always check with the appropriate state employment commission before instituting the use of instruments that may be perceived as being discriminatory toward any protected group. For example, some states prohibit the use of criminal record checks without permission. Potential perpetrators may apply for jobs or pose as employees, customers, or visitors. If the perimeter can't be made "bullet proof," consider what can be done to protect the premises once the intruder is on the inside. Airports, arenas, and any public gathering places could be targets. Inspection of bags and other devices that could conceal bombs or weapons, along with screening devices should be considered. If it is impossible to screen all entrants to the property, consider screening all entrants to given buildings or venues. The ability to lock down specific parts of the establishment or to visually monitor and tape key areas may also serve as protection and deterrence. Common arguments by management are that "this is poor public relations" or "this may damage our business."

Clear Area Around the Premises

Shrubbery, walls, and sculptures may all become potential hiding places for intruders or explosives. Depending on your desire to maintain a field of view, consider keeping a clear area extending to a specific distance around each building. Don't give intruders a place to hide themselves or explosives.

Doors, Windows, and Other Openings

Doors, windows, and other openings into the property should be secured and possibly alarmed. Technology permits us to know anytime a building is penetrated. This is critical when no one is on the premises. If there is a way into your property, it is likely to be through an existing opening. Check them all, including those on the roof. Guard or alarm them to repel access by outsiders. Bullet-resistive film is available to be installed on curtain glass panels, doors and windows of facilities. There are varying levels of protection available with some film able to stop high-powered rifle shots. The cost is relatively inexpensive and the material is highly resistive to the elements when properly installed.

Ventilation and HVAC Systems

Ventilation and HVAC systems can be used as points of penetration or as delivery mechanisms for chemicals, including gases. All parts of these systems should be located behind fences or out of reach and secured.

Lighting Systems

Lighting by itself is a deterrent to some intruders. Poor planning with lighting can cause shaded areas that provide more cover for intruders than no lighting at all. Consider removing or lowering obstructions that cast enough shade for intruders to hide. If they cannot be removed or lowered, add lighting to illuminate the shaded areas they provide. Parking lots provide a special problem with lighting so consider isolating them behind fencing with secure entrances to and from the premises.

Communication Systems

Any communication systems used should be tamperproof to the extent possible. Backup systems and secondary backup systems help ensure continuity of operations during an emergency. Discuss communications with outside emergency responders to help them meet their needs during an adverse event. Utilize all communication systems in day-to-day activities to maximize their effectiveness, ensure systems will operate in times of emergency, and familiarize employees with their use.

Continuity of Operations (COOP)

Some organizations are including continuity of operations (COOP) plans as supplements to their safety plans to ensure their companies continue to operate, albeit with more limited resources. The ability to continue to operate, even on a less than optimal basis, may enable a company to maintain all or a large part of its existing customer base. A Federal Emergency Management Agency (FEMA) template for COOP plans is available at http:// brgov.com/ Dept/OEP/pdf/COOPPlan.pdf.

Security Systems

Technology enables unguarded areas to become virtually impossible to enter undetected. Even moderately sophisticated security systems cannot be disabled or disarmed without sending out an alarm. All openings to the property can be protected. Fences can be enabled to send an alert if any intruder attempts to get close to, cut, climb, or get through them in any way. Pressure systems determine an intruder's presence and detectors prevent unnoticed cutting or climbing.

Keypads can be located near common entry doors to control access. The systems may also include other devices:

- Magnetic contacts to protect doors and windows that open
- Glass breakage sensors to protect windows and glass doors
- Motion sensors to detect movement within a defined area
- Panic switches that can be manually operated to summon the authorities and/or sound an alarm siren
- Window alarm screens that allow alarm protection with windows wide open
- A siren to alert employees and neighbors when an alarm is activated
- Smoke and heat detectors to alert employees and the fire department of a fire
- Environmental sensors such as freeze alarms, water leakage detectors, and carbon monoxide detectors

The typical security system can be designed in a number of ways with several levels of protection. For example, when the business is closed and no one is on the premises, all sensors and windows are activated. If a few employees are working in the evening, they may choose to have the door and window magnetic contacts activated and the motion sensors off. The magnetic contact is a two-part device that protects doors that open. One part of the contact is a switch, installed in the door or window jamb or frame; the other is installed in the door or window itself and contains a magnet that signals an alarm when it moves "out of contact" (range) with the switch.

There are two basic types of glass breakage sensors: an "acoustic" sensor that is mounted on a wall or ceiling and listens for the sound of breaking glass in a window or door and a "shock" sensor that is mounted on the window and "feels" the shock of breaking glass. Optimally, a glass breakage sensor could detect an intruder while he is still outside. A passive infrared motion detector (PIR) is a sensor that detects an intruder by sensing his body heat as he passes through the area "covered" by the PIR. Since PIRs are motion sensors, they are designed to be used when no one is present or not moving through the protected areas.

Surveillance Systems

When combined with an appropriate security system, surveillance systems can be triggered anytime there is movement at any specified level. Sensitivity settings can be adjusted to ignore dogs or other small creatures but to detect human entrance. Cameras record all movements of people on the premises and can easily be connected to a server for 24-hour Internet access. The motion detector can also send an alert via pager or e-mail when there is a disturbance. When door/window sensor is triggered, it will take snapshots and upload to a remote server for later investigation.

The system can automatically link events on one system to actions on others. For example, if it detects an intrusion, it calls up live video in the command center and shows the alarm location on a graphic map. This reduces false alarms and helps security staff respond more quickly.

Motion detectors, security codes, and cameras provide information to a smart security system, allowing it to determine whether an individual is an employee, a cleared visitor, or an intruder. Motion detectors trigger an alert, letting the artificial intelligence program know that there is someone or something to be evaluated. Facial recognition software and security codes allow the security system to permit employees to enter the facilities while restricting access to other individuals based on pre-programmed information.

In the event that the system detects someone who is unknown, it can provide video of the visitor to security personnel. Visitors that are welcome can be given clearance and allowed in remotely. Unwelcome visitors can be ignored, and individuals attempting to break in will trigger a security response and a call to the police.

Intruders and fires are not the only dangers. A smart security system can also protect and issue an alert to unanticipated health problems. Using the same cameras and motion detectors that protect the outside, the smart system can learn about the habits and normal movements of the employees. When the employee does something unexpected and does not resume normal activities, the system can alert security and emergency services.

Guards

Guards may be needed when mechanical or electronic surveillance devices are inadequate. This may require 24-hour, 7-day-per-week coverage or a minimum of four full-time personnel. Depending on needs, animals may provide an adequate level of security, and systems can be adjusted so animals do not trigger them. Guards must be continually trained and notified of specific threats. Guards are also the most expensive means of security, usually only justified following a cost-benefit analysis.

Entry and Movement of Vehicles and Visitors

A pass system may be initiated to check all entrants to the premises. Visitors can be screened by checking identification cards or prearranged passes. Employees may also be required to carry identification cards or to wear badges. Electronic badges or biometric identification points are used to monitor movement inside the premises and to limit access to certain locations.

Entry and Movement of Employees

Carefully check the credentials and backgrounds of all new employees. This is standard practice under any circumstances but even more so if the facility is subject to terrorist attack. Background and criminal checks where legal minimize the possibility of employing individuals who propose to destroy the organization. Carefully review records and references. Long, unaccounted-for time periods are especially important, so find out why potential employees have large gaps in their employment history and work to verify their accounts of them. Another way of accomplishing this—and one that airport operators have long urged FAA/TSA to implement—is a requirement to re-badge all employees on a regular (say, two-year) basis (Robert Raffel, personal communication, August 11, 2009). Excessive cost has been cited as the major reason this hasn't happened.

Computers and Networks

Sensitive data, vulnerabilities, access points, and other critical information are frequently stored on accessible computers. Consider implementing the use of a firewall to help prevent outsiders from roaming through your system. If that isn't feasible, computers with sensitive information can be physically and electronically isolated from others in the operation. When not in use they can be made more inaccessible by disconnecting them from any Internet or intranet system.

Threats

Terrorist threats may come indirectly via alerts received in the news media as disseminated by the FBI or Homeland Security. They may also come directly as bomb threats or other threatened hostilities aimed at disrupting company operations. The goal is often to disrupt operations, whether or not a bomb exists or other hostilities are planned. Threats need to be taken seriously, but decisions concerning their handling should be made in advance. Cool thinking prevails as you play the "What if?" game. In the heat of a terrorist disaster or even a threat, it may be difficult to make wise decisions. Consider the possibilities that may be faced and, as with any other potential emergency, develop standard operating procedures (SOPs) for each. Regularly scheduled tabletop and full-scale exercises can be extremely worthwhile approaches to hypothetical, but potentially real situations. Once the threat of disaster no longer fits the mold around which the SOPs were developed, then make the necessary subsequent decisions. By following SOPs, the whole organization operates in a more rational, cohesive, and likely intelligent manner.

Terrorist threats are often designed to disrupt the organization. A threat of a bomb or some other terrorist action may only be designed to slow or halt production. Yielding to such demands may be the only response the terrorist seeks and may encourage the action to be repeated. Attempting to identify the terrorist or the person delivering the threat is helpful. Receptionists and others receiving the call are instructed to learn all they can about the caller. It is important not only to ask questions about the type of bomb and where it may be hidden, but also to ask the caller for name and address. They may respond to the question without thinking, and you will thus learn the source of the call. Their responses may also give insight into the true nature of the threat.

Whether or not a threat is perceived to be legitimate, decisions must be made as to evacuate or protect-in-place. It may be safer to stay inside than to move outside the building. A threat may, in fact, be designed to lure workers or a particular employee outside the building for malevolent reasons.

Mail and Packages

Procedures for handling of mail and packages help lessen the possibility that bombs or toxic substances can be delivered to the mailroom and disseminated. Hand-addressed envelopes and packages are automatically suspect, as are those of unknown origin or with no return address. Mailrooms should always be located on the perimeter of the building so, in the unlikely event of an explosion or contamination, damage can be limited.

YOUR ROLE

Once you have considered the possibilities, advise management of the risks your company faces. Prepare a plan to deal with threats and acts of terrorism. Consider employees at every level and their appropriate responses. In most cases they are instructed to evacuate. Thought must be given as to how and where they will move.

Employees with loved ones will likely want to unite with them. In major cities evacuation routes may be jammed and chaos will prevail if companies and municipalities have not worked together to determine routes and destinations. A written plan to be implemented in conjunction with municipal plans is essential. Following the disaster at the World Trade Center, bridges and tunnels were closed for security purposes. Many employees were trapped in Manhattan and unable to communicate with loved ones. In some cases days or months passed before relatives learned whether loved ones had been injured or killed. Appropriate preplanning with governmental authorities can help minimize the effects of transportation and communication problems.

INSURANCE

The last approach to any loss control problem is insurance. Companies purchase insurance in case all efforts at loss prevention or control fail. Historically, insurance companies have refused to pay losses resulting from acts of war. Some major insurance companies are also excluding acts of terrorism. The potential for loss from such acts can be catastrophic for an individual carrier and it may not be willing to run the risk. The federal government has promised to help but it has capped its own liability. Companies may be forced to negotiate with insurance carriers in agreeing to take certain precautionary measures resulting in higher standards of prevention efforts in return for coverage. The alternative may be no coverage or very limited coverage.

CONCLUSION

There is no way to absolutely eliminate the risks faced with terrorism. As with all forms of safety, the goal is to minimize exposure. Huge sums of funds can be spent in pursuit of this task; however, one thwarted terrorist attempt may make this expenditure a worthwhile investment. A thorough risk analysis of the potential problems and the resulting costs should be incorporated into the overall safety analysis.

QUESTIONS

1. Talk to representatives from your local fire and police departments. Ask what preparations for terrorism they are helping local companies to make.
2. Why has the safety and health professional become the primary individual responsible for corporate terrorism response preparation? What is his or her role relative to that of line management?
3. From an insurance perspective, why would carriers be unwilling to underwrite companies against terrorist losses? What alternatives are available to the companies?
4. What companies are vulnerable to terrorist attack in your community? Why? Do they appear to be taking efforts to thwart those attacks? What efforts are they taking?
5. What steps can the government take that companies are unable to take in order to lessen the probability of terrorist attack?

REFERENCE

U.S. Department of State. 2002. "September 11, 2001: Basic Facts." Washington, DC: U.S. Department of State. Available at http://2001-2009.state.gov/coalition/cr/fs/12701.htm.

15

Hazardous Materials

Tracy L. Zontek, PhD, CIH, CSP,
and Burton R. Ogle, PhD, CIH, CSP

CHAPTER OBJECTIVES

After completing this chapter, you will be able to

- Outline basic recent environmental legal history
- Understand the Resource Conservation and Recovery Act (RCRA)
- Identify the types of hazardous waste
- Determine hazardous waste generator status
- Know how to obtain an EPA identification number
- Understand the Comprehensive Environmental Response, Compensation, and Liability Act (CERCLA)
- Understand the Superfund Amendments and Reauthorization Act (SARA)
- Explain the Hazard Communications Standard (HAZCOM)
- Understand how the Globally Harmonized System for Hazard Communication of Classification and Labeling of Chemicals is implemented in the United States

CASE STUDY

Two employees of a North Carolina company were helping to unload a truck on a warm sunny day when they inadvertently punctured a container of an unknown liquid chemical. A quick check of the contents revealed they had spilled a concentrated pesticide, intended to be mixed with water before application. Their first response was to grab a garden hose and use the water pressure to push it to one side. In a short time, they began to feel faint and short of breath, so they decided to ask for help. Since no one at the facility had specialized training to deal with the spill, the manager on duty called the local fire department and an ambulance.

When the fire department arrived, it soon became evident their personnel lacked the training to deal with spills of hazardous chemicals. They formulated a plan to wash the remainder into the storm drain and use a vacuum truck to intercept the waste as it flowed through the system. A few minutes after the flushing began, the crew at the spill site received a call from the group at the truck asking how soon the flushing would begin. As it turns out, the truck was stationed at the wrong storm sewer opening and it missed capturing the contents. By the time all the mistakes were remedied, cleanup costs amounted to over $50,000. The employees were held overnight at the hospital for observation and released with no lasting health effects.

INTRODUCTION

The increase in environmental regulation may be the most significant legal development in America in the last 50 years. Safety practitioners do not need a detailed knowledge of the fine points of these laws and regulations. They do, however, need a basic understanding of the structure, scope, and framework of these laws in order to effectively measure and evaluate the nature of the compliance task faced.

Many of the environmental problems faced by the safety practitioner are covered or touched on by the Resource Conservation and Recovery Act (RCRA). When this act was passed in 1976, Congress was attempting to complete a series of legislative acts designed to protect the environment. A brief overview of some of the legislation that led to the passage of RCRA will be useful.

The Hazardous Materials Transportation Act (HMTA) was passed in 1975 to give the Department of Transportation (DOT) the authority to regulate the transportation of hazardous materials by air, waterways, rail, or highway. To avoid exposing large populated areas to significant risks from exposure to hazardous materials, the DOT was given authority to establish routes for hazardous materials shipping.

In 1976, Congress passed the Toxic Substances Control Act (TSCA), giving broad powers to the Environmental Protection Agency (EPA) to regulate the production and use of potentially hazardous chemicals and to ensure that new chemicals do not pose unreasonable hazards. Under the TSCA, the EPA may require a company to submit test data on specified chemicals. Anyone who intends to manufacture or import a chemical not listed on the TSCA Chemical Substance Inventory must submit a pre-manufacture notice (PMN) at least 90 days in advance of the manufacture or import of a new chemical. The EPA may then prohibit the manufacture, process, or distribution of the chemical. It may also place limits on the chemical's production or ban its production.

After the passage of the HMTA and the TSCA, Congress put the finishing touches on its hazardous material legislation with the passage of RCRA. RCRA controls the chemicals at their source during transportation and disposal. The RCRA requirements often fall within the scope of the safety practitioner's responsibilities, not only because of the natural relationship that management perceives between environmental and safety responsibilities, but also because of subsequent legislation, which will be discussed later.

The Globally Harmonized System of Classification and Labeling of Chemicals (GHS) was adopted by the United Nations in 2003 to reduce trade barriers and make hazard information uniform, thus increasing productivity, reducing confusion across country borders and reducing incidents. The GHS, referred to as the Purple Book, provides an updated method to classify health and physical hazards; labels to delineate a signal word, pictogram and hazard statement for each hazard class and category; a 16-section format to Safety Data Sheets (SDS), formally known as Material Safety Data Sheets (SDS); and information and training requirements for the system. Although the GHS is not an international law, the United States is currently changing its regulations to align with the GHS. In each of the regulations discussed below, the impact of GHS implementation will be evaluated.

BACKGROUND

The Resource Conservation and Recovery Act (RCRA) was passed by Congress to direct the EPA to implement a program that would "protect human health and the environment from improper hazardous waste management" (U.S. Environmental Protection Agency [US EPA], 2011). RCRA can be administered by the EPA or under state EPA plans. State-administered EPA programs must meet and/or exceed the standards set by the federal EPA. The first set of RCRA regulations was published in the Code of Federal Regulations under Title 40 that deals with Protection of the Environment.

Early regulations focused on large companies producing more than 1,000 kilograms of hazardous waste per month. Subsequent amendments directed the EPA to also include producers of between 100 kilograms and 1,000 kilograms per month of hazardous waste.

WHAT IS A HAZARDOUS WASTE?

A waste is something unwanted, usually a solid, liquid or a contained gas. It is the responsibility of the generator to determine if the waste is non-hazardous, hazardous, or acutely hazardous. The EPA classifies a waste as hazardous if, through improper handling, it can cause injury or death or can damage or pollute the environment (US EPA, 2011). To determine if a waste is hazardous, it may be necessary to refer to the Code of Federal Regulations at 40 CFR Part 261, "Identification and Listing of Hazardous Waste." In this section, *a waste is considered hazardous if it is (1) a listed waste or (2) a characteristic waste.*

The regulations include four lists identified by the letters F, K, P, and U that together describe about 400 hazardous wastes. The F-list includes hazardous waste from nonspecific sources or generic waste streams, whereas the K-list covers hazardous wastes from specific sources. F-list wastes are more general in nature and K-list wastes are generated by a specific manufacturing process. Those on the P-list are considered acutely toxic hazardous wastes. The U-list identifies commercial chemical products, chemical intermediates, and off-specification chemical products. Some states may include additional lists in this section, such as Kentucky's N-list, which covers nerve and blister agents (State of Kentucky, 1992, chap. 31). Wastes that appear on any of these sections are referred to as "listed wastes."

If a waste cannot be found on one of the RCRA lists, this does not mean it is not hazardous. Unlisted waste must still be tested to determine if it has certain properties or characteristics that render it hazardous. A waste is hazardous if it has one or more of the following properties: ignitability, corrosivity, reactivity, or toxicity.

Ignitable wastes have a flash point lower than 140 degrees Fahrenheit. *Corrosive wastes* are those that are acidic (at or below a pH of 2) or caustic (at or above a pH of 12.5). *Reactive wastes* produce violent results when mixed with water, air, or other chemicals. Results include explosions and the generation of toxic gases. *Toxic characteristic wastes* contain specified percentages of specific metals, pesticides, or organic chemicals. They are discovered by testing the waste stream with the *toxicity characteristic leaching procedure* (TCLP). This test is used to determine the amount of chemical that would leach out into the groundwater under specific conditions. Wastes that exhibit any one of the above characteristics are referred to as *characteristic wastes* or *TCLP wastes* and are coded as *D wastes*.

DETERMINING GENERATOR STATUS

Once it has been determined the waste produced by a plant is hazardous, the facility becomes a hazardous waste generator. To determine the generator status of a facility, the total hazardous waste generated per calendar month must be determined. The correct generator status is important because different regulations apply to each different status.

A facility is a *conditionally exempt, small-quantity generator* (CESQG) if it produces less than 100 kilograms of hazardous waste per month. If, however, a CESQG stores more than 1,000 kilograms of hazardous waste on-site at any time, it becomes a small-quantity generator. A CESQG is not required to obtain an Environmental Protection Agency identification number, but most licensed waste haulers will require an EPA identification number to ship the waste off-site. Some states do not recognize this generator status.

A facility is a *small-quantity generator* (SQG) if it produces between 100 kilograms and 1000 kilograms of hazardous waste per month. Small-quantity generators are required to obtain an EPA identification number. They can accumulate up to 6,000 kilograms of hazardous waste in any 180-day period. If, however, the hazardous waste is to be transported more than 200 miles, the generator is allowed to accumulate hazardous waste up to 270 days.

A *large-quantity generator* (LQG) is a facility that produces 1,000 kilograms of hazardous waste or one kilogram of acutely hazardous waste per month. Large-quantity generators are allowed to accumulate hazardous waste for 90 days. If the hazardous waste must be transported more than

200 miles, the generator is allowed to accumulate hazardous waste for 180 days (US EPA, 2011).

Only conditionally exempt, small-quantity generators are exempt from most RCRA regulations. Other generators must have an EPA identification number. Some states may have regulations more stringent than those of the federal government. The practitioner should check with the state EPA if there are any questions about a given facility.

OBTAINING AN EPA IDENTIFICATION NUMBER

To obtain an EPA identification (ID) number, a Notification of Hazardous Waste Activity (Form 8700-12) may be downloaded from the U.S. EPA or state EPA. If the facility is a treatment, storage, and disposal (TSD) facility, additional forms must be completed. Once the EPA has reviewed the form, a 12-character EPA ID number will be assigned to the facility (Kaufman, 1990). This form must be modified each time additional waste streams are generated or deleted or the generator status changes.

MANAGING HAZARDOUS WASTE ON-SITE

Hazardous waste can be stored on-site in containers or tanks. Proper storage during the accumulation period is necessary to prevent accidents and/or spills. Some of the basic requirements for temporary storage include the label *Hazardous Waste*, proper identification of the waste, and the accumulation start dates for each container or tank. Containers or tanks must be kept closed except when filled or emptied, and operational logbooks must be kept to record any activity. Containers must be inspected weekly for leaks. Safety equipment must also be inspected weekly. Inspection reports and a written emergency plan must be kept on file. Incompatible chemicals must be segregated and three feet of aisle space must be maintained in the accumulation area. Signs must be posted identifying the area as a hazardous materials storage area. When hazardous waste is transported off-site, the package must comply with the U.S. Department of Transportation (DOT) regulations. DOT has implemented the Globally Harmonized System of Classification and Labeling of Chemicals.

CERCLA

Once RCRA was passed in 1976, Congress no doubt believed most of the environmental problems with hazardous chemicals had been addressed. It wasn't until the discovery of Love Canal and other similar problems in the late 1970s that Congress realized there was a significant hole in its plan.

Love Canal was built near Niagara Falls, New York, in the 1800s to link waterways. Although it was never completed, the canal, in the form of a half-mile ditch, remained until the Hooker Chemical Company purchased it in the early part of the twentieth century as a dumpsite for hazardous waste. Once full, the canal was covered and eventually wound up as the building site for a school with residential properties nearby.

Figure 15-1. Example of hazardous material accumulation/dispensing center with spill containment

Figure 15-2.   Example of a single drum hazardous material collection station

When this and other abandoned sites were exposed as significant threats to the environment and residents, Congress realized it had not addressed the problems associated with *existing* waste sites. Thus, in 1980, the Comprehensive Environmental Response, Compensation, and Liability Act (CERCLA) was passed. CERCLA, or Superfund, as it is sometimes called, gives the government authority to pursue potentially responsible parties (PRPs) for past and future cleanup costs at abandoned waste dumps and other places where hazardous substances have been released. PRPs include existing owners and operators, former owners and operators, transporters to the site, generators of the waste, and others. Any person found to be even partially responsible for creating the hazardous waste site could be held completely responsible for its cleanup. This law affects sites that were in existence before the legislation was passed, as well as sites that may have been legal at the time of their creation. EPA cast a wide net to encompass as many potentially responsible parties as possible under this regulation. Unknowing parties could be held responsible for thousands or even millions of dollars of cleanup costs through CERCLA.

SARA

One of the most prodigious tasks an individual who is responsible for environmental regulations must do is gather data, file it, and in some cases, submit it to the government. These responsibilities were increased with the passage of the Superfund Amendments and Reauthorization Act (SARA) in 1986.

SARA was proposed as a more stringent response to hazardous waste releases. SARA and CERCLA address hazardous substance releases into the environment and the cleanup of inactive hazardous waste disposal sites (Noll et al., 1995, p. 5).

SARA is broken down into five titles:

I.   Provisions Relating Primarily To Response and Liability
II. Miscellaneous Provisions
III. Emergency Planning and Community Right-to-Know
IV. Radon Gas and Indoor Air Quality Research
V.   Amendments of the Internal Revenue Code of 1986

Figure 15-3.   Example of a hazardous material four-drum pallet with spill containment

Title I and II feature the major changes to sections of CERCLA, while Title V discusses Superfund and its revenues. Under SARA, any company with threshold quantities of extremely hazardous substances (EHSs) must contact state and local emergency planners and the fire department. Threshold quantities refer to specified amounts of certain chemicals. If the company has these chemicals, it must also submit copies of its Safety Data Sheets (SDSs) to the fire department and emergency planners.

If a company exceeds certain thresholds, it will have to submit chemical inventory forms to state and local firefighters. EPA refers to these forms as Tier I and Tier II. Tier I forms are submitted by March 1 each year and Tier II forms are submitted when requested by emergency response officials.

Superfund sites (known or threatened releases of hazardous substances) are found on the National Priorities List (NPL). These sites are areas where the land is significantly contaminated and warrants further study and/or remediation through EPA. EPA's website has an interactive "Where You Live" function that allows citizens to learn about Superfund sites in their geographical location.

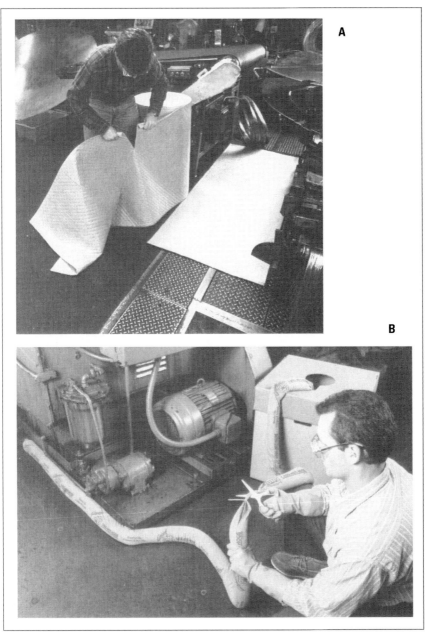

Figure 15-4.　(A) Example of hazardous material perforated absorption mat (B) Example of spill containment "sock"

The Occupational Safety and Health Administration (OSHA) sets requirements for the health and safety of hazardous waste workers, RCRA/TSD site
workers, and emergency responders. These requirements are found in the Code
of Federal Regulations at 29 CFR 1910.120, Hazardous Waste Operations
and Emergency Response, which is an outgrowth of Title I of the Superfund
Amendment and Reauthorization Act (SARA). This regulation is also known
as HAZWOPER (Hazardous Waste Operations and Emergency Response).

WORKER PROTECTION STANDARDS

Employees who work in hazardous waste operations want to be able to
perform their duties safely and be assured their health is not at risk. The
HAZWOPER law proposed standards that facilities must follow to protect the
health and safety of their employees in a hazardous waste operation or emergency response involving hazardous substances. ***Regulations were issued on
these standards that include but are not limited to the following provisions:***

- ***Site Analysis.*** Requirements for a formal hazard analysis of the site and
  development of a site-specific plan for worker protection
- ***Training Requirements.*** Requirements for contractors to provide initial
  and routine training of workers before they are permitted to engage in
  hazardous waste operations that could expose them to toxic substances
- ***Medical Surveillance.*** Requirements for a program of regular medical
  examination, monitoring, and surveillance of workers engaged in hazardous waste operations which could expose them to toxic substances
- ***Protective Equipment.*** Requirements for appropriate personal protective equipment, clothing, and respirators for individuals working in
  hazardous waste operations
- ***Engineering Controls.*** Requirements for engineering controls (e.g.
  ventilation, isolation) concerning the use of equipment and exposure of
  workers engaged in hazardous waste operations
- ***Air Monitoring.*** Requirements for ensuring air concentrations do
  not exceed the OSHA Permissible Exposure Limits or Immediately
  Dangerous to Life and Health (IDLH) limits for workers engaged in
  hazardous waste operations, including necessary air monitoring and assessment procedures

- *Informational Programs.* Requirements for a program to inform workers engaged in hazardous waste operations of the nature and degree of toxic exposure likely as a result of such hazardous waste operations
- *Handling.* Requirements for the handling, transporting, labeling, and disposing of hazardous wastes
- *New Technology Program.* Requirements for a program that introduces new equipment or technologies that will maintain worker protection
- *Decontamination Procedures.* Requirements for decontamination procedures
- *Emergency Response.* Requirements for emergency response and protection of workers engaged in hazardous waste operations

HAZWOPER sets standards on the amount of training that employees at hazardous waste sites must have in order to protect their health and safety. Everyone who is going to be involved with the hazardous waste operations must be trained for a minimum of 40 hours of initial instruction off-site, including three days of actual field experience under trained supervision. The employees must have additional training if the hazards they are working with are unique or special. Supervisors must have the same training as the other employees, plus an additional eight hours of specialized training on managing a hazardous waste operation.

Workers at treatment, storage, and disposal (TSD) facilities and emergency responders must have a minimum of 24 hours of initial training. Workers in all categories must complete an eight-hour refresher course annually.

Workers at sites where hazardous waste is generated must also be trained in 29 CFR 1910.1200, the Hazard Communication Standard (also referred to as HAZCOM). Although SDSs and some other parts of the standard are not required for hazardous waste generation, they must be covered relative to the chemicals that may turn into hazardous waste. HAZCOM requires companies maintain a file of all SDSs, label all hazardous chemicals, and train workers in the handling of hazardous chemicals.

CONTINGENCY PLANS

Contingency plans must be drawn to include emergency procedures, emergency phone numbers, the name of the emergency coordinator, and containment measures available. Training records for all emergency responders must also be maintained.

HAZARDOUS WASTE DISPOSAL

Hazardous waste may not be disposed of on-site unless a disposal permit has been obtained. Obtaining a permit to store, treat, or dispose of hazardous waste on site (40 CFR Part 270) can be costly and time consuming.

The first step in the proper disposal of hazardous waste off-site is to select a licensed hazardous waste hauler and disposal facility. Companies must ship waste with qualified haulers and to facilities with EPA ID numbers. Either the waste generator or the hauler must package and label wastes for shipping and prepare a hazardous waste manifest (Kaufman, 1990).

When hazardous waste is shipped off-site, it is still the responsibility of the generator. Incineration of the waste, by a licensed treatment facility, is usually the selection of choice due to the fact that the waste is destroyed and no future liability can be incurred. Land disposal requires land ban exemption forms to be filed.

Uniform Hazardous Waste Manifest and DOT Regulations

The Department of Transportation (DOT) regulations, 49 CFR Part 172, require hazardous waste to be properly packaged in containers that are acceptable for transportation on public roads. Proper labeling of containers includes the identification of the waste, the words "Hazardous Waste," and the DOT waste codes and placards. DOT has implemented the Globally Harmonized System of Classification and Labeling. The Department of Transportation also requires a uniform hazardous waste manifest. This is a six-part shipping document that must accompany all hazardous waste shipments. It is designed so that hazardous waste can be tracked from the generator to the final destination or disposal, or cradle-to-grave. The waste in the shipment must be the same as the waste listed on the manifest.

Additional Reporting Requirements

Hazardous Waste Annual Report Forms must be submitted annually to the USEPA or state EPA (State of Kentucky, 1992, chap. 31). The annual reporting form includes information on all the hazardous waste generated and shipped during the report year. All facilities that treat, dispose, or recycle hazardous waste on site must complete the form; companies must maintain meticulous records of all hazardous waste. Waste minimization activities must be identified on the form.

EPA WEARS MANY HATS

As seen in the plethora of regulations administered by EPA, the EPA wears many hats: developing standards, conducting inspections, and implementing enforcement actions. The EPA is divided into multiple offices at EPA headquarters and ten regional offices around the nation.

The first role of the EPA is to write regulations for Congress to authorize, thus becoming legally enforceable. EPA regulations are found in the U.S. Code of Federal Regulations (CFR), Title 40: "Protection of Environment." Each state can create more stringent regulations than EPA. State-level agencies administer how EPA regulations are implemented, including compliance assistance, permits, and training.

The EPA also acts as a policeman conducting inspections and investigating incidents. The Office of Enforcement and Compliance Assurance (OECA) conducts inspections to ensure compliance with environmental requirements, such as hazardous waste generation facilities and TSDs. The compliance officer will inspect the facility and fill out an inspection report. During the inspection, the compliance officer will do a walk-through tour of the facility, take pictures of any violations, and review the necessary paperwork. The compliance officer will have a closing interview with the operator of the facility and point out any violations. A notice of violation (NOV) will be sent by certified mail. Any violations must be corrected within a time period specified on the NOV. The compliance officer will conduct a follow-up inspection to make sure all the violations are corrected.

At a certain level, the EPA acts as judge and jury when a violation is not corrected within the given time period or when penalties for a violation are assessed. Enforcement actions can include civil penalties or criminal charges, which can lead to fines and/or imprisonment. If an EPA inspection does result in a citation, the facility may choose to contest the citation through an appeals process within the EPA instead of paying the fine. The facility has 30 days after receiving the NOV to contest the citation.

The vast majority of federal environmental issues are handled at the EPA level. There are, however, some critical cases that do wind up in the federal courts. Many environmental statutes contain provisions for criminal penalties. Violations of the criminal provisions are prosecuted in the courts rather than through administrative actions. Under certain, rather narrow circumstances, companies may appeal adverse EPA decisions on regulatory violations to the courts.

HAZARD COMMUNICATION STANDARD (29 CFR 1910.1200)

OSHA estimates tens of millions of U.S. workers are potentially exposed to one or more hazardous chemicals in their workplaces. There are hundreds of thousands of hazardous chemicals found in the workplace, with numerous new ones being introduced each year. The chemicals can be health hazards, physical hazards, or, in some cases, both. Simple rashes or more serious conditions such as burns, sterility, lung damage, heart ailments, kidney damage, or cancer can be associated with exposures to hazardous chemicals (State of Kentucky, 1992, p. 1).

Protecting the employee from the harmful effects of hazardous chemical exposures is a formidable task for the employer. In order for a chemical to harm an individual's health, it must come in contact with or enter the body. *The four routes of entry or pathways by which a chemical can enter the body are:*

1. Inhalation
2. Absorption
3. Ingestion
4. Injection

For the employer to develop and implement control measures to block these routes of entry, all pertinent information about a chemical is essential.

In 1983, the Occupational Safety and Health Administration enacted the Hazard Communication Standard to confront the seriousness of this health and safety problem. The purpose is to ensure that the hazards from all chemicals produced or imported are evaluated and the hazard information transmitted downstream to all affected parties. The rule also guarantees the workers' right to know about the hazardous chemicals in their workplaces; therefore, it was often referred to as the right-to-know standard. After the passage of SARA, it has been typically referred to as the HAZCOM Standard.

Global Harmonization

Recently, the OSHA HAZCOM standard has been updated to align with the Globally Harmonized System of Classification and Labeling of Chemicals (GHS). The most significant change from HAZOM to GHS is the use of hazard classes and categories. GHS provides more quantitative measures to assess the hazard class and category of a chemical or mixture. Health hazard classification guidance is given for the following: acute toxicity; skin corrosion/irritation; serious eye damage/eye irritation; respiratory or skin sensitization; germ cell mutagenicity; carcinogenicity; reproductive toxicity; specific target organ toxicity and aspiration hazard. Physical hazard classification guidance is given for explosives; flammable gases; flammable aerosols; oxidizing gases; gases under pressure; flammable liquids; flammable solids; self-reactive chemicals; pyrophoric liquids; pyrophoric solids; self-heating chemicals; chemicals which, in contact with water, emit flammable gases; oxidizing liquids; oxidizing solids; organic peroxides; and corrosive to metals.

The following chemicals are excluded from the provisions of the standard:

- Hazardous waste as defined by the Resource Conservation and Recovery Act (RCRA)
- Hazardous waste as defined by the Comprehensive Environmental Response, Compensation, and Liability Act (CERCLA) when it is the focus of remedial or removal action under CERCLA
- Tobacco and tobacco products
- Wood or wood products
- Articles that do not release or otherwise result in exposure to a hazardous chemical under normal conditions of use
- Food or alcoholic beverages in a retail establishment
- Any drug in solid, final form for patient administration or packaged by the manufacturer for sale in a retail establishment, as well as drugs for personal consumption
- Cosmetics packaged for sale in a retail operation or for personal consumption by employees
- Nuisance particulates established by manufacturer or importer not to pose any physical or health hazard

- Ionizing and nonionizing radiation
- Biological hazards

There are four basic requirements under the HCS:

1. Written hazard communication program
2. Labeling
3. SDS
4. Employee training

Written Program

The employer must develop, implement, and maintain at the workplace a written hazard communication program that contains at least the following information:

- Information explaining labels and other forms of warning
- Information on location and availability of SDSs
- Details of how the employee training requirements are to be satisfied
- A list of all the hazardous chemicals known to be present in the workplace
- The methods the employer will use to inform employees of the hazards involved in non-routine tasks and the hazards associated with chemicals contained in unlabeled pipes in their workplace areas

Organizations requiring contracted services in locations of their facility where hazardous materials are used or stored must inform those contracted employees of all hazards present. This must be part of the written program. This program should include methods and procedures the employer will use to provide all contractors with SDSs, information on labeling and other forms of warning, and information concerning precautions needed to protect their employees while on-site.

The employer shall make the written program available, on request, to any employee or his designated representatives. The Occupational Safety and Health Administration (OSHA), the director of the National Institute of Occupational Safety and Health (NIOSH), or their designees may also request the program.

Labeling

Chemical manufacturers, distributors, or importers must ensure that each container of hazardous chemicals leaving their locale is labeled, tagged, or marked with the following information:

- Identity of the hazardous chemical
- Signal word
- Hazard statement(s)
- Pictogram(s)
- Precautionary statement(s)
- Name, address, and telephone number of the manufacturer, distributor, importer, or other responsible party

It should be noted that other information is often required to comply with other regulations and standards beyond the scope of HAZCOM, such as DOT packaging requirements.

A.

B.

Figure 15-5. Examples of required labels under the Globally Harmonized System of Classifying and Labeling Chemicals: A. Flame; B. Corrosion; C. Health Hazard

C.

The employer is required to:

- Ensure that all incoming shipments are properly labeled and the information is in English.
- Ensure that each container of hazardous chemicals in the workplace is labeled, tagged, or marked with the GHS shipping label OR product identifier and words, pictures, symbols, or a combination thereof that provides chemical hazard information.
- Ensure that all labels on existing stock are in place, legible, and not removed or defaced.
- Ensure chemical labels are revised within six months of new information.
- Ensure that all portable containers are labeled unless it is for the immediate use of the employee that makes the transfer.

Safety Data Sheets (SDSs)

Chemical manufacturers or importers shall obtain or develop an SDS for each hazardous chemical. These SDSs should be sent to the distributor or user prior to or with the initial chemical shipment. An SDS must be sent with each initial shipment after each subsequent SDS update—updates are required within three months upon new significant information regarding chemical hazards or controls. The employer shall maintain a current SDS for each hazardous chemical or mixture present in the workplace. These SDSs must be readily accessible to all affected employees on all shifts without the employees having to leave the work area.

The SDSs must know contain standard information. They must be in English and contain no blank spaces. Any SDS that contains blank spaces is considered unacceptable and should be returned to the vendor, or the missing information can be requisitioned from that vendor; if information is unknown, it should be marked as such and not left blank. ***The complete SDS contains the following information in English (the employer may also maintain copies in other languages):***

- The identity used on the label
- Hazard(s) identification
- Composition/information on ingredients
- First-aid measures
- Fire-fighting measures
- Accidental release measures

- Handling and storage
- Exposure controls/personal protection
- Physical and chemical properties
- Stability and reactivity
- Toxicological information
- Ecological information
- Disposal considerations
- Transport information
- Regulatory information
- Other information, including date of preparation or last revision

Training

Employee training must be conducted during the initial assignment to a work area containing one or more hazardous chemicals and each time a new chemical hazard is introduced into the workplace. ***The employee training shall include the following:***

- The requirements of 1910.1200
- Any operations in their work area where hazardous chemicals are present
- The location and availability of the written program, including lists of hazardous chemicals and SDSs
- Methods and observations used to detect the presence or release of hazardous chemicals in the work area
- The physical and health, simple asphyxiation, combustible dust and pyrophoric gas hazards, as well as hazards not otherwise classified of the chemicals present in the employees' work areas
- How employees can protect themselves from the chemical hazards
- Details of the hazard communication program developed by the employer, including an explanation of the labeling system on shipped containers and workplace labeling system if different, the SDSs including the order of information, and how employees can obtain and use the appropriate hazard information

CONCLUSION

Compliance with regulations that govern the handling, storage, transportation, and disposal of hazardous materials is a critical component of any safety practitioner's job. The mishandling of hazardous materials can be costly in a number of ways. Monetary losses and human suffering can result from poorly planned actions on the part of untrained personnel. Violators can find themselves at the mercy of the EPA, OSHA, and the courts. The proactive safety practitioner should become familiar with existing regulations at both the federal and state levels, and make every effort to stay apprised of changes in the law as they affect the workforce.

QUESTIONS

1. What hazardous chemicals are located in your area? Contact your local fire department and tell them you wish to learn about hazardous chemicals in your area. Ask them if copies of SDSs are available and where they can be found.
2. What are the responsibilities of environmental managers? Find someone who has environmental responsibilities for his company and ask what his major responsibilities are. Ask specifically about filing SDSs with the fire department and emergency planning officials. Also ask about what training is required for workers to respond to spills of any hazardous chemicals on the property. Ask this person which CFRs are available on-site and if you can take a look at them. Some companies now purchase these on CD-ROM, so it's hard to get a feel for the scope of the regulations.
3. Why do you think many companies complain about the difficult legal framework and red tape they face in trying to comply with all applicable regulations? Take a look at some of the individual CFRs in your library or on the OSHA and EPA websites.

REFERENCES

Freeman, H. 1990. *Hazardous Waste Minimization*. New York: McGraw-Hill.

Kaufman, J. (Ed.). 1990. *Waste Disposal in Academic Institutions*, 2nd ed. Chelsea, MI: Lewis.

Noll, G. G., Hildebrand, M. S., and Yvorra, J. G. 1995. *Hazardous Material: Managing the Incident.* Stillwater, OK: Fire Protection Publications.

State of Kentucky. 1992. Waste Management—Identification and Listing of Hazardous Wastes: Title 401. *Kentucky Waste Management Regulations*.

U.S. Department of Labor. 2002. *Hazardous Waste Operations and Emergency Response: 29 CFR §1910.120*. Retrieved from http://www.osha.gov.

U.S. Department of Labor. 2012. *Red Line Strikeout HCS Final Regulatory Text 2012*. Retrieved from https://www.osha.gov/dsg/hazcom/redline.html.

U.S. Environmental Protection Agency. 2002. *Protection of the Environment: Title 40 CFR*. Retrieved from http://www.osha.gov.

U.S. Environmental Protection Agency. 2011. *RCRA Orientation Manual 2011: Resource Conservation and Recovery Act*. Retrieved from http://www.epa.gov/osw/inforesources/pubs/orientat.

U.S. Environmental Protection Agency. 2012. *Hazardous Waste Generator Regulations: A User Friendly Reference Document*. Retrieved from http://www.epa.gov/waste/hazard/downloads/tool2012.pdf.

U.S. Environmental Protection Agency. 2012. *Pesticides: International Activities. GHS Implementation in EPA's Office of Pesticide Programs*. Retrieved from http://www.epa.gov/oppfead1/international/ghs/implement.htm.

U.S. Environmental Protection Agency. 2013. *Superfund*. Retrieved from http://www.epa.gov/superfund/index.htm.

BIBLIOGRAPHY

Bowman, V. 1988. *Checklist for Environmental Compliance*. 4th ed. Newton, MA: Cahners.

Chandler, R. L., Iannaccone, M., and Toki, A. (Eds.). 1995. *Best's Safety Directory: Industrial Safety*, vol. 2. Oldwick, NJ: A. M. Best.

16

Construction Safety and the Multiemployer Worksite Doctrine

Mark A. Friend and Dan Nelson,
with additions by William Walker

CHAPTER OBJECTIVES

After completing this chapter, you will be able to

- List the major hazards typically found on a construction worksite
- Know how to deal with the major hazards found on construction worksites
- Explain the doctrine of a multiemployer worksite as viewed by OSHA
- List the components of a company's effective multiemployer worksite policy

CASE STUDY

An 18-year-old male worker contacted electrical energy when he kneeled to plug a portable appliance into a 100–120 V/20 amp floor outlet. After a scream was heard, the victim was found convulsing on the damp floor, with one hand on the plug and the other on the receptacle box. A supervisor went to the electrical panel but was unable to locate the appropriate circuit breaker. A coworker attempted to take the victim's pulse and received an electric shock, but was uninjured. After contacting medical help, the supervisor returned to the panel and de-energized all circuits (three to five minutes after the worker contacted the electrical energy). After five minutes, a second call was placed to the emergency squad, and the supervisor yelled for another employee, who came and performed CPR. It was over six minutes from the time the employee was injured. Emergency services were on the scene within 10 minutes of the first call and began treating the employee. He was pronounced dead at the hospital.

INTRODUCTION

Although many safety practitioners in general industry have few direct responsibilities regarding construction safety, they often find themselves in situations where knowledge of construction and safe practices is useful, if not imperative. The federal regulations regarding construction safety are found in 29 CFR 1926. They provide essential guidance relative to safety for professionals desiring to steer clear of government-imposed sanctions. Knowledge of those regulations can also help in the management of engineering problems on a construction site. Even though a company may not be in the business of construction, it is important to know whether others engaged in construction practices on the site are following appropriate safety procedures in their work.

OSHA's regulations only provide a starting point for a quality safety program, but its experiences have caused the government to approach the construction problem in a manner that is important to discuss here. OSHA learned that their compliance officers at construction sites were spending too much time on issues of little importance and too little time inspecting for the hazards most likely to cause fatalities. This resulted in a change in inspection policy. When OSHA visits a construction site today, the compliance officer (CO) determines whether there is project coordination by the general contractor and whether that contractor has an adequate safety and health program. COs use OSHA guidelines to determine whether the program is adequate. The inspector also considers whether a designated competent person responsible for and capable of implementing the program plan is on-site. If so, the CO will proceed with a focused inspection, which zeroes in on a few key areas. Otherwise, the employer can expect a general inspection.

The CO determines adequacy of a safety and health program by looking at the comprehensiveness of the plan, its degree of implementation, whether there is a competent person designated (if required by standards relevant to that particular jobsite), and how the plan is enforced. Plan enforcement looks at management policies and activities, employee involvement, and training.

During the course of a focused inspection, citations will be issued for violations of the four leading hazards and any other serious hazards observed. The four leading hazards include:

- Falls from elevations
- Struck by
- Caught in-between
- Electrical shock

If a worksite has multiple employers, citations are issued to any employers whose employees are exposed to hazards. In addition, the employer creating the hazard, the employer responsible for jobsite safety and health conditions, and the employer who has responsibility for correcting the hazard are also cited.

The conditions that employers typically expose their employees to include not only the above, but also those commonly found in general industry violations. The key concerns of OSHA are those mentioned above.

CONSTRUCTION SAFETY RECOMMENDATIONS

Prevention is the primary goal of any occupational safety program. Prevention isn't always possible, so workers on construction sites should never be alone. In the event of an incident, someone needs to be able to call for help. All workers should be trained in the hazards they face on a construction site, and they should also be equipped with the appropriate personal protective equipment. When hazards can be engineered out, they should be. The safety professional needs an in-depth knowledge of potential construction hazards so that he or she knows how to take an engineering approach. Worker attitude and behavior must be addressed. Horseplay, shortcuts, and lack of attention can and do kill on a construction site. New workers and experienced employees lulled into a routine after months or even years on the job need reminders to stay alert while working around construction sites.

Falls from Elevations

Construction employees often work from heights, which can be hazardous. In observing a worker at height, ask yourself if you would be comfortable in that same job. If your answer is no, you may have a problem. The employer should ensure that any worker performing his or her job on any walking/working surface at a height of six feet or more above a lower level is adequately protected from falls. This isn't simply a matter of complying with OSHA. This is a matter of protecting employees.

After years of watching construction workers perform death-defying feats, many employers have become complacent and even nonchalant about worker safety. This is true on commercial construction sites as well as on many residential construction sites. (See figure 16-1 for an example of a residential construction site where no fall protection measures have been taken and no personal protective equipment has been issued to workers.) If the worker is doing a job at a height where a fall could injure or kill, the employee must be protected from a fall. Protection can come in the form of guardrails, fall protection systems (such as safety nets), fall arrest systems (such as body harnesses and lanyards), or monitored separation from the fall hazard by distance from the edge. It is up to the employer to ensure that the appropriate steps have been taken to protect the worker regardless of the regulations.

Workers should typically be equipped with hard hats and steel-toed shoes. Other personal protective equipment (such as safety glasses and hearing

Figure 16-1. Lax safety practices in residential construction

protection) is also necessary in appropriate circumstances. It is up to the employer to ensure that appropriate PPE is provided.

Engineering of the jobsite is critical. Stairways or ladders are needed whenever there is a break in elevation of 19 inches or more. Two or more ladders are needed if there are 25 or more employees on the site. OSHA estimates there are as many as 36 fatalities per year due to falls from stairways and ladders used in construction. (See figure 16-2 for an example of a construction worker not observing OSHA standards for ladder use.)

Figure 16-2. Ladder hazard

Scaffolding is another critical area requiring special attention to protect workers from falls. There are many different types of scaffolds and each has its own requirements. Certain guidelines apply to all types. Among others, there must be:

- Firm footing or anchorage to support the intended load
- A competent person to erect, move, dismantle, or alter the scaffold (a ***competent person*** is one who is capable of identifying existing and predictable hazards in the surroundings or working conditions which are unsanitary, hazardous, or dangerous to employees, and who has authorization to eliminate them)
- Standard guardrail systems or personal fall arrest systems (harnesses), which are required at a working height of more than 10 feet above a lower working level
- Capability of supporting four times the intended load
- Planking of scaffold grade (it will have markings to show it is appropriate)
- "Where platforms are overlapped to create a long platform, the overlap shall occur only over supports, and shall not be less than 12 inches unless the platforms are nailed together or otherwise restrained to prevent movement"; each platform unit (e.g., scaffold plank, fabricated plank, fabricated deck, or fabricated platform) shall be installed so that the space between adjacent units and the space between the platform and the uprights is no more than 1 inch wide, except where the employer can demonstrate that a wider space is necessary (for example, to fit around uprights when side brackets are used to extend the width of the platform)
- Unsecured scaffold planking that extends a minimum of 6 inches beyond the ends of the scaffold support
- Flat planking that extends no more than 12 inches past the scaffold support if the planking is 10 feet or less in length, and no more than 18 inches if the planking is over 10 feet in length (both with exceptions)
- Overhead protection
- No slippery conditions
- No work during high winds or storms

These and other conditions are found in 1926 Subpart L—Scaffolding.

Struck by and Caught In-between

A construction site is loaded with moving objects. It is up to the employer to ensure that the moving objects don't strike the employee in such a way as to cause injury or death. This can be accomplished in a number of ways. Safety professionals should not automatically resort to PPE to solve construction worksite problems. Behavior modification and engineering are better approaches for protecting workers from fast-moving objects than relying on PPE.

There are many recorded cases of workers being struck by moving equipment or machinery on a construction site. The employer should implement procedures to ensure that equipment is never moved without someone watching where the equipment is moving and making certain that everyone is out of the way. Backup alarms on heavy trucks can be useful, but on a major construction site, alarms are sounding all the time. Employees soon tune them out. Some companies rely on spotters to make certain the way is clear and to physically keep people out of their paths. The same should be done for any moving equipment where people might be in the path and the operator cannot see where the equipment is moving.

Unsecured or improperly secured loads can also be a problem. Improper instruction of workers regarding movement about the worksite and around machinery can also cause accidents. Workers can be caught between moving equipment or a shifting load and stationary objects. Proper training on worksite procedures can help prevent caught-between incidents. Supervisors should be trained to watch out for the welfare of employees and correct them when they see unsafe behaviors. A worker getting caught between the superstructure of equipment and a wall, a worker being mashed when a piece of hydraulically operated equipment unexpectedly falls, and a worker caught under a shifting load are only a few of the incidents reported on the OSHA website.

As already mentioned, hard hats and steel-toed shoes are essential. Relatively few workers who sustain head injuries wear hard hats, although many are required to wear them for at least some of the tasks they perform. Less than one in four workers with foot injuries wear steel toes and less than half of workers with eye injuries wear eye protection. Protective equipment works, but only if employees use it. The same is true for ear, respiratory, and other forms of protection.

Electrical Shock

Hundreds of occupational electrocutions occur each year. One of the primary goals of any safety program should be to prevent workers from contacting electrical energy. A secondary goal must be to provide appropriate medical care to workers who come in contact with electrical energy. The National Electrical Code (NEC) divides voltage into two categories: 600 volts or less (low voltage) and greater than 600 volts (high voltage). Live parts of 50 volts or more must be guarded against accidental contact. Brief contact with low voltage may not cause a burn, but it could cause *ventricular fibrillation* (a rapid, ineffective heartbeat). High-voltage contacts can cause the heart to stop completely. When the circuit breaks, the heart may resume beating normally. But if there are also extensive burns, death may result anyway.

There is an old saying: "It ain't the volts that kill you; it's the amps." One milliampere is 1/1000 of an amp. It provides just a faint tingle. Muscular control can be lost from 6–30 milliamperes. Somewhere between 50 and 150 milliamperes, extreme pain, respiratory arrest, and severe muscular contractions will occur. If an employee is holding on to something when the shock occurs, he can't let go. Death is possible at this point. Between 1 and 4 amps, death is likely. A 100-watt light bulb operates at 833 milliamperes of current. Environmental conditions can greatly affect how severely the body is shocked. An employee performing heavy labor on a hot, humid day could come into contact with a level of electrical energy exactly the same as six months before and have completely different results. On a cold, dry day with little moisture on the skin, the energy may result in a relatively mild shock. On a hot, humid day, injury or death may result.

For obvious reasons, an employee on the construction site should have a working knowledge of basic electricity. All work is to be done in compliance with the National Electrical Code, unless the regulations provide otherwise. Electrical energy needs to be kept separate from all employees. This means that any electrical equipment or tools not double insulated must be grounded and all cords should be of the three-wire type with grounding attached. It is a good idea to replace worn, frayed, or cut cords. Avoid the use of splices on the construction site. Replace cords; don't patch them. Keep all electrical boxes closed and make certain there is no way accidental contact with a bare wire could occur. All openings to electrical panels must be kept covered.

Electricity and water don't mix. Use ground-fault circuit interrupters (GFCIs) for tools and any equipment or appliances that might be used in a damp location. The **GFCI** is a fast-acting circuit breaker that senses small imbalances in the circuit caused by current leakage to ground and, in a fraction of a second, shuts off the electricity. The GFCI continually matches the amount of current going to an electrical device against the amount of current returning. Whenever the amount going differs from the amount returning by about 5 milliamps, the GFCI interrupts the power within as little as 1/40 of a second. All temporary power branch circuits used for construction-related activities are required to be GFCI protected (unless an assured equipment grounding conductor program is in place, where then only 120-volt, single-phase, 15- and 20-amp branch circuits are required to be GFCI protected). GFCIs don't last forever, so they should be tested on a regular basis to ensure that they are in proper working condition.

Ensure that each disconnecting means for motors and appliances and each service feeder or branch circuit is clearly marked to indicate its purpose. Don't permit employees to look for a particular circuit by switching them individually. A shutoff and restart could seriously injure someone who stops to inspect an electric tool to determine the problem. Every disconnect switch and breaker must be clearly marked to indicate its purpose.

Any temporary electrical systems should be inspected regularly (at least weekly) for proper connections, polarity, and so on. A polarity tester should be used to insure that all ground wires are properly connected. Follow all regulations regarding lockout/tagout. This is particularly true regarding electricity.

Other Hazards

Construction sites often have many of the same hazards that are present in manufacturing operations and other non-construction workplaces. Other hazards commonly found on construction worksites include those related to confined space, trenching, and equipment. (See figures 16-3, 16-4, and 16-5 for examples of these hazards at construction sites.)

Case Study

An employee was installing a small-diameter pipe in a trench 3 feet wide, 12–15 feet deep, and 90 feet long. The trench was not shored or sloped, nor was there a box or shield to protect the employee. Further, there was evi-

Figure 16-3.   Confined space hazard

Figure 16-4.   Potential for collapse

Figure 16-5.   Saw without guard

dence of a previous cave-in. The worker may not have been aware of that. The employee apparently entered the trench and a second cave-in occurred, burying him. He was found face down in the bottom of the trench.

Trenching

Trenching is a particularly hazardous area on construction sites. Even relatively shallow trenches are capable of trapping and suffocating a worker within their walls. Many accidents occur because no one took into account the specific conditions associated with the site of the trench. Traffic, heavy structures nearby, a high water table, weather, and soil type are among factors contributing to weakened trenches. Before beginning the job, a thorough analysis of the area to be trenched must be performed by a competent person.

The trench also requires ongoing inspection by a competent person on at least a daily basis or whenever conditions change (such as rain). Large or complex trenching operations may require more; employees must be protected from cave-ins. A small person in a shallow trench might be bending to perform work when the trench collapses. Even though the worker's head is above the ground, the weight of the earth can prevent breathing and suffocate the worker before he is rescued. OSHA *requires* protection via sloping or shoring/trench box at a depth of 5 feet. However, the competent person may determine that unstable conditions mandate stabilization and/or protection prior to reaching that depth.

OSHA requires adequate means of egress in the form of ladders, steps, ramps, or the like from any trench 4 or more feet deep. These need to be placed within 25 feet of travel for any worker.

Many companies are able to take care of their own construction safety problems, but they run into problems when interfacing with other companies. They may have employees who work with employees of other companies. They may hire companies who bring employees onto their site. These issues all create problems of their own.

MULTIEMPLOYER WORKSITE POLICY

Employers around the country are finding that hiring an expert, independent contractor is riskier than anyone (except, perhaps, government regulators) ever imagined. The Occupational Safety and Health Administration (OSHA) is citing host employers for safety violations committed by independent contractors and their employees. Why? Host employers have an obligation to provide safe working conditions to any person working at their facilities, even the employees of some other employer.

This *multiemployer worksite doctrine* first arose in the construction industry, where the presence of numerous contractors and subcontractors on one site sometimes made it difficult to determine who was responsible for safety violations. In response to this problem, OSHA established a rule that basically imposed responsibility on any employer who had control of the site. The rule has been and continues to be challenged in courts across the country, and rulings tend to support individual employer-employee responsibility.

Safety professionals need to analyze and define contractor safety requirements. Contractors are being used extensively on- and off-site by a large number of companies. There are some facilities that provide safety direction and supervision to contractors, while others provide little or no safety direction and/or supervision. Actions of contractors that work for any organization can have a profound impact on the hiring company's loss experience (dollar and injury). Contractors can cause injury to host employees, other contractors, the general public, and property owned by others.

The legal relationships presented by associations of these groups remain complex. Questions continually arise over responsibility and process ownership. If an employee creates a hazard while working with others, who is responsible? What if the other employers ignore the hazard and continue to work or even contribute to the situation? How does a compliance safety and health officer (CSHO) handle the audit and/or citation? Multiemployer worksites can be complicated by these differing work relationships.

Early in the development of the multiemployer doctrine, OSHA's attempt to clarify issues was the usual one-standard-fits-all approach. The Williams-Steiger Occupational Safety and Health Act of 1970 requires employers to provide their employees a safe workplace. In fact, the original multiemployer doctrine is entirely based on the following standard quotation:

(a) Each employer (1) shall furnish to each of his employees employment and a place of employment which are free from recognized hazards that are causing or are likely to cause death or serious physical harm to his employees; (2) shall comply with occupational safety and health standards promulgated under this Act. (OSHA ACT 1970, Section 5, Duties [29 USC 654(a)])

OSHA sees these two statements as imposing two distinct duties. First, (a)(1) requires employers to protect their own employees from hazards in the workplace. The employer's duty under (a)(1) owes only to its employees, as indicated by the language requiring a hazard-free workplace for "his employees." Second, (a)(2) requires employers to comply with the Act's safety standards. Unlike (a)(1), *it does not limit its compliance directives to the employer's own employees, but requires employers to implement the Act's safety standards for the benefit of all employees in a given workplace, even employees of another employer.* OSHA issues citations based on the multiemployer doctrine under (a)(2) (Briscoe; *Universal Construction Co., Inc. v. OSHRC*).

Doctrine History

The multiemployer doctrine is one of many processes outlined in OSHA's *Field Inspection Reference Manual* (FIRM). The FIRM is a manual that provides guidance regarding some of OSHA's operational processes. Position responsibilities within OSHA are also outlined here, along with inspection procedures. In 1974, OSHA's enforcement policy on multiemployer worksites was first stated in the Field Operations Manual (FOM). The policy provided for the citing of employers who exposed their employees to hazards. Over the years, this policy was changed to provide for the citing of controlling employers. In 1981, OSHA issued directive CPL 2.49, Multiemployer Citation Policy, which was incorporated into the FOM. The actual manual revision was eventually published in 1983.

An employer who has general supervisory authority over the worksite, including the power to correct safety and health violations itself or require others to correct them, can be said to have *control*. Control can be established by contract, or in the absence of explicit contractual provisions, by the exercise of control in practice. The courts accepted this interpretation in one form or another. (See *Marshall v. Knutson Construction Co.* [8th Cir., 1977] or *Brennan v. OSHRC* [2nd Cir., 1975] for more information on this.) Eventually, OSHA's policy evolved to also provide for the citing of *correcting* and *creating* employers as well.

Currently the doctrine is not only intended for construction sites, but is equally applicable to temporary employees and professional contractors within the workplace. That is not to say that the policy has not had its legal bumps and bruises along the way. The changing workplace and legal opinions have prompted yet another change in the doctrine.

The multiemployer directive, CPL 2-0.124 (11/99), is a result of court opinions recognizing individual company and employee behavior on multiemployer worksites (e.g., Silberman, *IBP, Inc. v. Herman* [DC Cir., 1998]). This revision continues OSHA's existing policy for issuing citations on multiemployer worksites. Court opinion holds that the following can be held liable:

- An employer who is engaged in a common undertaking, on the same worksite, as the exposing employer and is responsible for correcting a hazard. This usually occurs where an employer is given the responsibility of installing and/or maintaining particular safety/health equipment or devices.
- The employer that caused a hazardous condition that violates an OSHA standard.

A subcontractor whose employees are threatened by a hazard created and controlled by another subcontractor has only two options: request the offending subcontractor to abate the hazard or request the general contractor to correct or direct correction of the condition. As a practical matter, the general contractor may be the only party on-site with authority to compel compliance with OSHA safety standards (decision, Briscoe). (Also see *Anning-Johnson Co. v. OSHRC* [7th Cir., 1975].)

The multiemployer doctrine is particularly applicable to multiemployer construction worksites. The nature of construction requires that subcontractors work in close proximity with one another and with the general contractor at the same worksite. In these situations, a hazard created by one employer could be seen as reasonably affecting the safety of other employers (*Bratton Corp. v. OSHRC* [8th Cir., 1979]). Specific areas of expertise or job area responsibility may limit a subcontractor's ability to abate hazards posed to its own employees that may be created by another subcontractor, general contractor, or host employer (*IBP, Inc. v. Herman* [DC Cir., 1998]).

J. Anthony Penry, an attorney in Raleigh, North Carolina, states, "Every serious accident and even some not-so-serious accidents, will result in a failure-to-inspect citation, because it is now very easy to prove. Expect an increase in third-party claims. Because courts often confuse the OSHA standard for liability and liability for negligence, and because the cases are less than clear on this issue, expect injured subcontractor employees to assert more claims against general contractors" (*Risky Business*, Summer 2005, p. 1).

If outside contractors are working at a facility, the agreement should clearly set forth the contractor's safety responsibilities. It's also a good idea to adopt a specific outside contractor safety program.

Contractor Qualifications and Programs

Contractor programs are administered in one of two ways: hands on and hands off. Both methods incorporate varying degrees of the following recommendations. Whichever approach is chosen, consistent application and thorough documentation are necessary. Each of these approaches requires an equal amount of initial paperwork. The first things that must be reviewed are the contract and the safety qualifications of the prospective contractor, otherwise known as contractor qualifications. Three sources of information provide ways for an employer to evaluate the probable safety performance of a prospective contractor. Review:

- Experience modification rates for workers' compensation insurance
- OSHA incident rates for recordable injuries and illnesses
- Contractor safety practices and written procedures

The insurance industry has developed experience-rating systems as an equitable means of determining premiums for workers' compensation insurance (see chapter 3 on workers' compensation). These rating systems consider the average worker's compensation losses for a given firm's type of work and payroll and predict the dollar amount of expected losses to be paid by that employer in a designated rating period (usually three years). The rating is based on comparison of firms doing similar types of work and in each work classification. Losses incurred by the employer for the rating period are then compared to the expected losses to develop the ratio factor.

The second suggested analysis should take the contractors' OSHA incident rates into consideration. OSHA requires employers to report and record accident information on occupational injuries and illnesses on the annual OSHA 300 log. The employer must retain completed forms for five years and the following information can be obtained from them:

- Number of fatalities and injuries
- Number of days involving lost or restricted time
- Incident rates based on the employer's annual hours-worked calculation

The third performance measurement is very important in determining a company's safety record and attitude. Companies that hold their project management accountable for accidents along with productivity, schedules, and quality, and so on are usually the ones that carry the best safety records. As with any safety program, a few of the things to be considered are:

- Management commitment
- Written safety program
- Hazard assessments
- Training programs and employee qualifications
- Emergency plans and procedures
- Accident reporting protocol
- Regular safety meetings and inspections

Once the prospective contractor qualifies, an agreement, a written contract, must be prepared. This contract, at a minimum, should state the host company's safety beliefs and expectations and a statement extending these to the contractor. Thus, the contract would include statements similar to the following:

- Contractor employees are required to adhere to all applicable federal and state occupational safety and health laws as they apply to this contract.
- The contractor shall enforce the host company's safety rules and practices as they apply to the contractor's employees, in addition to the contractor's own safety rules and procedures.
- The contractor shall provide all of their subcontractors with copies of all safe working procedures and shall ensure their enforcement.

Without relieving the contractor of full responsibility to comply with all appropriate safety requirements, the host employer should ensure that a project manager is assigned and advised to keep management apprised of all activity and work progress. With regard to construction managers, host employers should be aware of the following OSHA interpretation:

To the extent that a construction manager has a role in directing the manner or timing of the work, it may be cited as a "creating" employer if a violation occurs as a result of its direction. Depending on the circumstances, including contractual responsibility or the assumption of a safety-monitoring role, a construction manager may also be a *controlling employer*. A controlling employer is one having the responsibility or authority to have violative conditions corrected. General (or prime) contractors are controlling employers for many types of violations that occur on construction sites, but they may choose to carry out their safety role in whole or in part through a construction manager. (OSHA, August 5, 1993)

Contractor safety orientation is another critical step in ensuring that contractor employees understand hazards related to the job, know site safety practices and rules, and are familiar with local emergency practices related to the site. Contractor safety orientation methods involving classroom sessions or video presentations can be very effective in preventing needless losses. The duty of a contractor in a host employer's workplace should be, at a minimum, to maintain the same level of safety and compliance that the host employer practices. These requests and requirements should be outlined in the scope of work and contract documentation. The host employer should also remember that the contractor's responsibility (liability) only applies to the extent to which the contractor has control of or can reasonably be expected to have control of the site. Those actions by the host employer that may create or expose employees to hazards remain the responsibility of that employer.

In summary, OSHA has determined that employers at a multiemployer worksite fall into four basic categories: controlling, creating, correcting, or exposing. The *controlling employer* is the employer who, by contract or actual practice, has the responsibility and authority for ensuring that hazardous work conditions are corrected. This employer is usually the general contractor, or GC. When a company acts as the general contractor for a construction project, it is considered the controlling employer and would be responsible for the safety and health of all workers at the site.

The *creating employer* is the employer whose activities actually create a hazardous condition, while the *correcting employer* is the employer that has the responsibility for correcting the hazardous condition. An *exposing employer* is any employer whose workers are exposed to the hazardous condition. Depending upon the situation, any employer at a construction site could fall into one or more of these classifications and could be issued a citation by OSHA.

Employers should consider the multiemployer worksite rules whenever their workers are working with other employers, or whenever they are acting as the project manager for such an activity. In situations where the host company is acting as the general contractor, the burden of providing a safe worksite rests with the project manager and each of its supervisors involved with the project. However, even on those projects where an outside contractor is acting as the general contractor, subcontractors or departments are still responsible for their own workers' safety. Any hazardous condition should be brought to the attention of the general contractor and/or host employer. If the condition is so hazardous as to be imminently dangerous, supervisors should remove their workers from the worksite and contact their assigned safety representative and/or management.

CONCLUSION

Even companies operating under general industry standards find themselves dealing with construction practices and standards. While OSHA standards provide a basis for practices, safety practitioners not dealing with construction on a regular basis may not be familiar with even basic construction safety principles. This lack of knowledge can create a triple-barreled shock to the employer. An injured or killed worker who may not be employed directly by the host company can create problems in terms of dealing with the tragedy, paying costs associated with the tragedy directly and subsequently in litigation and increased insurance premiums, and paying OSHA fines. Safety practitioners need to familiarize themselves with basic construction safety practices and with the multiemployer worksite doctrine when dealing with outside contractors.

QUESTIONS

1. Why is it important for the safety professional to be familiar with construction safety practices?
2. What are the major construction safety problems the safety practitioner is likely to encounter on the job?
3. Interview a safety professional in general industry. Ask about this person's familiarity with construction safety practices. Ask if the professional's employer hires construction contractors and what the company policies are regarding these contractors. Are both the company and the

safety professional prepared to deal with the multiemployer workplace problem?

4. How has the multiemployer worksite doctrine changed since this chapter was written? Review the OSHA website at www.osha.gov and determine what changes have been made.

REFERENCES

Anning-Johnson Co., Petitioner, v. OSHRC, Respondents. 516 F.2d 1081 (7th Cir., 1975).

Bratton Corporation, Petitioner, v. Occupational Safety and Health Review Commission, and Ray Marshall, Secretary of Labor, Respondents. 590 F.2d 273 (8th Cir., 1979).

IBP, Inc., Petitioner/Cross Respondent v. Alexis M. Herman, and DOL, Respondents/Cross-Petitioners. 144 F.3d 861 (DC Cir., 1998).

OSHA ACT 1970, Section 5, Duties. [29 USC 654(a)(1)(2)]

OSHA Directive, Multiemployer Citation Policy. CPL 2-0.124, 12/10/99.

OSHA Field Inspection Reference Manual (FIRM). Instruction CPL 2.103, 9/26/94.

OSHA Letter of Interpretation. Duties of a "Construction Manager" on a Multiemployer Worksite. 8/5/99.

Peter J. Brennan, Secretary of Labor, Petitioner, v. OSHRC and Underhill Construction Corp., Respondents. 513 F.2d 1032 (2nd Cir., 1975).

Ray Marshall, Secretary of Labor, Petitioner, v. Knutson Construction Company and Occupational Safety and Health Review Commission, Respondents. 566 F.2d 596 (8th Cir., 1977).

Risky Business. Summer 2005. *Outcome of New Lawsuit May Create More Exposure than GC.* Raleigh, NC: Builders' Mutual.

Universal Construction Co., Inc., Petitioner, v. OSHRC, Respondent. 182 F.3d 726 (10th Cir., 1999).

United States, Plaintiff-Appellee, v. Pitt-Des Moines, Inc., Defendant-Appellent. 168 F.3d 976 (7th Cir., 1999).

17

Transportation Safety

Scotty Dunlap, EdD, CSP

CHAPTER OBJECTIVES

After completing this chapter, you will be able to

- Define what a hazardous material is according to DOT regulations
- Explain the safety implications of transporting hazardous materials
- Describe the components of commercial motor vehicle driver qualifications
- Differentiate between the two DOT regulations that address drug and alcohol testing
- Describe how the components of DOT safety can be applied to fleet safety

CASE STUDY

After receiving her degree in agriculture, Ellen entered the grain industry, where she was able to successfully work her way up the ladder. She began her career as a production supervisor at a grain elevator that purchased, stored, and shipped corn and soybeans from local farmers. Through numerous promotions within her company, she decided to take the bold step of accepting a position as vice president of operations with another grain company. Within the new organization, Ellen learned that they not only handled grain but also sold agricultural chemicals to farmers.

On a tour of one of the facilities Ellen saw many familiar sites. There were the towering concrete grain bins, a maintenance shop, and an office building; however, the agricultural chemical area was new to her. She soon realized the complexity of the operation. Plant employees delivered anhydrous ammonia in tanks that were pulled behind company-owned pickup trucks. There was also a fleet of five mid-sized cars used by plant sales managers. Ellen's previous company was very safety conscious, so she immediately began inquiring about the safety implications of the operation. A production supervisor sufficiently answered her questions regarding the safety precautions used at the plant when anhydrous ammonia was being received into the large storage tanks and when it was later transferred into the smaller transport tanks. She was also concerned about what happened once the rubber wheels of the company trucks and anhydrous ammonia tanks touched the public road. In normal driving activity she remembered seeing large trucks on the road, hauling chemicals that had placards on each side of the trailers, but she did not notice such placards on the anhydrous ammonia trailers. When she asked the production supervisor about it, he simply stated those rules did not apply to their operation. She did not receive information that was much better when asking about the maintenance of the cars used by the sales managers or the process by which drivers were approved to operate company vehicles.

Several days later when Ellen was back at her office, she felt uneasy about what was being done in the transportation of anhydrous ammonia, but she had never experienced issues related to the Department of Transportation in her career, so she was not certain of where to start. The phone rang and when she answered it she was greeted with a panicked plant manager. A company truck pulling an anhydrous ammonia tank had been involved in an accident. The driver was injured and was being transported to a hospital. The fire department, local police, and a DOT enforcement officer were on the scene. Ellen had a sickening feeling in the pit of her stomach.

HAZARDOUS MATERIALS TRANSPORTATION

Safety issues related to transportation are complex and large in number. In this chapter key issues will be addressed for vehicles regulated by the Department of Transportation (DOT), and also for those unregulated, but included as part of an organization's fleet. One of the greatest areas of motor carrier exposure to legal liability is in hazardous material handling, and it also presents some of the most complex compliance issues.

What Is a Hazardous Material?

As defined in 49 CFR 171.8, a *hazardous material* is

> a substance or material that the Secretary of Transportation has determined is capable of posing an unreasonable risk to health, safety, and property when transported in commerce, and has designated as hazardous under section 5103 of Federal hazardous materials transportation law (49 U.S.C. 5103). The term includes hazardous substances, hazardous wastes, marine pollutants, elevated temperature materials, materials designated as hazardous in the Hazardous Materials Table (see 49 CFR 172.101), and materials that meet the defining criteria for hazard classes and divisions in part 173 of subchapter C of this chapter.

This definition addresses both what hazardous materials are and what they include. A "hazardous material" is any material so designated by the DOT. The three key considerations are

1. unreasonable risk to health, safety, or property;
2. the health and safety of individuals including drivers, warehouse workers, emergency response personnel, motorists, and pedestrians; and
3. concern for conserving the environment and maintaining the integrity of privately owned property.

Anyone offering the service or performing the act of transporting hazardous materials is covered by the requirements of this standard. The standard also provides a number of features for transporters to help minimize the risk, including a hazardous materials table; shipping papers; emergency response information; placarding; and provisions for driver training, emergency equipment, spill prevention and response training, and an emergency response plan.

Hazardous Materials Table

Information and governing guidelines for truck drivers and emergency response personnel are addressed in the Hazardous Materials Table in 49

CFR 172. This table identifies all materials considered hazardous during the transportation process, listing them alphabetically, by their proper chemical name, and providing other relevant information in columns across the table, including hazard class or division, identification numbers, and packing group.

Shipping Papers

One of the byproducts of the Hazardous Materials Table, **shipping papers**, are required in 49 CFR Subpart C, to be carried by truck drivers at all times during the delivery of hazardous materials. They list hazardous materials in the load and serve as a resource for emergency response personnel in the event of an accident or spill by providing proper shipping name, identification number, hazard class or division, and packing group. These four items together are known as the *proper description*. The proper shipping name and identification number are used to find emergency response information as shown later. The class or division identifies the type of hazard. The packing group provides the degree of danger the hazardous material presents. Altogether, this information supplies the first few pieces of the puzzle for responders to a hazardous material incident.

Emergency Response Information

Emergency response information must be available in the event of an emergency; primary concerns are that:

- Information is written in English for emergency response personnel (fire, police, EMS).
- It is available to employees.
- It contains an emergency contact phone number in case of an incident.

Forms of emergency response information include shipping papers, the DOT Emergency Response Guidebook, and Safety Data Sheets (SDSs). In the event a tractor-trailer is involved in an accident where hazardous materials have been spilled, shipping papers carried by the truck driver will be the first source of information. The proper description will be listed on the shipping papers, including the name of the hazardous material, its identification number, its hazard class or division, and the packing group to which it belongs. This information will first alert emergency response personnel to the type and danger of the involved hazardous material(s). The *Emergency Response Guidebook* will be the next piece of information to be referenced. This publication of the Department of Transportation (DOT) lists directs the user to useful information for handling a hazardous material leak, spill, or fire. The book is divided into five sections, each having its own color of paper:

- White—contains introductory remarks and a depiction of the various placards used by motor carriers
- Yellow—lists all hazardous materials numerically by their identification number
- Blue—lists all hazardous materials alphabetically by their proper name
- Orange—provides emergency response information for the individual hazardous materials
- Green—provides initial isolation and protective action distances

Placarding

Placarding requirements are given in the .500 series of regulations in 49 CFR 172. Part 172.504 states that with certain exceptions, placards must be placed on both sides and both ends of hazardous materials transport containers. Weight calculation is also addressed in .504 and tables are provided to assist in the proper selection of placards. The tables include three columns of information:

- Hazard class or division or other appropriate description
- Placard name
- Placard design section reference

The hazard class or division must first be found by using the Hazardous Material Table. Once it is located on the table, simply scan across for the placard name, such as "EXPLOSIVES 1.1," and the placard design section reference, such as "172.522." "EXPLOSIVES 1.1" provides the name of the placard to be used and "172.522" provides the reference for the regulation that gives the specific requirements related to that placard. Requirements for placarding subsidiary hazards are discussed in .505. In .506, the burden to provide placards for motor carriers is placed on the shipper, who must provide the driver with the appropriate placards when presenting a hazardous material for transportation. Motor carriers are prohibited from transporting hazardous materials when the load is not properly placarded.

Driver Training

Drivers and other workers associated with the safe transportation of hazardous materials must be properly trained, as indicated in 172.704 include:

- General Awareness—basic understanding of hazardous materials regulations
- Function Specific—issues that are directly related to each person's responsibilities
- Safety—personal safety and emergency response information
- Security—recognize and respond to security risks
- In-Depth Security—understanding of the organization's specific security plan

Emergency Response Planning

Planning for possible emergencies by qualified personnel, to include breakdown, violence (road rage), vehicle accidents, hazardous material release or spill, or ignition of a hazardous material, is critical, so that appropriate response measures can be generated to adequately and safely address each. Actions may range from the driver directly handling the situation to his or her simply calling 911. In any case, the safety of the driver and others are the primary consideration.

Emergency Response Equipment

Each driver should be provided with emergency response equipment to include portable fire extinguishers; reflective triangles; a spill response kit; and, at the discretion of the motor carrier, possibly other equipment, such as first aid kits. Drivers must be properly trained in the use of all equipment provided.

Spill Prevention and Response

Regulations published the Department of Transportation and the Environmental Protection Agency also address spill prevention and response. For example, DOT regulation 49 CFR 130.11 requires a means of communication stating that the shipment contains oil. This can be accomplished through the use of properly completed shipping papers. Prior to shipments of oil falling under the requirements of this regulation, a plan must be developed to address various areas of emergency response. As stated in 130.31, the emergency response plan must include:

- Description of how emergency response will occur
- Consideration of the worst possible scenario by considering the maximum amount of oil that may be released from the packaging
- Delineation of resources for emergency response
- Delineation of who must be notified in the event of a spill, to include individuals and governmental agencies

In the event an incident causes the release of a hazardous material during the transportation process, there may be applicable reporting requirements governed by the Environmental Protection Agency (EPA). Information is provided in 40 CFR 355 informing the user of spill quantities requiring immediate reporting to the EPA. The information is listed in Appendix A of the regulation in alphabetical order and in Appendix B, by CAS number of the various hazardous materials.

Hazardous materials incident reporting requirements are detailed in 171.15 and .16. Incidents requiring immediate reporting to the Department of Transportation as listed in 171.15 include those where:

- A fatality has occurred
- Injuries have occurred that result in at least one person requiring off-site medical attention
- An evacuation of the general public lasts one hour or more
- At least one major transportation artery or facility is closed for one hour or more
- The operational flight pattern or routine of an aircraft is altered

Part 171.16 expands the reporting requirement to include any other unintentional release of a hazardous material during loading, transport, unloading or while in temporary storage.

Security Plan

The events surrounding 9/11 gave rise to the development of security plans to effectively protect hazardous materials shipments. Part 172.802 indicates that such plans must consider:

- Personnel security addressing the hiring process of those who will transport hazardous materials
- Unauthorized access establishing who may or may not be authorized to access the material
- En route security addressing the risk analysis to protect the hazardous material while in transit

FEDERAL MOTOR CARRIER SAFETY REGULATIONS

Commercial Driver's License

Obtaining a Commercial Driver's License (CDL) is necessary for driving a commercial motor vehicle on public roads. This regulation states the purpose of the CDL process as follows: "to help reduce or prevent truck and bus accidents, fatalities, and injuries by requiring drivers to have a single commercial motor vehicle driver's license and by disqualifying drivers who operate commercial motor vehicles in an unsafe manner" (49 CFR 383.1(a)).

Although the CDL process provides measurement for a level of expertise and a standard of performance, the main purpose is to provide for the safety of human life. Applicants are responsible for reporting certain convictions, in addition to issues related to their standard driving license. Though a standard driver's license and a CDL are separate, offenses made under a standard driver's license may influence the ability to obtain a CDL.

In applying for a CDL, an applicant must be tested according to the type of vehicle(s) that he or she wishes to operate. Regulation 49 CFR 383 Subpart F contains the divisions for the different groups of vehicles. In table 17-1, GCVR refers to gross combination vehicle weight rating and GVWR refers to gross vehicle weight rating.

Provided a driver has the appropriate endorsement, he or she may operate a vehicle in any group below that tested. For example, if a driver has

Table 17-1. Divisions for different groups of vehicles

Group	Type	Configuration	Specifications
A	Combination Vehicle	Combination	GCWR > 26,000 lbs. (vehicle(s) towed > 10,000 lbs.)
B	Heavy Straight Vehicle	Single Vehicle	GVWR > 26,000 lbs. (trailers must be < 10,000 lbs.)
C	Small Vehicle	Single Vehicle or Combination	Does not satisfy Group A or B definitions and is intended to carry 16 or more passengers or hazardous materials

tested and passed in group A and has received the proper endorsements on their license, he or she may also operate commercial motor vehicles in either group B or C.

Driver Qualifications

In order to operate a commercial motor vehicle, prospective drivers must meet minimum qualifications, according to 391.11:

- Must be at least 21 years of age
- Must be fluent in English to the extent that he or she can communicate with the public, understand road signs, and make appropriate log entries
- Can safely operate their motor vehicle
- Satisfies DOT requirements for being physically fit
- Has a current CDL
- Has provided their employer with a list of appropriate driving violations, if any
- Has not been disqualified to drive a motor vehicle under DOT regulations
- Has passed a road test given by their employer

Some exemptions are given, as is the case in some farm-use applications.

All appropriate documentation must be maintained in a "driver's qualification file." Once each year, the employer is responsible for reviewing the file to ensure the driver has not committed any infraction disqualifying him or her from operating a motor vehicle. The employer must also conduct a road test to allow the driver to demonstrate the ability to safely operate the motor vehicle to be driven. Prior to operating a vehicle for a motor carrier, drivers must also pass an exam considering physical condition, health issues, vision, and hearing.

Drivers are not permitted to operate their vehicle while under the influence of drugs or alcohol. In addition, when their condition affects the safe operation of the vehicle, ill or fatigued drivers are not permitted to drive. The Department of Transportation established "hours of service," meaning how long a driver may drive in addition to time for rest—and each driver is required to keep accurate records of service. In the event a driver is pulled over, inspected, and found to be in violation of the hours-of-service requirements, the officer can immediately place him or her out of service for a 24-hour period.

Emergency Equipment

Just as in facility buildings, tractors must be equipped with equipment for routine emergencies, and though not specified in this section, drivers should be properly trained in its use and application. Requirements in 393.95 specify including the following:

- Fire extinguishers
- Spare fuses
- Warning devices for stopped vehicles (reflective triangles; red emergency reflectors)
- Fuses and liquid burning flares
- Red flags

Maintenance of Accident Registers and Reports

Similar to OSHA injury and illness recordkeeping, motor carriers are also responsible for maintaining records associated with vehicle accidents. The following information must be maintained on accident registers:

- Date of the accident
- City and state where the accident occurred
- Name of the driver
- Number of injuries
- Number of fatalities
- Whether or not hazardous materials being carried were released

DRUG AND ALCOHOL PROGRAMS

Background Checks

Employers are responsible for ascertaining the previous two-year history of drug and alcohol tests from prospective drivers. Following the consent of the applicant, the employer must contact previous employers covering that period, no later than 14 days after the driver has begun work performing safety-sensitive functions, but preferably prior to the individual beginning work.

Testing Programs

Employers with commercial motor vehicle operations in the United States, whether domestically or foreign-owned, must establish and maintain drug and alcohol testing programs, as required in 49 CFR 382. It is imperative that drivers remain free of the influence of drugs and alcohol while performing "safety-sensitive functions" to include all time

- spent on duty at the place of employment,
- spent performing vehicle inspections or maintenance,
- spent operating a commercial motor vehicle,
- spent in a commercial motor vehicle except while resting in a sleeper berth,
- associated with loading or unloading the vehicle, and
- associated with a vehicle breakdown.

In addition, alcohol consumption is restricted

- any time a driver is performing a safety-sensitive function,
- within four hours of performing a safety-sensitive function, and
- within eight hours following an accident requiring an alcohol test or until the test is conducted, whichever comes first.

Unless a doctor has prescribed a controlled substance, and the drug will not adversely affect the driver while performing his or her duty, employers cannot allow drivers to perform safety-sensitive functions while under its influence. Employers must inform drivers about the drug and alcohol program prior to testing, and drivers must sign a statement that they have received the information. The employer keeps the original and the driver is given a copy. Requirements for the drug and alcohol testing process include:

- **Test Preparation:** the employee has no ability to contaminate or provide a false sample and that the container is clean and cannot have been tampered with prior to the collection of the specimen
- **Specimen Collection:** chain of custody must be documented as the procedure is followed to completion
- **Medical Review Officer:** communications are channeled through the Medical Review Officer (MRO) with regard to test results and findings from examinations; the MRO is in contact with testing laboratories, examining physicians, the employer, and the employee
- **Laboratory:** test results are reported to the employer through the MRO

Drug tests are required:

- pre-employment
- post-accident
- randomly
- for reasonable suspicion
- upon return-to-duty
- as follow-up

In all but the last two of the above instances, the employer is responsible for informing the individual of test results, but should always inform of any controlled substance(s) showing positive and ask the person to contact the MRO. The actual drugs for which employers are permitted to test can be found in 49 CFR 40.85 and include marijuana, cocaine, amphetamines, opiates, and phencyclidine (PCP).

Rehabilitation

Drivers found to be using drugs or alcohol, as prohibited by this regulation, must be provided with information regarding access to a substance abuse program. The substance abuse professional will establish a treatment plan in accordance with 49 CFR 40 Subpart O. Due to the scope and application of some regulations in very specific situations, there are many topics covered within the Federal Motor Carrier Regulations not addressed in this chapter.

Readers with specific responsibilities in this area should seek additional information at www.fmcsa.dot.gov/rules-regulations/administration/fmcsr/fmcsrguide.aspx.

FLEET SAFETY

Even though some company vehicles may fall outside the DOT regulations, other company cars or trucks can be addressed through a fleet-safety program using these rules as a template. The application of safety practices to the operation of any company vehicle is very similar to those of operating a commercial motor vehicle.

Driver Authorization

Minimum standards may be established for individuals who wish to operate a company vehicle. Similar to DOT requirements, these may include parameters such as:

- Is at least 21 years of age
- Is fluent in English to the extent that he or she can communicate with the public and understand road signs
- Can safely operate their motor vehicle
- Can ensure cargo is properly loaded and secured
- Knows methods of securing load with regard to the motor vehicle
- Is physically fit
- Has current driver's license
- Has not had driver's license previously revoked
- Has passed a drug and alcohol screening

Since some of these are not required by law, the human resources (HR) department should provide guidance to ensure requirements are consistent with employment policies and existing regulations. Work will also need to be done to explore unique applications to each organization. For example, improperly securing loads presents liability for transporting them in a small pickup truck. A loose bucket or tool in the back of a truck can become a projectile that could possibly damage another vehicle.

The process of authorizing drivers may also include a driver-training component. Various organizations, including local schools and contractors may provide these services. The goal is to determine key performance objectives and compare course contents. If an appropriate resource cannot be identified, the organization may choose to create an internal program.

Establishing a process to obtain a Motor Vehicle Report (MVR) will provide information regarding the employee's driving history. In consultation with HR, a policy can be established that delineates what offenses will prohibit an employee from operating a company vehicle.

Preventive Maintenance

The failure of a fleet vehicle to function properly can be very costly. Incidents could range from a flat tire causing loss of control to a burned-out brake light causing a rear-end crash. These accident scenarios are often avoidable through a simple preventive maintenance program. Cycles of maintenance can be established to include *periodically*:

- Checking and changing fluids, such as oil, windshield wiper fluid, and transmission fluid
- Measuring brake shoe thickness and replacing as needed
- Measuring tire tread depth and replace

The organization will need to determine if internally trained resources are available or if maintenance will need to be performed by a third party. Pre-trip and post-trip inspections can be integrated into the process to ensure problems are identified proactively. A typical inspection might include a check of the following:

- Brake lights are working
- Headlights are working
- Turn signals are working
- Mirrors are functioning and are not damaged
- Windshield is not damaged
- Tire pressure is sufficient
- Body of the vehicle is in good condition
- Oil level is sufficient

This list can go into greater detail depending upon organizational needs. The program will need to specify what issues found during a pre-trip inspection would cause the vehicle to be taken out of service versus those that will be noted for later repair. For example, a cracked windshield may cause the vehicle to be taken out of service, while a dent in a fender may be identified for later repair.

Employee training must be conducted in an organized fashion to insure all fleet drivers know how to perform inspections. Content may include how to complete and submit inspection forms, what damage will cause a fleet vehicle to be taken out of service, and how to initiate a work order for repairs to be made by the maintenance department or a third party. Checking the oil level may appear to be a simple task, but it cannot be assumed that all drivers know how to do so.

Vehicle Operation

Fleet vehicle operation is saturated with behavioral issues that need to be identified and appropriate policies must be developed and training conducted. Examples might include:

- Distractive behavior (eating, texting, talking on a cell phone)
- Wearing seatbelts
- Observing vehicle weight ratings
- Observing towing restrictions
- Carrying passengers
- Reporting accidents

One distractive behavior receiving a great deal of attention is the use of cell phones while driving. Initially, the problem was driving and talking, but it has now evolved to include using cell phones to read and send text messages, surf the web, and read and send e-mails. Not only has cell phone use been identified as a significant risk in safely operating a vehicle, but in many geographical areas it is now illegal. Distractive behavior can also include a number of other behavioral issues. Eating while driving may require unwrapping a sandwich with both hands, thus altering how the vehicle is steered, which in turn could cause an accident.

A behavior that affects the safety of both the driver and passengers is the use of seatbelts. An individual not wearing a seatbelt involved in a low-speed collision is exposed to significant injuries. The use of seatbelts has evolved from simply being the right thing to do by "buckling up" to jurisdictional laws requiring use to now being a primary offense by which a driver may be pulled over by law enforcement. In light of the personal and legal risk involved, the use of seatbelts should be mandatory among all fleet vehicle drivers.

Weight ratings are provided for each vehicle prescribing loads at which the vehicle can be safely operated. Such ratings need to be identified and communicated to fleet vehicle drivers to avoid vehicles from becoming over-loaded. Related to this issue is towing capacity. A load that is too heavy may also cause the vehicle to not stop properly due to the momentum caused by the weight pushing the vehicle. Knowing and understanding how to manage weight and towing capacities can help to avoid such incidents.

The presence of passengers while operating a fleet vehicle can present risk to an organization. An evaluation should be performed to determine acceptable standards for passengers in a fleet vehicle. For example, it may be determined that it is acceptable for a fleet vehicle to carry any employee of the organization on company business. However, a fleet vehicle permanently assigned to a given employee may have the prohibition of carrying family members as passengers on personal travel.

Risk Management

Risk management and insurance are key considerations with any motor vehicle fleet. Insurance coverage can cover a broad spectrum of risk based on what risk an organization is willing to assume and what it wants to be insured. Understanding uses of fleet vehicles will help determine the risk presented by each. Such use may include vehicles operated:

- Only on private company property
- Locally on public streets
- On an interstate basis
- Only by company employees carrying only company employees
- Only by company employees carrying non-company employees
- Solely for company use as compared to those where some personal use may occur

Each of these situations and others pertinent to your environment will help determine the risk and subsequent insurance coverage needed. For example, it may be determined that a fleet vehicle operated only on company property does not carry the same level of risk as one operated on a public road.

In the event of an accident, fleet vehicle drivers must know accident management and reporting procedures. Accident management may include camera use to immediately take pictures of the accident scene and knowledge of how to communicate with others involved in the accident, law enforcement, and emergency responders. Accident reporting may include immediate notification to a contact person within the organization who can initiate an insurance claim and manage the situation from a larger perspective. The fleet vehicle driver would then need to comply with different aspects of the accident investigation process from law enforcement and organizational perspectives.

CONCLUSION

While Ellen may not have understood her responsibilities regarding fleet safety, the information was available to her. Not only should she have been concerned with hazardous materials, their identification, and the appropriate precautions taken during shipment, she also should have addressed the need to properly license the drivers and assure that each had appropriate training. Drug testing and periodic reviews of fleet drivers should have occurred. In addition, a training program could have been implemented for non-fleet drivers. All of these steps would have helped ensure a safer fleet and compliance with the applicable laws.

QUESTIONS

1. According to DOT, what is a hazardous material and where do you go in the regulations to find information regarding various hazardous materials?
2. There are a few components that are of critical importance when responding to a commercial motor vehicle accident that involves hazardous materials. List each component and describe its importance.
3. What are the different triggers for conducting a drug and alcohol test?
4. What are the different groups of commercial motor vehicles? How do these groups affect the Commercial Driver's License process?
5. What things might you consider when developing a transportation safety program that addresses the qualification of DOT and non-DOT regulated drivers to operate company vehicles?

BIBLIOGRAPHY

Department of Transportation. 2013. *Hazardous Materials and Oil Transportation.* Retrieved from http://ecfr.gpoaccess.gov/cgi/t/text/text-idx?sid=c4fb6180a7fd7554239f970035a610c c&c=ecfr&tpl=/ecfrbrowse/Title49/49cfrv2_02.tpl.

Dunlap, E. S. 2000. *Motor Carrier Safety: A Guide to Regulatory Compliance.* Boca Raton, FL: CRC Press.

Federal Motor Carrier Safety Administration. 2013. *Federal Motor Carrier Safety Regulations.* Retrieved from http://www.fmcsa.dot.gov/rules-regulations/rules-regulations.htm.

Pipeline and Hazardous Materials Safety Administration. 2013. *DOT Emergency Response Guidebook.* Washington, DC: Government Printing Office.

18

Introduction to Extreme Weather

Randell J. Barry, PhD

CHAPTER OBJECTIVES

After completing this chapter, you will be able to

- Describe basic concepts and phenomenon in meteorology
- Describe a variety of extreme weather phenomenon and their associated hazards
- Define severe weather watch, severe weather warning, weather advisory, and hazardous weather outlook
- Identify sources to monitor and forecast extreme weather conditions

CASE STUDIES

On Sunday August 28, 2005, at 10 a.m. CDT Hurricane Katrina, a strong Category 5 storm, sat over the Gulf of Mexico heading north toward the city of New Orleans. Maximum wind speeds within the storm were an impressive and dangerous 150 knots. The National Hurricane Center (NHC) had issued hurricane warnings along the Gulf Coast from Morgan City, Louisiana, to the Alabama-Florida border. Hurricane watches and tropical storm warnings were also in place westward of Morgan City and eastward into the Florida panhandle. NHC's forecast storm track was calling for the center of Katrina's circulation to continue to move northward, make landfall south-southeast of New Orleans, and then continue northward just to the east of the city. The storm intensity forecast was calling for slight weakening before the storm made landfall and then more significant weakening after that. Approximately 20 hours later Katrina did, in fact, make landfall just to the southeast of New Orleans, but as a weaker than expected but still dangerous Category 3 storm. Wind speeds at landfall were estimated to be 110 knots. It then continued to move northward and weaken, passing just to the east of New Orleans. Despite this weakening, however, Katrina would turn out to be the costliest and one of the five deadliest hurricanes in American history. When all was said and done, the total estimated cost of damage from this storm was $133.8 billion. In addition, approximately 1,800 people lost their lives.

Early on the morning of February 5, 2008, a surface frontal boundary extended from the Great Lakes region southwestward toward Texas. This front separated warmer, moister air to the southeast from colder, drier air to the northwest. Surface lows were located along the front near the Texas-Oklahoma border and over southern Ontario. Strong low-level flow out of the south-southwest on the warm side of the front continued to pump warm, moist, and unstable air into eastern Texas. This air continued toward the northeast into the Ohio River Valley. Thunderstorms were occurring in the warm air along the front but, generally, none were of the severe variety. But that would change later in the day. As a result of the meteorological conditions that morning, the Storm Prediction Center (SPC) was, in fact, calling for a slight to moderate chance of severe thunderstorms developing later that day. And that's what would occur. On that early February day, numerous severe thunderstorms would develop over a broad area running from northeast Texas to southern Ohio. Over 500 reports of severe weather were made in that region on that day including 131 tornado reports. This severe weather would result in over a billion dollars in damage, cause 57 fatalities, and come to be known as the "Super Tuesday" outbreak.

INTRODUCTION

The events presented in the case study section of this chapter were just two examples of extreme weather events that led to both extensive economic damage and significant loss of life. Both of these events are included in an ongoing study conducted by the National Climatic Data Center (NOAA/NCDC 2009). This study catalogs extreme weather events since 1980 that have resulted in a billion or more dollars in damage. Ninety such events have been identified (see figure 18-1). And while the study focused only on economic damage, these events often led to a loss of life.

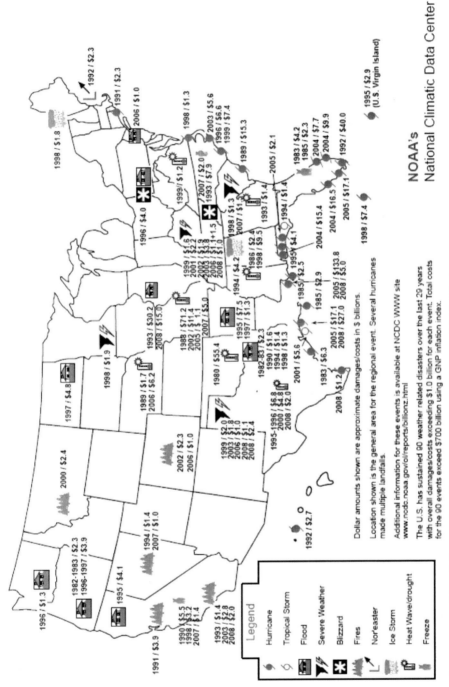

Figure 18-1. Billion dollar weather disasters, 1980–2008 (Source: NOAA/NCDC)

The NCDC study clearly shows the threat the United States faces from extreme weather. It is often the responsibility of the safety professional within an organization to take actions to protect people and resources when extreme weather threatens. It is also true that the safety professional who has a strong understanding of the nature of these weather events and the resources that can be used to monitor and forecast these events is best able to respond to this threat.

In this chapter we will provide the reader with an introduction to the science of meteorology, take a more detailed look at extreme weather phenomenon, and then present some resources that can be used to monitor and forecast extreme weather.

BASIC METEOROLOGY

In order to better understand extreme weather, one must first look at some introductory concepts from *meteorology*—the study of the atmosphere and, in particular, weather. The *atmosphere*, for the most part, can be defined as the mixture of gases that surround the earth and are held in place by the earth's gravitational attraction. This mixture includes gases such as nitrogen and oxygen. But when trying to understand the behavior of the atmosphere, it is a gas that exists in relatively small amounts that turns out to be quite important—that gas is water vapor. The term *humidity* is often used when referring to how much water vapor is in the air.

In addition to the gases of the atmosphere, there also are solid and liquid particles suspended in the atmosphere. These particles include the clouds and precipitation we observe. A *cloud* is simply suspended liquid water drops or ice particles in the atmosphere, while *precipitation* is just cloud particles that have become too large to remain suspended in the air and therefore fall to the ground. Clouds form when water vapor, a gas, condenses forming liquid drops or deposits forming ice crystals. Generally, it is the cooling of air that leads to cloud formation (i.e., the condensation or deposition of water vapor) and this cooling is often the result of upward-moving air. It is for this reason that meteorologists often look for where the air is moving upward to determine where clouds and subsequently precipitation will form.

Another important meteorological variable is atmospheric pressure. The *atmospheric pressure* at any given point in space is determined by the amount, or mass, of atmosphere above that point. Why is that? Since the atmosphere is held in place by earth's gravity, the atmosphere has weight. The weight of the atmosphere, a force, exerted over a given area is what we experience as atmospheric pressure. Therefore, the more atmosphere (i.e., mass) there is above a location, the greater the weight and, in turn, the greater the pressure. And the opposite is also true—less mass, therefore less weight, therefore less pressure. This is why pressure decreases as you move upward through the atmosphere.

But the amount of atmosphere above a given level not only varies as you go up (i.e., in the vertical) but also varies in the horizontal. This, therefore, leads to the situation where high and low pressure areas can also exist at a given level (e.g., mean sea level) as you move in the horizontal. These are the high and low pressure features (i.e., **highs** and **lows**) that meteorologists often refer to. For example, figure 18-2 shows a surface weather map that includes an analysis of atmospheric pressure at mean sea level on June 6, 2008. Lines of constant pressure known as *isobars* have been drawn on this map to determine where the pressure is relatively high or relatively low. High pressure centers have been labeled with an "H" and low pressure centers with an "L." This analysis shows that, in general, there is high pressure in the eastern United States and Canada. A high pressure center is located near New England and the Maritime Provinces and another is located over the Atlantic Ocean off the southeastern United States. A broad area of low pressure is centered along the Canadian border from Minnesota to Washington. A deep, strong low pressure center is seen within that area over Minnesota and the Dakotas. An area of weaker high pressure is also observed over the Southwest.

Why is this horizontal pressure distribution important? It is this change in pressure in the horizontal that, for the most part, determines the motion of the atmosphere (i.e., the **winds**) and ultimately the weather that is experienced. In the Northern Hemisphere we often observe that air circulates in a counter-clockwise motion around low pressure centers. This counter-clockwise motion around a low is also referred to as cyclonic motion and therefore low pressure centers are also known as *cyclones*. We tend to observe the opposite circulation around high pressure centers (i.e., clockwise motion). This is known as anticyclonic motion. For this reason high pressure centers can also be called *anticyclones*.

To illustrate how the pressure pattern determines the winds that one experiences, recall the pattern depicted in figure 18-2. A low pressure center was located near Minnesota and a high pressure center was located off of

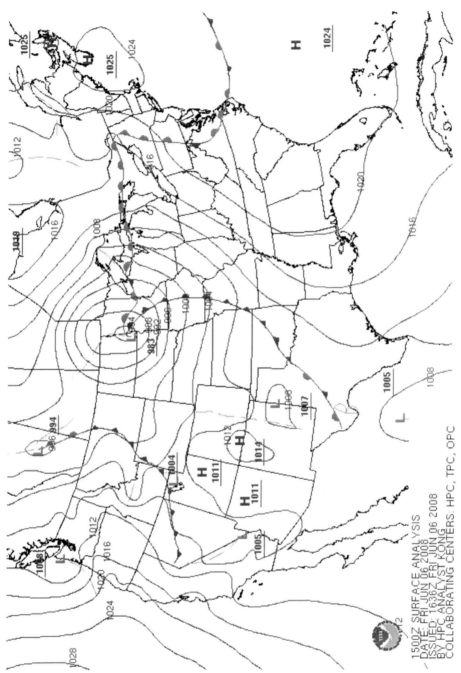

Figure 18-2. Surface weather map on June 6, 2008 (Source: NOAA/HPC)

the Southeast. Although not shown in the figure, the winds in the area of the Midwest between these two features were generally blowing from the southwest toward the northeast at that time. And this, of course, is consistent with the counterclockwise motion typically observed around lows and the clockwise motion around highs.

The direction is not the only aspect of wind that is affected by the horizontal pressure distribution, the speed is also affected. Wind speeds tend to be faster when the pressure changes more rapidly in the horizontal. This would be seen in an area where the isobars on the weather map are packed more closely together. And, of course, the opposite is true. Slower wind speeds are typically observed when the isobars are spread further apart. Once again, referring back to figure 18-2, where the isobars are closer together around the low near Minnesota, the winds were relatively strong. In the area of the high pressure center affecting the Southeast, where the isobars are spread further apart, the winds were weaker.

As mentioned earlier, the horizontal pressure distribution is not only important in determining the wind that is experienced but is also important in determining the overall weather. Generally, low pressure regions bring "poorer" weather (i.e., clouds and precipitation) and, in the extreme, often lead to the storms we experience. High pressure regions generally bring "better" weather (i.e., clear skies). The reason for this is that the air typically tends to move upward in areas of low pressure while air tends to sink with highs. Recalling that clouds form through cooling air and this cooling is brought about by air moving upward, it is easy to see the association of clouds with lows. And furthermore, the upward motion that leads to clouds also helps to bring about the precipitation that is often observed with low pressure centers (i.e., cyclones). The sinking air in highs (i.e., anticyclones) strongly hinders the formation of clouds and therefore limits precipitation in those areas.

Another important concept in meteorology is that of the air mass. An *air mass* is a relatively large body of air that has similar characteristics in terms of temperature and humidity. For example, there are air masses that are cold and dry, warm and dry, cold and moist, and so forth. These air masses form when relatively large anticyclones stagnate over certain regions of the globe. Therefore, at the center of these air masses there is high pressure. The regions where they form include the poles, which produce the colder air masses, and the subtropics, which produce the warmer air masses. The moisture content or humidity of the air mass is determined by whether the anticyclone has persisted over land (dry) or water (moist). And while these air masses develop in the polar and subtropical regions, they do move, and therefore affect mid-latitude locations.

Another meteorological concept related to air masses is that of the front. And like air masses, this concept is quite important. A ***front*** can be defined as a transition zone between differing air masses and, in particular, between air masses of differing temperature. Examples of differing air masses and the resulting frontal boundaries can be seen on the map shown in figure 18-2. A cool high or air mass is located over the Northeast, a warm, moist air mass is affecting the Southeast, and a cooler, drier air mass has pushed southward behind the low in the upper-middle part of the United States. The frontal boundaries between the warm air mass and the two cooler air masses are represented as the line on the map with either the half circles or triangles. For example, the front running from the low near Minnesota southward toward Texas represents the boundary between cooler, drier air to the northwest and warmer, moister air to the southeast. In general, fronts are important because, like low pressure, they tend to bring clouds and precipitation. In fact, there is a strong relationship between fronts and low pressure centers or cyclones.

The pressure along a front tends to be lower as, once again, the differing air masses that come up against each other to form a front have high pressure at their centers. Furthermore, these frontal boundaries are a breeding ground for the low pressure centers or cyclones that can bring our more significant or stormy weather. In particular when a low forms and strengthens along a front it is termed an ***extra-tropical*** or ***mid-latitude cyclone***. These lows have a distinct center and clear frontal boundaries. Referring back to figure 18-2, the pressure feature near Minnesota is a nice example of an extra-tropical cyclone. A deep low pressure center is present and fronts extend southward and eastward from this system. Once again, these extra-tropical cyclones and their associated fronts are often responsible for the significant and stormy weather we experience in the mid-latitudes such as blizzards, Nor'easters, and ice storms.

Another atmospheric phenomenon that brings stormy weather is the thunderstorm. A ***thunderstorm*** is a convective type cloud that produces lightning and thunder. ***Lightning*** is a flow of electricity that occurs in the atmosphere due to the strong vertical motions and the subsequent charge separation that occurs within the storm. Other hazards that can occur with a thunderstorm include heavy precipitation, hail, strong surface wind gusts, and tornadoes. Individual thunderstorms, compared to extra-tropical cyclones, are relatively small. An individual thunderstorm is on the order on 10 km (6 miles) across while extra-tropical cyclones affect areas that are on the order of 1,000 km (600 miles) across. But note that when the conditions are right, numerous thunderstorms can form over a relatively large area at any given time.

All convective type clouds, including thunderstorms, form in an atmosphere that is relatively unstable. In an atmosphere that is unstable, a region of air that is lifted tends to become warmer than the air around it and therefore continues to easily rise on its own. This process (i.e., warm air rising) is known as convection, and the clouds that form due to this process are known as convective clouds. Unstable atmospheres usually occur when the air is warm and moist near the surface but cooler and somewhat drier higher up in the atmosphere. The more unstable the atmosphere, the stronger the vertical motions that result from the convection and the more likely that thunderstorms will form.

Convection in the atmosphere can also lead to the formation of another type of cyclone—a *tropical cyclone*. Recall that extra-tropical cyclones form in the mid-latitudes along frontal boundaries. In contrast, tropical cyclones form over the warm tropical oceans in regions of preexisting low pressure, well away from any mid-latitude frontal boundary and associated temperature contrast. In the case of tropical cyclone formation, the warm moist air over the tropical oceans leads to an unstable atmosphere that supports the formation of numerous convective clouds. When these convective clouds cluster over a certain area, the pressure decreases in that area forming a low pressure center and the cyclonic circulation that goes with it. Tropical cyclones may strengthen becoming potentially dangerous tropical storms or hurricanes.

In the opening paragraph of this section we defined meteorology as "the study of the atmosphere and, in particular, weather." We then went on to formally define the term "atmosphere" and discussed several terms and concepts from meteorology. But we never formally defined the term weather, although we alluded to it several times.

Weather, simply put, is the condition of the atmosphere at a particular time and location; its condition as described by the various meteorological variables we have discussed. Once again, these variables include things such as the temperature, humidity, atmospheric pressure, wind, cloudiness, and the presence of precipitation. This, of course, would also include extreme weather elements such as the high temperatures associated with a heat wave, the damaging winds associated with a hurricane, and significant precipitation associated with a flooding event. In the next section we will look in greater detail at the more extreme weather events and phenomena like the ones just mentioned.

EXTREME WEATHER

While there is no formal definition for "extreme weather," we have been and will continue to use this term for any weather condition or meteorological phenomenon that has the potential to produce some type of economic impact or that may cause injury or loss of life. Therefore, events like Hurricane Katrina in 2005 and the "Super Tuesday" severe thunderstorm outbreak in 2008 were clearly examples of extreme weather. In this section, we will continue to increase our knowledge of meteorological concepts. But now we will look in greater detail at extreme weather itself.

Tropical Storms/Hurricanes

In the previous section we stated that "tropical cyclones may strengthen, becoming dangerous tropical storms or hurricanes." Specifically, a ***tropical storm*** is a tropical cyclone that has wind speeds that are at least 34 knots (39 m.p.h.) within its distinct cyclonic circulation. Likewise, a ***hurricane*** is a tropical cyclone that has wind speeds of 64 knots (73 m.p.h.) or greater. Once a tropical cyclone reaches hurricane strength it is further classified according to the **Saffir-Simpson Hurricane Intensity Scale** shown in table 18-1. Note that Category 3 storms and greater are considered ***"major" hurricanes***.

When a hurricane becomes well organized, certain distinct structures are often observed within the storm. These include the storm's eye, eyewall, and spiral rain bands. The ***eye*** of the storm is the cloud-free, relatively calm area at the center of the storm's circulation. Figure 18-3(a) is a visible satellite

Table 18-1. Saffir-Simpson Hurricane Intensity Scale

Saffir-Simpson Category	Wind Speeds
1	64–82 knots (74–95 m.p.h.)
2	83–95 knots (96–110 m.p.h.)
3	96–113 knots (111–130 m.p.h.)
4	114–135 knots (130–155 m.p.h.)
5	greater than 135 knots (greater than 155 m.p.h.)

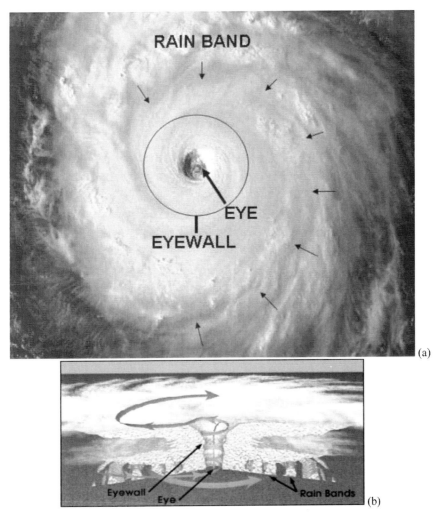

Figure 18-3. (a) Satellite image of Hurricane Katrina and (b) hurricane illustration depicting eye, eyewall, and spiral rain band structures (Source: NOAA/NWS)

image showing Hurricane Katrina as it was heading toward the Gulf Coast. The eye is clearly visible in this image. Note that the winds are rapidly rotating in counter-clockwise sense around the eye but are relatively weak within the eye itself.

The *eyewall* is the ring of "deep" rainshowers and thunderstorms that surround the eye. Figure 18-3(b) shows an illustration depicting a vertical slice through a hurricane. The "deep" clouds surrounding the eye associated with the eyewall and the eye itself can be seen in this figure. This denser ring of clouds that makes up the eyewall is also evident in the satellite image of Katrina (figure 18-3[a]). It is important to note that the most intense precipitation and the strongest winds of the storm typically occur within the eyewall.

Lastly, the **spiral rain bands** are organized lines of clouds, rainshowers, and thunderstorms that spiral inward toward the center of the storm. It is within the rain bands that the heaviest precipitation occurs outside of the eyewall. A single rain band is highlighted by arrows in the image of Katrina (figure 18-3[a]) and rain band structures are also evident in figure 18-3(b).

It has been pointed out that the strongest winds of a hurricane typically occur toward the center of the storm within the eyewall. Figure 18-4 presents a detailed analysis of wind speed near the center of Katrina's circulation at the time the storm was moving toward New Orleans. Note once again that the wind speeds are fastest toward the center within the eyewall, decreasing as one moves outward. Also note, however, that these wind speeds are not distributed evenly (i.e., symmetric) about the center of the storm, especially within the eyewall. The fastest winds are located on the east side of the storm only within an area exceeding 130 knots (150 m.p.h). When taking the perspective of the motion of the center of the storm (i.e., toward the north in this case), this area of intense wind would be to the right of the center of circulation. In general, this is where the fastest wind speeds are found in most hurricanes relative to the overall motion of their center.

The high winds contained within the cyclonic circulation of the storm are not the only hazard associated with a hurricane. Other hazards include the storm surge, flooding, and tornadoes. The **storm surge** is a "wall" of water that is pushed inland by the winds of the storm as the storm makes landfall. This depth of water can reach 15 feet or greater. While locations directly on the shore are most vulnerable, this water can be pushed miles inland. The depth and inland extent of the storm surge are affected by two factors—the strength of the storm and the slope of the shoreline. Generally speaking, the stronger the storm and the shallower the slope of the shoreline, the greater the depth and inland extent of the surge. Figure 18-5 depicts the storm surge phenomenon. This figure is drawn such that the storm is moving out of the page. The storm surge therefore typically occurs in the area of the storm where the winds are blowing onshore and where the wind speeds tend to be the fastest (i.e., to the right of the motion in the vicinity of the eyewall).

Tropical storms, hurricanes, and their remnants often produce very intense rainfall; therefore, **flooding** can also be an issue. This is true not only for the region affected by the storm at landfall but also for regions well inland

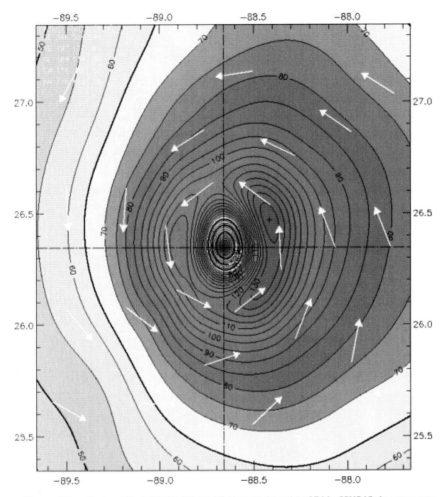

Observed Max. Surface Wind: 138 kts, 16 nm NE of center based on 1744 z SFMR43 sfc measurem
Analyzed Max. Wind: 138 kts, 18 nm NE of center

Figure 18-4. Analysis of surface winds associated with Hurricane Katrina (Source: NOAA/AOML HRD)

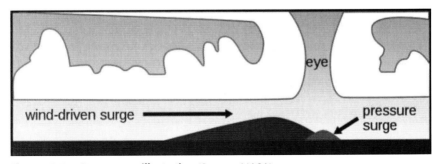

Figure 18-5. Storm surge illustration (Source: NASA)

long after the storm's winds have weakened. The two key factors affecting the amount of flooding is the rainfall rate (i.e., how much falls in a given time period) and the speed at which the storm is moving. These storms have very large rainfall rates and therefore a significant amount of water can accumulate in an area in a relatively short period of time. Furthermore, if a storm is moving slowly across a region, these relatively high rates will be experienced for a longer period of time and therefore increase the depth of water locally. A storm that stalls over a region can be especially problematic.

A *tornado* is a rapidly rotating column of air that descends from an individual thunderstorm. While the majority of tornadoes occur with thunderstorms not associated with a hurricane (i.e., severe thunderstorms), they can occur within the thunderstorms and convective clouds embedded in hurricanes. Speeds in a tornado can be over 175 knots (200 m.p.h.). But typically the tornadoes that occur within a hurricane are of the weaker variety, although still highly dangerous and destructive. These tornadoes, if they form, develop over land within the spiral rain bands. And it is typically within the spiral rain bands located within the right-front quadrant of the storm's circulation. Once again, this "right-front quadrant" is relative to the motion of the storm as it is moving on-shore and inland.

Severe Thunderstorms

Recall that earlier we defined a thunderstorm as "a convective type cloud that produces lightning and thunder." *Severe thunderstorms* are thunderstorms that contain frequent lightning, large hail, downbursts, and/or tornadoes. These hazards are, to a large extent, the result of the strong vertical motions that occur within the storm. As indicated earlier in the chapter, it is the strong upward motions (i.e., updrafts) that produce charge separation that subsequently leads to a flow of electricity in the atmosphere (i.e., lightning). Furthermore, this strong vertical motion forms in the most unstable of atmospheres—an atmosphere that has warm moist near the surface and cooler, drier air aloft.

Hail, relatively large particles of ice that fall to the ground from thunderstorms and like the thunderstorm itself, is also the result of the large upward motion in the storm. Hail ranges in size from pea size to softball size or greater. Hail forms by smaller ice particles staying suspended in regions of the atmosphere where additional liquid water can freeze on their surface. This allows the particles to grow bigger than they would be otherwise. This can only occur in areas where the air is rapidly moving upward. The strong updrafts do not allow the ice particles to fall to the ground; instead, they stay suspended and grow. After growing, the larger hail is then blown out of the region of the updraft and falls to the ground.

Downbursts are strong surface wind gusts that occur in response to sinking air within the thunderstorm (i.e., downdrafts). This air descends to the surface beneath the storm and then accelerates outward. Once again, this phenomenon is related to the strong vertical motions that can form within the unstable environment of the storm. In this case, a portion of the thunderstorm becomes colder than the air around it and rapidly sinks to the surface. This mixes faster wind speeds that are higher in the atmosphere to the surface and affects pressure differences near the surface that can accelerate the wind. These straight-line winds can be quite fast, with wind speeds greater than 90 knots (104 m.p.h.), and therefore are quite destructive.

In an earlier section we defined a tornado as "a rapidly rotating column of air that descends from an individual thunderstorm." We pointed out that tornadoes can contain wind speeds that exceed 175 knots (200 m.p.h.). There is, however, a range of wind speeds that can occur with a tornado. This range has been broken down into the specific categories that makeup the Enhanced Fujita (EF) Scale (see table 18-2). Ratings go from EF-0 to EF-5. EF-2 tornadoes and greater are sometimes labeled "significant" but all tornadoes, to some degree, are dangerous and can be destructive.

Tornadoes tend to form in thunderstorms with the strongest updrafts and that possess some rotation themselves—rotation that is not as narrow or rapid as the tornado, but some type of broad, precursor rotation is likely needed. While the exact tornado formation is not completely understood, it is believed that the strong vertical motions of the thunderstorms stretch in the vertical and contract in the horizontal slower rotating columns of air causing those columns to rotate faster. These faster rotating columns of air then ultimately lead to a tornado. This process is somewhat analogous to the

Table 18-2. Enhanced Fujita (EF) Scale

EF Scale	Wind Speeds
0	65–85 m.p.h.
1	86–110 m.p.h.
2	111–135 m.p.h.
3	136–165 m.p.h.
4	166–200 m.p.h.
5	greater than 200 m.p.h.

ice skater who slowly spins as they hold their arms out but then spins more quickly when pulling their arms in.

Severe thunderstorms, and the hazardous phenomenon that goes with them, typically develop near the boundary of contrasting air masses. In other words, they generally form along a front separating warm, moist, and unstable air toward the south from cold, dry, and stable air to the north. And it is to the south of the front within the warm, moist, and unstable air that the severe thunderstorms form and organize themselves.

The central portion of the United States, between the Rocky and Appalachian mountains, is a region where cold, dry air masses from the north often collide with warm, moist air masses to the south, especially in the spring of the year. Therefore, it is also a region where severe thunderstorms tend to form. This is seen in the climatology of hail with diameters greater than two inches (figure 18-6), the downburst climatology (figure 18-7), and the tornado climatology (figure 18-8). Note the tendency for these severe phenomena to maximize in the upper Midwest, central plains, and southeastern portions of the United States—regions of frequent frontal activity.

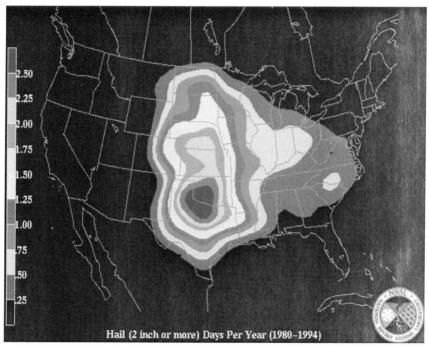

Hail (2 inch or more) Days Per Year (1980–1994)

Figure 18-6. Climatology of hail with diameters of two inches or greater (Source: NOAA/NSSL)

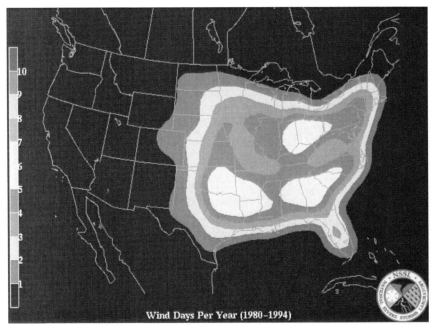

Figure 18-7.   Downburst climatology (Source: NOAA/NSSL)

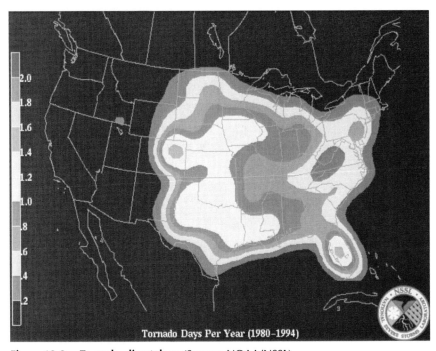

Figure 18-8.   Tornado climatology (Source: NOAA/NSSL)

Winter Storms

Extreme winter weather can have both economic and human impacts. Business can be brought to a halt, transportation is almost always affected, structures can be significantly damaged, and individuals can be harmed by the hazardous conditions that occur within winter storms. In this section we will examine the blizzard phenomenon, storm systems known as Nor'easters, major ice storms, and lake-effect snow. The first three of these generally occur with larger scale extra-tropical cyclones and/or active frontal boundaries, and therefore typically affect a broader area. The fourth phenomenon, lake-effect snow, tends to have a more localized affect but can be equally as dangerous.

A ***blizzard*** or blizzard conditions occur when low temperatures (e.g., less than 20°F) and high winds (i.e., greater than 30 knots [35 m.p.h.]) combine to produce blowing and drifting snow. Snow may or may not be falling from the sky at the time but, at minimum, the high winds must be blowing snow already present on the ground into the air, causing a significant reduction in visibility (i.e., the distance one can see). These conditions are extremely hazardous. They can make travel next to impossible, create conditions in which the onset of hypothermia and frostbite occurs within minutes, and cause significant property damage.

Blizzard conditions typically develop when cold air combines with a strong extra-tropical cyclone. In the United States and Canada, wintertime mid-latitude cyclones often develop near the eastern portion of the Rocky Mountains and then move eastward. Two regions are particularly favorable for development—near Colorado and in the vicinity of Alberta. Storms that originate in these areas are sometime referred to as "Colorado Lows" and "Alberta Clippers," respectively. If these storms significantly "deepen" (i.e., the pressure at the center of the storm drops) as they push east, strong pressure horizontal differences are generated. This, in turn, produces strong winds. With cold temperatures and snow, blizzard conditions can and will result.

A wintertime storm system that can have a significant impact on the eastern portion of the Unites States, especially the East Coast, is the Nor'easter. ***Nor'easters*** are extra-tropical cyclones that typically first form along the Gulf Coast or along the East Coast near the Carolinas or mid-Atlantic states. They then deepen as they move northeastward along the coast. These storms get their name from the strong, steady winds that blow out of

for an extended period of time can also lead to flooding. This is especially true when these features draw warm, moist air into the affected area. Furthermore, if this warm, moist air is also highly unstable, slow-moving rain showers and thunderstorms with higher than average rainfall rates will develop, rapidly leading to a flooding situation.

The amount of rainfall in a given area is not the only factor that determines whether a region will flood. Things such as local topography, the nature of the soil and degree to which it is already saturated with water, snow melt, and ice jam formations in rivers all affect flooding potential.

Heat Waves

Heat waves can create an extremely dangerous situation, especially for individuals who have to work outdoors. *Heat waves* occur when the already warm summertime temperatures and higher humidity values increase and remain significantly above their typical values for an extended period of time. A weather pattern that leads to heat waves in the central and eastern portion of the United States (i.e., east of the Rocky Mountains) involves a hot, humid air mass developing and stagnating over this region. The center of the surface anticyclone associated with this air mass sits over or just off the each coast putting the central and eastern region in an area of warm, moist flow out of the south. And although very warm and moist near the surface, this air mass tends to be relatively stable so that little relief in the form of rain and thundershowers can be expected across the area.

Cold Waves

Wintertime *cold waves* bring temperatures that are significantly colder than average. Like heat waves, the temperature experienced in cold waves can also make being in the outdoors quite dangerous. The sub-zero degree Fahrenheit temperatures that occur with these cold air outbreaks can quickly lead to hypothermia and frostbite. Furthermore, although this cold air gets warmed somewhat as it pushes south, it can still produce extended periods of sub-freezing temperatures that can have a significant effect on, for example, agriculture in the southern states (e.g., the citrus industry in Florida).

The strong cold waves that occur in the United States typically affect areas in the eastern Rocky Mountains and eastward. They occur when a cold, dry arctic air mass moves relatively rapidly, with a minimum of modification, from the polar region toward the south over North America. The center of this air mass, and the surface anticyclone that goes with it, ultimately sets up over the central United States. This puts the eastern part of the country in relatively strong and cold flow out of the north. And this flow can persist for an extended period of time. In addition to high in the center part of the country, a rapidly developing cyclone off the East Coast (e.g., a Nor'easter) may also be present helping to produce this cold northerly flow.

Fire Weather

The weather, particularly over an extended period of time, can have a significant effect on a region's susceptibility to forest and grass fires. Weather conditions that favor fire formation, sometime referred to *fire weather*, involve a combination of low humidity and higher temperatures, and typically stronger winds. An area where these conditions can occur is on the periphery (i.e., away from the center) of a warm, dry anticyclone. For example, a pattern favorable for forest fires in southern California has a warm, dry high located to the north of the region creating relatively fast winds out of the east across the area. These winds are hot and dry, not only due to the nature of the air mass from which they originate, but also due to the down slope condition produced by easterly winds in that area. When air sinks, as it does in this case, it warms and dries. This phenomenon is locally known as the Santa Ana winds.

RESOURCES TO MONITOR AND FORECAST EXTREME WEATHER

In this section we will discuss a sampling of resources for monitoring and obtaining forecast information on extreme weather. The focus will be on resources and products provided directly by the National Weather Service. Resources provided by commercial weather companies and television/radio sources, although also quite useful, will not be discussed.

National Weather Service

The mission of the United States' National Weather Service (NWS) is to provide "weather, hydrologic, and climate forecasts and warnings for the United States, its territories, adjacent waters and ocean areas, for the protection of life and property and the enhancement of the national economy. NWS data and products form a national information database and infrastructure which can be used by other governmental agencies, the private sector, the public, and the global community" (NOAA/NWS 2007). The NWS is made up of 122 local weather forecast offices (WFOs) whose responsibility it is to produce forecast products and warnings for their designated areas. In addition to the WFOs, there are national centers providing products that support both the WFOs and the public, in general. Two of these centers—the Storm Prediction Center (SPC) and the National Hurricane Center (NHC)—are particularly important with regard to extreme weather. SPC is responsible for forecasting severe thunderstorm development and fire weather conditions. NHC is responsible for monitoring and forecasting tropical cyclones. Both of these organizations will be discussed in more detail later.

Terminology

When monitoring and responding to extreme weather, there are certain types of NWS products that are particularly important and useful. Having immediate access to these products, in some cases, could be the difference between life and death for you and the people around you. These products are severe weather watches, severe weather warnings, weather advisories, and hazardous weather outlooks.

A *severe weather watch* "is used when the risk of a hazardous weather or hydrologic event has increased significantly, but its occurrence, location, and/or timing is still uncertain. It is intended to provide enough lead time so that those who need to set their plans in motion can do so" (NOAA/NWS 2009).

On the other hand, a *severe weather warning* "is issued when a hazardous weather or hydrologic event is occurring, is imminent, or has a very high probability of occurring. A warning is used for conditions posing a threat to life or property" (NOAA/NWS 2009).

Weather advisories highlight "special weather conditions that are less serious than a warning. They are for events that may cause significant inconvenience, and if caution is not exercised, it could lead to situations that may threaten life and/or property" (NOAA/NWS 2009).

And lastly, *hazardous weather outlooks* "indicate that a hazardous weather or hydrologic event may develop. It is intended to provide information to those who need considerable lead time to prepare for the event" (NOAA/NWS 2009).

Weather Radio

A simple and relatively inexpensive way to get NWS forecast products including watches, warnings, and advisories is through the NOAA weather radio. The NOAA weather radio is a nationwide network of radio transmitters that continuously broadcast weather information. The radios are designed not only to allow you to get general forecast information (i.e., what will the weather be today?), but will also sound an alarm followed by relevant information when, for example, severe weather watches and warnings are issued for your area. This, therefore, allows those within hearing distance of the radio to take immediate and appropriate action when extreme weather occurs or is imminent.

Internet Resources

In general, the Internet is an excellent means to obtain weather information with multiple sources providing products (e.g., the NWS, commercial weather providers, and television stations). The Internet can be especially useful for monitoring extreme weather conditions in real time. Note, however, Internet sites do not typically have an alarm feature like the weather radio does.

The suite of products provided by the NWS can be obtained on the web by first going to www.weather.gov—the NWS's homepage. A sample from July 22, 2013, for this site appears in figure 18-9. This webpage displays a map of the United States with current watches, warnings, advisories, and outlooks indicated by color coding in the affected areas. Note, for example, that there is a severe thunderstorm watch running from eastern Oklahoma into Arkansas and a flash flood warning has been issued to the east of that area in northern Mississippi.

Detailed information on local watches, warnings, advisories, and outlooks, as well as other forecast information, can be obtained by clicking on the location of interest on the map. Doing this will take you to the WFO responsible for that location where you can obtain, for example, specific text products for each watch, warning, advisory, or outlook that is in effect and, on a more general level, an extended weather forecast.

Links to national forecast centers such as SPC and NHC are also located on this page in the dropdown menu labeled "Forecast." Clicking on the link labeled "Severe Weather" will take you to SPC's website. A sample of SPC's homepage, also from July 22, 2013, appears in figure 18-10. This

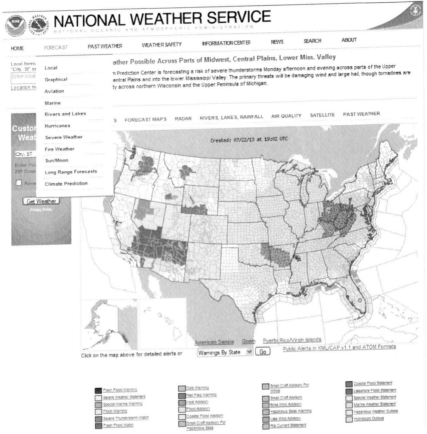

Figure 18-9. National Weather Service website (www.weather.gov) (Source: NOAA/NWS)

webpage displays SPC's current convective outlook for the United States. This product indicates where conditions are such that severe thunderstorms may develop and the relative probability for development. Note that the NWS defines a thunderstorm as being severe when it produces hail one inch or greater, surface wind gusts 50 knot of greater, and/or a tornado. Also displayed are current severe thunderstorm watches (note, for example, the severe thunderstorm watch in Oklahoma and Arkansas mentioned earlier) and real-time weather radar indicating thunderstorm locations.

NHC's homepage can be accessed by clicking on the link labeled "Hurricanes" in the aforementioned dropdown menu. You will find products identifying areas of potential tropical cyclone development and, when necessary, forecast information on tropical storms and hurricanes. As an example,

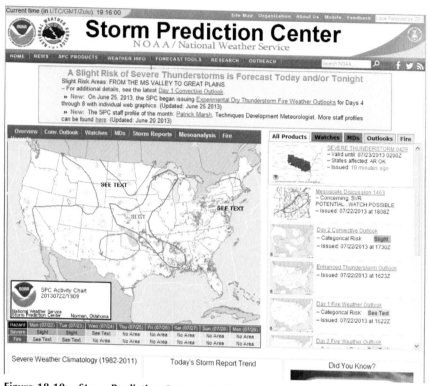

Figure 18-10.   Storm Prediction Center website (www.spc.noaa.gov) (Source: NOAA/ SPC)

figure 18-11 shows the five-day forecast track for Hurricane Katrina issued on August 28 at 10 a.m. CDT in 2005. This track is indicated by the dark heavy line connecting the current position with forecasted future positions. Also displayed is a representation of forecast uncertainty based on the historical forecast error made in previous tropical storm and hurricane track forecasts. This uncertainty is represented by the shaded cone, with the width of the cone indicating how much, on average, the NHC forecast position differed from a storm's actual position at a particular forecast time. Lastly, the location of any tropical storm and/or hurricane watches or warnings currently in effect are indicated by highlighting the affected coastline.

CONCLUSION

Extreme weather can have significant economic impacts and deadly conse-quences. The two case studies at the beginning of this chapter, Hurricane

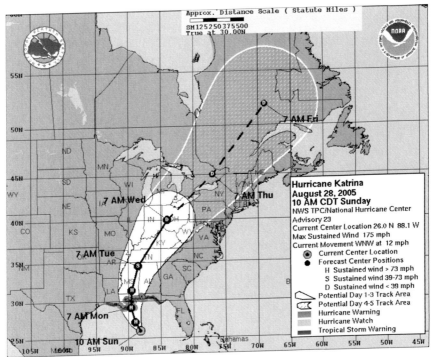

Figure 18-11. Sample hurricane forecast map (Source: NOAA/NHC)

Katrina and the "Super Tuesday" severe thunderstorm outbreak, clearly illustrate that. Both events produced significant damage and numerous fatalities, especially in the case of Katrina.

The purpose this chapter was to provide a variety of information on extreme weather so as to minimize the effects of such events. Once again, the safety professional who has a strong understanding of the nature of this threat, combined with knowledge of resources to monitor and forecast these events, is best able to respond in a manner that protects property and saves lives.

QUESTIONS

1. Describe the typical weather experienced with surface high and low pressure systems. For each system, explain why one usually experiences that type of weather.

2. Compare and contrast extra-tropical cyclones to tropical cyclones. Consider their basic structure and how they form.

3. Describe the hazards associated with a hurricane. Where is the most dangerous region of a hurricane and why?

4. Describe the general weather pattern that leads to severe thunderstorm development and where severe thunderstorms tend to occur in the United States.

5. Describe the differences between a severe weather watch, a severe weather warning, and a weather advisory.

6. Discuss different sources of weather information.

REFERENCES

Ahrens, C. D. 2009. *Meteorology Today: An Introduction to Weather, Climate, and the Environment.* Belmont, CA: Brooks/Cole.

Edwards, R. 2009. *The Online Tornado FAQ.* Posted at http://www.spc.noaa.gov/faq/tornado/.

Knabb, R. D., Rhome, J. R., and Brown, D. P. 2006. *Tropical Cyclone Report Hurricane Katrina 23–30 August 2005.* Posted at http://www.nhc.noaa.gov/2005atlan.shtml.

NOAA/AOML HRD. 2007. *Frequently Asked Questions.* Posted at http://www.aoml.noaa.gov/hrd/tcfaq/tcfaqHED.html.

NOAA/NCDC. 2009. *Billion Dollar U.S. Weather Disasters.* Posted at http://www.ncdc.noaa.gov/oa/reports/billionz.html.

NOAA/NWS. 2009. *National Weather Service Glossary.* Posted at http://www.nws.noaa.gov/glossary/.

NOAA/NWS. 2009. *The Rapid City Flood of 1972.* Posted at http://www.crh.noaa.gov/unr/?n=1972_Flood.

NOAA/NWS. 2008. *NOAA Weather Radio All Hazards.* Posted at http://www.nws.noaa.gov/nwr/.

NOAA/NWS. 2008. *Reference Sheet for Warning/Advisory Thresholds.* Posted at http://www.erh.noaa.gov/box/warningcriteria.shtml.

NOAA/NWS. 2007. *NWS Headquarters Organization and Structure.* Posted at http://www.weather.gov/hdqrtr.php.

NOAA/SPC. 2010. *Frequently Asked Questions.* Posted at http://www.spc.noaa.gov/faq/.

Rauber, R. M., Walsh, J. E., and Charlevoix, D. J. 2005. *Severe and Hazardous Weather An Introduction to High Impact Meteorology.* Dubuque, IA: Kendall/Hunt.

19

Required Written Programs

CHAPTER OBJECTIVES

After completing this chapter, you will be able to

- Understand the content of the major written programs for general industry
- Explain when each of the major written programs is required
- Know where to find an explanation of each of the major written programs

CASE STUDY

Jimmy Sanders manages a small motor machining shop. Much of his work involves reworking high-horsepower engines for competitive applications. Like many small business owners, Jimmy started in his garage, but he now owns a small building west of town with seven employees. Jimmy is aware he does not have to maintain an OSHA log, but when an employee complains he does not know what to do in the event of exposure to certain chemicals, Jimmy soon learns he doesn't know about the HAZCOM standard. Although the law exempts him from maintaining certain accident records, he finds he is not exempt from maintaining other required programs. After exploring the Hazard Communication requirements, Jimmy is surprised to learn he is required by OSHA to maintain a HAZCOM program with only one employee. He soon comes to believe it is in everyone's best interests to maintain written programs regardless of legal requirements. Jimmy carefully reviews regulations and learns what he must do to have a top-notch program wherever needed.

Most of the OSHA-required written programs are referenced in other portions of this text. As a review or summary, most are presented here in a format designed to help guide the reader in knowing what is required and where. Some companies falling under CFR 1926 or other safety provisions may have additional required programs. Only summaries of the requirements are provided here. Individual regulations should be accessed to learn more.

SAFETY AND HEALTH PROGRAM

Although OSHA does not require a written safety and health program, they strongly encourage it. In fact, they demonstrate a model program that emphasizes management commitment and employee involvement.

Management commitment provides the motivating force and the resources for organizing and controlling activities within an organization. In an effective program, management regards workers' safety and health as a fundamental value of the organization and applies its commitment to safety and health protection with as much vigor as to other organizational purposes.

- *The first parts of any comprehensive safety and health program should be a written policy from top management and visible support for the complete safety and health program.* The written policy is the cornerstone of the program. Responsibility is assigned and communicated for all aspects of the program and managers are held accountable for safety performance. Management makes itself visible and supportive in each stage of the program from the time of the first announcement and enlistment of employee support to the unveiling and ongoing operation of the program. Management sets the example in terms of obeying all safety rules and regulations. It also has a representative present at safety meetings, awards ceremonies, and selected safety activities. Management puts the teeth and backbone into any program, so managers are compelled to be present during all phases of its development and operation.
- *Employee involvement provides the means through which workers develop and/or express their own commitment to safety and health pro-*

tection, for themselves and for their fellow workers. They are included in the development and the implementation of the safety and health program and share responsibility for it. Their responsibilities are outlined in writing. Employee involvement includes employee input in the writing of the initial program. Employees have ideas and suggestions on how the program is shaped and on its components. Since they are closer to the job and the hazards of the job than management, they can provide insight that may never occur to someone who hasn't actually performed the job. By including employees in the process, management is more likely to have "buy-in" and active participation by them.

- *The program should also include a section on worksite analysis.* Worksite analysis involves a variety of worksite examinations, to identify not only existing hazards but also conditions and operations in which changes might occur to create hazards. Lack of awareness of a hazard, stemming from failure to examine the worksite, is a sign that safety and health policies and/or practices are ineffective. Effective management actively analyzes the work and worksite, to anticipate and prevent harmful occurrences. This requires periodic examination of the workplace through inspections, audits, and the application of other safety tools such as the job safety analysis.

- *Next, hazard prevention and controls are triggered by a determination that a hazard or potential hazard exists.* Where feasible, hazards are prevented by effective design of the jobsite or job. Where it is not feasible to eliminate them, they are controlled to prevent unsafe and unhealthful exposure. Once a hazard or potential hazard is recognized, elimination or controls are accomplished in a timely manner. Techniques for identifying and controlling hazards should be discussed in the program.

- *Lastly, training addresses the safety and health responsibilities of all personnel concerned with the site, whether salaried or hourly.* Incorporate this training into sessions concerning performance requirements and job practices. The intricacy of training depends on the size and complexity of the worksite and the nature of the hazards and potential hazards at the site.

A complete description and outline for this program are available online at http://www.osha.gov/pls/oshaweb/owadisp.show_document?p_table=FEDERAL_REGISTER&p_id=12909&p_text_version=FALSE.

HAZARD COMMUNICATION PROGRAM

The most common required program is the Hazard Communication Program (CFR 1910.1200). Nearly every company has hazardous materials on their premises and, therefore, needs a HAZCOM program. The question arises as to what is a hazardous material. If you can't safely wash your eyes with it, consider it hazardous. The standard is obviously more technical, giving the employer exceptions. The most common exception is that if the material used is only maintained in household quantities, it does not have to meet all the requirements of the standard. For most hazardous materials, which usually come in the form of chemicals, a few basic requirements exist. The Hazard Communication program is more thoroughly discussed in chapter 15 dealing with hazardous materials.

EMERGENCY ACTION PLAN

Employers should always develop a written emergency action plan as outlined in CFR 1910.38. Historical records of the company, the area, and similar companies should be reviewed to determine likely emergencies the employer might face. Address each emergency by developing standard operating procedures for employees in the event of fire, weather emergency, terrorist attack, civil insurrection, or other likely adverse situations. *Plans include at least the following elements:*

- Whom the employee notifies in the event of an emergency—that individual (typically a receptionist or dispatcher) will then notify local emergency response authorities, appropriate management, and in-plant response teams; every individual notified has standard operating procedures for each anticipated emergency situation
- Emergency escape routes and procedures for different types of emergencies for all employees—for example, with a tornado warning, employees may be instructed to assemble in an interior hallway, whereas a fire would require escape to a certain location via a particular route
- Standard operating procedures for employees who remain in the facility for critical operations
- Standard operating procedures for reporting fires and other emergencies
- Standard operating procedures to account for all employees following evacuation or emergency notification

- Standard operating procedures for in-house emergency responders
- List of managers or plant personnel to be contacted for additional information in an emergency
- The alarm system and how employees will be notified for different types of emergencies
- Training

Employers with fewer than ten employees need not maintain a written plan, but they must communicate to each employee responsibilities in the event of an emergency.

FIRE PREVENTION PLAN

Closely related to the emergency evacuation plan, but under a different section and for a different purpose, is the fire prevention plan found in 1910.38. This is simply a written plan of what fire prevention procedures your company follows and what you will do to protect your workplace from fire. *The essential elements of the plan include:*

- A list of the major fire hazards in your workplace
- Proper storage and handling procedures for those hazards
- List of ignition sources
- Fire protection equipment to protect your workplace from the fire hazards
- List of personnel responsible for maintaining fire control equipment and systems
- List of personnel responsible for control of fuel source hazards

Employers with fewer than ten employees need not maintain a written plan, but they communicate to each employee responsibilities in the event of a fire.

EMERGENCY RESPONSE PLAN

Companies involved in emergency response operations to the release of or substantial threat of release of hazardous substances must develop an emergency response plan (1910.120). In other words, if workers respond to chemical spills or releases, this plan is required. The only other option in the event of an emergency is to evacuate all employees and arrange for an outside agency to respond to anything more than an incidental spill or release. ***The plan is detailed, requiring at least the following elements:***

- Pre-emergency planning and coordination with outside parties
- Personnel roles, lines of authority, training, and communication
- Emergency recognition and prevention
- Safe distances and places of refuge
- Site security and control
- Evacuation routes and procedures
- Decontamination procedures not covered by the site safety and health plan
- Emergency medical treatment and first aid
- Emergency alerting and response procedures
- Critique of response and follow-up
- Personal protection equipment (PPE) and emergency equipment

Emergency response plans must also include site topography, layout, and prevailing weather conditions, as well as procedures for reporting incidents. Training requirements for responding personnel are extensive. For liability and expense reasons, many companies choose to simply depend on outside agencies for response help. Employers deciding not to implement an emergency response team under this regulation should contact local authorities to learn of their options and know exactly what to do before the incident occurs.

PERMIT-REQUIRED CONFINED SPACE PLAN

Under this section companies are required to identify confined spaces throughout their facilities as detailed in 1910.146. Although the standard specifically exempts the construction industry, OSHA can cite under the general duty clause. ***A confined space:***

- Is large enough and so configured that an employee can bodily enter and perform assigned work
- Has limited or restricted means for entry or exit (for example, tanks, vessels, silos, storage bins, hoppers, vaults, and pits are spaces that may have limited means of entry)
- Is not designed for continuous employee occupancy

In order for the confined space to be considered permit-required, it:

- Contains or has a potential to contain a hazardous atmosphere
- Contains a material that has the potential for engulfing an entrant
- Has an internal configuration such that an entrant could be trapped or asphyxiated by inwardly converging walls or by a floor that slopes downward and tapers to a smaller cross section
- Contains any other recognized serious safety or health hazard

If an employer decides employees will not enter permit spaces, he posts an appropriate danger sign and prevents employees from entering the spaces. If he decides employees will enter permit spaces, he must develop and implement a written program. There is a major exception to this. If the employer can demonstrate that the only hazard is atmospheric, and that it is alleviated through continuous forced air ventilation, all procedures do not have to be followed.

Major components of the written program typically include the following:

- Policy regarding permit-required confined space entry
- Potential hazards and the hazard evaluation process
- Safe entry procedures
- Surveillance procedures
- Evaluation procedures prior to entry
- Certification verifying that spaces are safe to enter and that appropriate procedures have been taken for all spaces
- Duties for entrants, attendants, and supervisors
- Training program
- Rescue and emergency services
- Review procedures of the program and how often this occurs
- Enforcement of the procedures

Examples of permit-required confined space programs appear in the appendixes of 1910.146.

LOCKOUT TAGOUT

Companies are required to comply with the lockout tagout standard under 1910.147 if:

- Servicing and/or maintenance takes place during normal production operations in which
  —An employee is required to remove or bypass a guard or other safety device, or
  —An employee is required to place any part of his or her body into an area on a machine or piece of equipment where work is actually performed upon the material being processed (point of operation) or where an associated danger zone exists during a machine operating cycle.

Exceptions to the standard include minor tool changes and adjustments and other minor servicing activities which take place during normal production operations. These are not included if they are routine, repetitive, and integral to the use of the equipment for production, provided that the work is performed, using alternative measures which provide effective protection.

A typical, minimal lockout procedure is included in the appendix to 1910.147. *The program consists of:*

- Energy control procedures
- Employee training
- Periodic inspections to ensure that before any employee performs any servicing or maintenance on a machine or equipment where the unexpected energizing, start-up, or release of stored energy could occur and cause injury, the machine or equipment shall be isolated from the energy source, and rendered inoperative

PERSONAL PROTECTIVE EQUIPMENT (PPE)

Your company is required to have a written PPE program if your employees use personal protective equipment in conjunction with their duties. Of course, PPE is only to be used when engineering controls and administrative controls cannot be used to protect the employee. Under 1910.132, "PPE" refers to equipment to protect eyes, face, head, and extremities; protective clothing; respiratory devices; and protective shields and barriers. If employees provide their own protective equipment, the employer is also responsible for it. *The employer responsibilities include:*

- Hazard assessment and equipment selection. Once engineering and administrative controls have been exhausted, it is the employer's responsibility to determine if hazards are present necessitating the use of PPE. The employer must select and have each affected employee use the types of PPE necessary for protection.

- Verification by the employer that the hazard assessment has been performed through a written certification which identifies:
  —The workplace evaluated
  —The person certifying the evaluation has been performed
  —The dates of the hazard assessment
  It also provides identification of the given document as certification of a hazard assessment.
- Training for each affected employee on
  —When PPE is necessary
  —What PPE is necessary
  —How to properly don, doff, adjust, and wear PPE
  —Limitations of the PPE
  —Proper care, maintenance, useful life, and disposal of the PPE
- A demonstration by each employee of an understanding of the training specified before being permitted to use the PPE.

RESPIRATORY PROTECTION

If you have any of the following hazards and you cannot prevent atmospheric contamination, you are required under 1910.134 to have a written respiratory program that accounts for:

- Presence of harmful dust
- Fogs
- Fumes
- Mists
- Gases
- Smokes
- Sprays
- Vapors

Of course, lack of oxygen or the necessity of respirator use to protect the health of the employee will mean the company must maintain a worksite-specific program consisting of the following:

- Procedures for selecting respirators for use in the workplace
- Medical evaluations of employees required to use respirators
- Fit-testing procedures for tight-fitting respirators
- Procedures for proper use of respirators in routine and reasonably fore-seeable emergency situations
- Procedures and schedules for cleaning, disinfecting, storing, inspecting, repairing, discarding, and otherwise maintaining respirators
- Procedures to ensure adequate air quality, quantity, and flow of breathing air for atmosphere-supplying respirators
- Training of employees in the respiratory hazards to which they are potentially exposed during routine and emergency situations
- Training of employees in the proper use of respirators, including putting on and removing them, any limitations on their use, and their maintenance, and procedures for regularly evaluating the effectiveness of the program

A suitably trained program administrator must administer the respiratory program. Employers are not required to include in the written program employees whose only use of respirators involves the voluntary use of dust masks. The employer's responsibilities include providing respirators, training, and medical evaluations at no cost to the employee.

PROCESS SAFETY MANAGEMENT

Companies involved in processing certain hazardous chemicals are required to take additional precautions to protect the employees. The purpose is to prevent or minimize the consequences of catastrophic releases of toxic, reactive, flammable, or explosive chemicals that may result in toxicity, fire, or explosion hazards. ***They are required to document what the company has done in regard to process safety if***

- a process involves a chemical at or above the specified threshold quantities listed in Appendix A of 1910.119; or
- a process involves a flammable liquid or gas (as defined in 1910.1200[c]) on-site in one location in a quantity of 10,000 pounds or more at one time, except for hydrocarbon fuels (such as oil, gasoline, heating oil, or liquid propane) used solely for workplace consumption as a fuel (if they

are not part of a process containing another highly hazardous chemical covered by this standard), and flammable liquids stored or transferred in atmospheric tanks which are kept below their boiling point without chilling or refrigeration; and

- the company is not a retail operation, a normally unoccupied remote facility, or an oil or gas drilling operation.

Requirements

The employer must compile written process safety information before conducting any process hazard analysis to enable the employer and the employees to identify and understand hazards posed by processes involving highly hazardous chemicals. *This shall include information pertaining to:*

- Hazards of the highly hazardous chemicals used or produced by the process
- Technology of the process
- Equipment in the process

The employer shall perform an initial hazard evaluation on processes covered by the standard to identify, evaluate, and control the hazards involved in the process.

Written Procedures

The employer shall develop and implement written operating procedures that provide clear instructions for safely conducting activities involved in each covered process consistent with the process safety information. Address at least the following elements:

A) Steps for each operating phase:
- Initial start-up
- Normal operations
- Temporary operations
- Emergency shutdown, including:
  —conditions under which emergency shutdown is required
  —assignment of shutdown responsibility to qualified operators to ensure that emergency shutdown is executed in a safe and timely manner

- Emergency operations
- Normal shutdown
- Start-up following a turnaround, or after an emergency shutdown

B) Operating limits
C) Safety and health considerations
D) Safety systems and their functions

Written procedures must be established to maintain the ongoing integrity of process equipment. Procedures should address the mechanical integrity of:

- pressure vessels and storage tanks
- piping systems (including piping components such as valves)
- relief and vent systems and devices
- emergency shutdown systems
- controls (including monitoring devices and sensors, alarms, and interlocks)
- pumps

The employer shall establish and implement written procedures to manage changes (except for "replacements in kind") to process chemicals, technology, equipment, and procedures and changes to facilities that affect a covered process.

Procedures shall ensure that the following considerations are addressed prior to any change:

- the technical basis for the proposed change;
- impact of change on safety and health;
- modifications to operating procedures;
- necessary time period for the change; and
- authorization requirements for the proposed change.

Employees involved in operating a process and maintenance and contract employees whose job tasks will be affected by a change in the process shall be informed of, and trained in, the change prior to start-up of the process or affected part of the process.

INCIDENT INVESTIGATION

Any incident which resulted in, or could reasonably have resulted in, a catastrophic release of a highly hazardous chemical in the workplace must be investigated as promptly as possible, but not later than 48 hours following the incident. *A report shall be prepared at the conclusion of the investigation which includes at a minimum:*

- The date of the incident
- The date the investigation began
- A description of the incident
- The factors that contributed to the incident
- Any recommendations resulting from the investigation

Emergency Action Plan

The employer shall establish and implement an emergency action plan for the entire plant in accordance with the provisions of 29 CFR 1910.38 (as outlined above). In addition, the emergency action plan shall include procedures for handling small releases. Employers covered under this standard may also be subject to the hazardous waste and emergency response provisions contained in 29 CFR 1910.120(a), (p), and (q).

COMPLIANCE AUDITS

Employers shall certify that they have evaluated compliance with the provisions of this section at least every three years to verify that the procedures and practices developed under the standard are adequate and are being followed. A report of the findings of the audit shall be developed. The employer shall promptly determine and document an appropriate response to each of the findings of the compliance audit and document that deficiencies have been corrected.

CONCLUSION

Nearly every company needs some written programs. The above are the most common ones required by OSHA for general industry. In addition, companies must post the OSHA poster, and those with eleven or more employees are also required to maintain accident records as outlined in chapter 3 under recordkeeping.

QUESTIONS

1. Why are small companies exempt from maintaining some of the programs in writing?
2. What programs are required of all companies regardless of size? What company activities cause companies to fall under provisions requiring written programs?
3. Talk to an owner of a small business of fewer than 11 employees. Ask what safety programs he or she has to protect the employees. Ask the same question of the owner of a small business of 11–25 employees. Compare their responses and comment on the similarities and differences.

Resources on Safety and Health

The following is a noncomprehensive list of resources available to the safety professional seeking assistance in selected areas of occupational safety and health. There are numerous resources available that have not been included in this list because of space limitations. Inclusion in this list is not intended to be an endorsement of individuals, organizations, products, or services. Resources are listed under the broad topic areas that comprise the area of occupational safety and health. Additional electronic resources can be found by using search engines (examples: Google, Dogpile, Yahoo!, Bing, MSN) to locate Internet resources using relevant keywords (examples: "industrial safety" and "hazardous waste").

AGENCIES AND ASSOCIATIONS

Organized alphabetically by acronym. (See also listings for Associations and Organizations under other topics.)

ABIH (American Board of Industrial Hygiene)
6015 West St. Joseph, Suite 102
Lansing, MI 48917-3980
(517) 321-2638
Fax: (517) 321-4624
Website: www.abih.org

The board is responsible for the certification of industrial hygienists. It oversees the examination process and credentialing of industrial hygiene practitioners.

ABT (American Board of Toxicology)
PO Box 30054
Raleigh, NC 27622
(919) 841-5022
Fax: (919) 841-5042
Website: www.abtox.org
This association is the certifying body for toxicologists.

American Chemistry Council (ACC)
1300 Wilson Boulevard
Arlington, VA 22209
(703) 741-5000
Fax: (703) 741-6050
Website: www.americanchemistry.com
Formerly known as the Chemical Manufacturers Association (CMA), this group provides information on chemical industry environmental health and safety issues.

ACGIH (American Conference of Governmental Industrial Hygienists)
1330 Kemper Meadow Drive
Cincinnati, OH 45240
(531) 742-2020
Fax: (531) 742-3355
Website: www.acgih.org
This is a professional society of persons employed by official governmental units responsible for full-time programs in industrial hygiene. It is devoted to the development of administrative and technical aspects of worker health protection.

AIHA (American Industrial Hygiene Association)
2700 Prosperity Avenue, Suite 250
Fairfax, VA 22031
(703) 849-8888

Fax: (703) 207-3561
Website: www.aiha.org
E-mail: infonet@aiha.org
This is the professional society for industrial hygienists. It promotes the study and control of environmental factors affecting the health and well-being of industrial workers. Their website carries a listing of jobs in the safety and health profession.

ANSI (American National Standards Institute)
1819 L Street, NW
Washington, DC 20036
(202) 293-8020
Fax: (202) 293-9287
Website: www.ansi.org
ANSI serves as a clearinghouse for nationally coordinated voluntary safety, engineering, and industrial standards.

APPA (Association of Physical Plant Administrators)
1643 Prince Street
Alexandria, VA 22314-2818
(703) 684-1446
Website: www.appa.org
Also known as the Association of Higher Education Facilities Officers, this organization provides information on environmental, workplace safety, and emergency management topics.

ARR (Association for Radiation Research)
Mount Vernon Hospital
PO Box 100
Northwood
Middlesex, Greater London HA6 2JR
United Kingdom
44-1923 828 611
Website: www.le.ac.uk/cm/arr/ijrb.html
This is a group of scientific, medical, and similar professionals who are interested in ionizing radiation. It organizes research and grants money to students who wish to attend the AAR conference.

ASME (American Society of Mechanical Engineers)

ASME International
Three Park Avenue
New York, NY 10016-5990
(800) 843-2763
Fax: (973) 882-1717
Website: www.asme.org
ASME conducts research and develops boiler, pressure vessel, and power test codes. It sponsors the American National Standards Institute in developing safety codes and standards for equipment.

ASSE (American Society of Safety Engineers)

1800 East Oakton Street
Des Plaines, IL 60018
(847) 699-2929
Fax: (847) 768-3434
Website: www.asse.org
ASSE is the professional society for safety engineers, safety directors, safety professionals, and others concerned with accident prevention and safety programs. ASSE provides guidance and support to local chapters, legislative bodies, and institutions of higher education regarding safety concerns.

ASTM (American Society for Testing and Materials)

100 Barr Harbor Drive
PO Box C700
West Conshohocken, PA 19428-2959
(610) 832-9585
Fax: (610) 832-9555
Website: www.astm.org
E-mail: service@astm.org
Establishes voluntary consensus standards for materials, products, systems, and services.

AWS (American Welding Society)

PO Box 351040
550 NW LeJeune Road
Miami, FL 33126
(305) 443-9353 or (800) 443-9353

Fax: (305) 443-7559
Website: www.aws.org
E-mail: info@aws.org
Professional engineering society in the field of welding. Sponsors seminars and conferences on welding.

BCPE (Board of Certification in Professional Ergonomics)
PO Box 2811
Bellingham, WA 98227-2811
(888) 856-4685
Fax: (866) 266-8003
Website: www.bcpe.org
E-mail: bcpehq@bcpe.org
The BCPE was established to provide a formal organization and procedures for examining and certifying qualified practitioners of ergonomics.

BCSP (Board of Certified Safety Professionals)
208 Burwash Avenue
Savoy, IL 61874
(217) 359-9263
Fax: (217) 359-0055
Website: www.bcsp.org
E-mail: bcsp@bcsp.org
The board is responsible for the certification of safety professionals. It oversees the examination process and credentialing of practitioners in the safety field.

BLS (Bureau of Labor Statistics)
Division of Information Services
2 Massachusetts Avenue NE,
Room 2860
Washington, DC 20212
(202) 691-5200
Fax: (202) 691-6325
Website: www.bls.gov
E-mail: blsstaff@bls.gov
This federal agency gathers workplace-related statistical information.

BOSTI (Buffalo Organization for Social and Technological Innovation)
1479 Hertel Avenue
Buffalo, NY 14216
(716) 316-6377
Website: www.bosti.com
E-mail: sweidemann@bosti.com
BOSTI is involved in design, research, and consulting associated with the behavior, performance, and satisfaction of people.

CDC (Centers for Disease Control)
U.S. Department of Health and Human Services
1600 Clifton Road
Atlanta, GA 30333
(800) 232-4636
Website: www.cdc.gov
The CDC surveys national disease trends and epidemics and environmental health problems. It promotes national health education programs. It administers block grants to states for preventive medicine and health services programs.

CGA (Compressed Gas Association)
4221 Walney Road, 5th Floor
Chantilly, VA 20151
(703) 788-2700
Fax : (703) 961-1931
Website: www.cganet.com
The CGA submits recommendations to appropriate government agencies to improve safety standards and methods of handling, transporting, and storing gases. It acts as an advisor to regulatory authorities and other agencies concerned with safe handling of compressed gases.

CHEJ (Center for Health, Environment, and Justice)
150 South Washington Street, Suite 300
PO Box 6806
Falls Church, VA 22040
(703) 237-2249
Website: www.chej.org
E-mail: chej@chej.org

This organization is concerned with the physical effects of toxic chemicals that come into contact with people. It also provides information and guidance in dealing with hazardous problems.

DOT (U.S. Department of Transportation)

1200 New Jersey Ave, SE
Washington, DC 20590
(address for correspondence)
(866) 377-8642
Website: www.dot.gov
E-mail: dot.comments@ost.dot.gov
The Department of Transportation is a federal agency responsible for establishing regulations for the transportation of hazardous materials via air, land, and sea. It also sets policy for the nation's transportation system.

EPA (U.S. Environmental Protection Agency)

Ariel Rios Building
1200 Pennsylvania Avenue NW
Washington, DC 20460
(202) 272-0167
Website: www.epa.gov
The EPA administers federal environmental policies, research, and regulations. It provides information on many environmental subjects, including water pollution, hazardous and solid waste disposal, air and noise pollution, pesticides, and radiation.

FBI (Federal Bureau of Investigation)

US Department of Justice
J. Edgar Hoover Building
935 Pennsylvania Avenue NW
Washington, DC 20535
(800) 324-3000
Website: www.fbi.gov
Federal agency that participates in accident investigations, terrorist attacks, and so forth.

FCC (Federal Communications Commission)

445 12th Street SW
Washington, DC 20554

(888) 225-5322
Fax: (202) 418-0232
Website: www.fcc.gov
E-mail: fccinfo@fcc.gov
The FCC is the federal agency responsible for the Emergency Broadcast System.

FEMA (Federal Emergency Management Agency)
500 C Street SW
Washington, DC 20472
(202) 646-2500
Website: www.fema.gov
E-mail: opa@fema.gov
FEMA is the federal agency responsible for coordinating emergency response to natural and manmade disasters. It provides information on preparing for and surviving fires, floods, hurricanes, earthquakes, terrorism, and other hazards.

GPO (U.S. Government Printing Office)
Superintendent of Documents
U.S. Government Printing Office
732 North Capital St. NW
Washington, DC 20401
(202) 512-1800
Website: www.access.gpo.gov
E-mail: gpoaccess@gpo.gov
The GPO prints, distributes, and sells selected publications of the U.S. Congress, government agencies, and executive departments.

HFES (Human Factors and Ergonomics Society)
PO Box 1369
Santa Monica, CA 90406-1369
(310) 394-1811
Fax: (310) 394-2410
Website: www.hfes.org
E-mail: info@hfes.org
The HFES is a professional association concerned with the use of human factors and ergonomics in the development of systems and devices.

IATUR (International Association for Time Use Research)
Saint Mary's University
HaliFax, NS
Canada B3H 3C3
(902) 496-8728
Fax: (902) 420-5129
Website: www.smu.ca/partners/iatur/iatur.htm
IATUR facilitates communication and exchange of information in time and motion studies, ergonomics, and industrial efficiency.

ICRU (International Commission on Radiation Units and Measurements)
7910 Woodmont Avenue, Suite 400
Bethesda, MD 20814-3095
(301) 657-2652
Website: Website: www.icru.org/
ICRU develops recommendations on the dosages needed and allowed during radioactive procedures on or near the human body.

INCE-USA (Institute of Noise Control Engineering–USA)
PO Box 220
Saddle River, NJ 07458
(201) 760-1101
Website: Website: www.inceusa.org
Members consist of noise-control professionals who develop engineering solutions to environmental noise problems.

IRI (Industrial Risk Insurer)
Swiss Reinsurance Company Ltd.
PO Box
8022 Zurich
Switzerland
41 43 285 2121
Fax: 41 43 285 2999
Website: www.swissre.com
IRI provides reinsurance products to business and industry to help companies handle risk.

ISMA (International Stress Management Association)
Health Assessment Program, Inc.
638 St. Lawrence Avenue
Reno, NV 89509
Website: www.stress-management-isma.org
They conduct research on tension control and incorporating relaxation into everyday life.

ISTSS (International Society for Traumatic Stress Studies)
111 Deer Lake Road, Suite 100
Deerfield, IL 60015 USA
Phone: (847) 480-9028
Fax: (847) 480-9282
E-mail: istss@istss.org
Website: www.istss.org
The ISTSS is made up of professionals who treat individuals suffering from stress and conduct research.

MSHA (Mine Safety and Health Administration)
1100 Wilson Boulevard, 21st Floor
Arlington, VA 22209
(202) 693-9400
Fax: (202) 693-9401
Website: www.msha.gov
E-mail: The website offers e-mail addresses for the appropriate contact.
The MSHA administers and enforces the health and safety provisions of the Federal Mine Safety and Health Act of 1977. Its training facility is located in Beckley, West Virginia.

MTM (MTM Association for Standards and Research)
1111 East Touhy Avenue
Des Plaines, IL 60018
(847) 299-1111
Fax: (847) 299-3509
Website: www.mtm.org
The MTM Association for Standards and Research has provided business and industry with innovative methodologies designed to help them improve profitability, productivity, and quality.

NFPA (National Fire Protection Association)

1 Batterymarch Park
Quincy, MA 02269-9101
(617) 770-3000
Fax: (617) 770-0700
Website: www.nfpa.org
The NFPA develops, publishes, and disseminates standards prepared by technical committees, intended to minimize the possibility and effects of fire and explosion.

NIH (National Institutes of Health)

9000 Rockville Pike
Bethesda, MD 20892
(301) 496-4000
Website: www.nih.gov
E-mail: NIHinfo@od.nih.gov
Supports and conducts biomedical research into the causes and prevention of diseases and furnishes information to health professionals and the public.

NIOSH (National Institute for Occupational Safety and Health)

U.S. Department of Health and Human Services
Publications Dissemination
4676 Columbia Parkway, Mail Stop C-13
Cincinnati, OH 45226-1998
(513) 533-8328 or (800) 232-4636
Website: www.cdc.gov/niosh
E-mail: cdcinfo@cdc.gov
NIOSH, within the Centers for Disease Control, supports and conducts research on occupational safety and health issues. It provides technical assistance and training and develops recommendations for OSHA.

NIST (National Institute of Standards and Technology)

U.S. Department of Commerce
100 Bureau Drive, Stop 1070
Gaithersburg, MD 20899-3460
(301) 975-6478
Website: www.nist.gov
E-mail: inquires@nist.gov

The NIST develops engineering measurements, data, and test methods. It produces the technical base for proposed engineering standards and code change and generates new engineering practices. It also aids the international competitiveness of small and medium-size companies and consortia through technology development and transfer programs.

NOISE (National Organization to Insure a Sound-Controlled Environment)
415 Second Street NE, Suite 210
Washington, DC 20002
(202) 544-9844
Fax: (202) 544-9850
Website: www.aviationnoise.org
E-mail: contact@aviation-noise.org
They seek to eliminate jet noise from the environment by confronting legislatures to ensure that rules are being followed.

NSC (National Safety Council)
1121 Spring Lake Drive
Itasca, IL 60143-3201
(800) 621-7615
Fax: (630) 285-1315
Website: www.nsc.org
E-mail: info@nsc.org
The NSC is a voluntary, nongovernmental organization that promotes accident reduction by providing a forum for the exchange of safety and health ideas, techniques, experiences, and accident-prevention methods.

NTIS (National Technical Information Services) (ALSO see NAC)
U.S. Department of Commerce
5285 Port Royal Road
Springfield, VA 22161
(703) 605-6000
Website: www.nsc.org
E-mail: helpdesk@fedworld.gov
The NTIS is a distribution center selling to the public government-funded research and development reports prepared by the federal agencies, their

contractors, or grantees. It offers microfiche and computerized bibliography search services.

NTSB (National Transportation Safety Board)
U.S. Department of Transportation
490 L'Enfant Plaza SW
Washington, DC 20594
(202) 314-6000
Website: www.ntsb.gov
NTSB is a federal agency responsible for conducting accident investigations into major incidents involving trains, planes, buses, and so on.

OSHA (Occupational Safety and Health Administration, National Office)
U.S. Department of Labor
200 Constitution Avenue, NW
Washington, DC 20210
(800) 321-6742
Website: www.osha.gov
OSHA sets policy, develops programs, and implements the Occupational Safety and Health Act of 1970.

OSHRC (Occupational Safety and Health Review Commission)
One Lafayette Center
1120 20th Street NW, 9th Floor
Washington, DC 20036-3419
(202) 606-5400
Fax: (202) 606-5050
Website: www.oshrc.gov
E-mail: It_gpo@oshrc.gov
The OSHRC is an independent executive agency that adjudicates disputes between private employers and OSHA, arising from citations of occupational safety and health standards.

SOT (Society of Toxicology)
1821 Michael Faraday Drive, Suite 300,
Reston, VA 20190
(703) 438-3115

Fax (703) 438-3113
Website: www.toxicology.org
E-mail: sothg@toxicology.org
The SOT is a resource for advancing the science and applications of toxicology.

MANUFACTURERS AND SUPPLIERS

American Health and Safety, Inc.
325 Industrial Circle
Stoughton, WI 53589
(800) 522-7554
Website: www.ahsafety.com
This group manufactures a disposable mask that filters out unpleasant or nontoxic odors. This mask does not restrict breathing or hold in heat.

ANSUL Incorporated
One Stanton Street
Marinette, WI 54143-2542
(715) 735-7411
Website: www.ansul.com
ANSUL manufactures wheeled and handheld portable fire extinguishers.

Broner Glove and Safety
1750 Harmon Road
Auburn Hills, MI 48326
(800) 521-1318
Fax: (800) 276-6372
Website: www.broner.com
Broner makes custom, ergonomically designed work stations for assembly line workers.

Bruel and Kjaer Instruments
2815-A Colonnades Court
Norcross, Georgia 30071-1588
(770) 209-6907
Fax: (770) 448-3246
Website: www.bkhome.com

This company manufactures a device that measures the various energies of sounds on several levels.

CEL Instruments, Ltd.
17 Old Nashua Road
Amherst, NH 03031
(800) 366-2966
Website: www.casellausa.com
CEL manufactures noise dosimeters that can be worn by personnel to obtain individual exposure samples.

Drager USA
101 Technology Drive
Pittsburgh, PA 15275
(412) 787-8383
Fax: (412) 787-2207
Website: Website: www.draeger.com
Drager USA manufactures industrial hygiene and personal protective equipment.

Elvex Corporation
7 Trowbridge Drive
Bethel, CT 06801-0850
(203) 743-2488
Website: www.elvex.com
Elvex manufactures a wide range of personal protective equipment, including vision and hearing protection.

J. J. Keller & Associates, Inc.
3003 West Breezewood Lane
Neenah, WI 54957
(877)564-2333
Fax: (800) 727-7516
Website: www.jjkeller.com
Keller offers compliance audits. They offer an extensive list of training and operating aids for the safety professional.

Justrite Manufacturing Co.
2454 Dempster Street
Des Plaines, IL 60016-5344
(847) 298-9250
Fax: (847) 298-9261
Website: www.justritemfg.com
Justrite manufactures flammable-liquid storage cabinets, safety cans, and flammable waste disposal containers.

Lab Safety Supply, Inc.
PO Box 1368
Janesville, WI 53547-1368
(800) 356-0783
Website: www.labsafety.com
Lab Safety offers reference manual sheets that provide useful background information on toxic substances that can be used to compile documentation related to toxic substances. Their catalog is a must-read for safety students. It shows many types of safety equipment and explains each.

Marcom Group, Ltd.
20 Creek Parkway
Garnet Valley, PA 19061
(800) 654-2448
Fax: (610) 859-8106
Website: www.marcomltd.com
Marcom's audiovisual, computer programs, and literature detail ways the workplace might be altered to alleviate fatigue-related pain for workers.

New Pig
One Pork Avenue
Tipton, PA 16684-0304
(800) 468-4647
Fax: (814) 684-0972
Website: www.newpig.com
New Pig manufactures hazardous material spill cleanup equipment.

Orr Safety Corporation

PO Box 198029
Louisville, KY 40259-8029
(800) 669-1677
Fax: (800) 800-6774
Website: www.orrsafety.com
E-mail: custserv@orrcorp.com

Orr Safety is a supplier of high-friction thermoplastic material applied to tool handles and then molded to the shape of the user's hand to provide an ergonomic grip.

Quest Technologies

510 South Worthington
Oconomowoc, WI 53066
(800) 245-0779
Fax: (262) 567-4047
Website: www.quest-technologies.com

Quest manufactures noise-monitoring equipment, including octave band analyzers, dosimeters, and sound-level meters.

Sensidyne, Inc.

16333 Bay Vista Drive
Clearwater, FL 33760
(727) 530-3602 or (800) 451-9444
Fax: (727) 539-0550
Website: www.sensidyne.com

Sensidyne, Inc. manufactures of a wide range of air monitoring equipment.

SKC, Inc.

863 Valley View Road
Eighty Four, PA 15330
(800) 752-8472
Fax: (724) 941-1369
Website: www.skcinc.com

SKC manufactures a wide range of air sampling supplies and equipment.

FEDERAL REGIONAL OFFICES

States marked with an asterisk (*) have a state safety and health plan under section 18(b) of the OSHAct. Those states' agencies may be contacted directly for specific information regarding regulations in their jurisdictions.

OSHA Region I
(CT*, MA, ME, NH, RI, VT*)
JFK Federal Building, Room E340
Boston, MA 02203
(617) 565-9860

OSHA Region II
(NJ, NY*, PR*, VI*)
201 Varick Street
Room 670
New York, NY 10014
(212) 337-2378

OSHA Region III
(DC, DE, MD*, PA, VA*, WV)
The Curtis Center-Suite 740 West
170 South Independence Mall West
Philadelphia, PA 19106-3309
(215) 861-4900

OSHA Region IV
(AL, FL, GA, KY*, MS, NC*, SC*, TN*)
61 Forsyth Street, Suite 6T50
Atlanta, Georgia 30303
(404) 562-2300

OSHA Region V
(IL, IN*, MI*, MN*, OH, WI)
230 South Dearborn, Room 3244
Chicago, IL 60604
(312) 353-2220

OSHA Region VI
(AR, LA, NM*, OK, TX)
525 Griffin Street
Room 602
Dallas, TX 75202
(972) 850-4145

OSHA Region VII
(IA*, KS, MO, NE)
Two Pershing Square Building
2300 Main Street, Suite 1010
Kansas City, MO 64105
(816) 283-8745

OSHA Region VIII
(CO, MT, ND, SD, UT*, WY)
1999 Broadway, Suite 1690
Denver, Colorado 80202-5716
(702) 264-6550

OSHA Region IX
(AZ*, CA*, HI*, NV*)
90 7th Street,
Suite 18100
San Francisco, CA 94103
(415) 625-2547

OSHA Region X
(AK*, ID, OR*, WA*)
1111 Third Avenue
Suite 715
Seattle, WA 98101-3212
(206) 553-5930

STATE COMPLIANCE OFFICES

Alaska Department of Labor
PO Box 111149
Juneau, AK 99801-1149
(907) 465-2700

Industrial Commission of Arizona
800 West Washington St. Second Floor
Phoenix, AZ 85007-2922
(602) 542-5795

California Department of Industrial Relations
455 Golden Gate Avenue, 10th Floor
San Francisco, CA 94102
(415) 703-5050

Connecticut Department of Labor
200 Folly Brook Boulevard
Wethersfield, CT 06109
(860) 263-6505

State of Hawaii
Department of Labor and Industrial Relations
830 Punchbowl Street
Honolulu, HI 96813
(808) 586-8842

Illinois Department of Labor
1 W. Old State Capitol Plaza, Room 300
Springfield, IL 62701
(217) 782-6206

Indiana Department of Labor
State Office Building
402 West Washington Street, Room W195
Indianapolis, IN 46204-2751
(317) 232-2378

Iowa Division of Labor Services
1000 East Grand Avenue
Des Moines, IA 50319-0209
(515) 281-8067

Kentucky Labor Cabinet
Occupational Safety and Health Program
1047 US 127 South, Suite 4
Frankfort, KY 40601
(502) 564-3070

Maryland Department of Licensing and Regulation
Maryland Occupational Safety and Health
1100 North Eutaw Street, Room 613
Baltimore, MD 21201-2206
(410) 767-2241

Michigan Department of Consumer and Industry Services
Bureau of Safety and Regulation
PO Box 30643
Lansing, MI 48909-8143
(517) 322-1817

Minnesota Department of Labor and Industry
443 Lafayette Road North
St. Paul, MN 55155
(651) 284-5010

Nevada Division of Industrial Relations
Occupational Safety and Health Enforcement Section
400 West King Street, Suite 400
Carson City, NV 89703
(775) 684-7260

New Jersey Department of Labor and Workforce Development
Office of Public Employees
Occupational Safety & Health (PEOSH)

1 John Fitch Plaza
PO Box 386
Trenton, NJ 08625-0386
(609) 292-2975

New Mexico Environment Department

Occupational Safety and Health Bureau
525 Camino de Los Marquez, Suite 3
PO Box 26110
Santa Fe, NM 87502
(505) 827-2855

New York Department of Labor

Division of Safety and Health
State Office Campus Building 12, Room 130
Albany, NY 12240
(518) 457-2238

North Carolina Department of Labor

1101 Mail Service Center
Raleigh, NC 27699-1101
(919) 733-0359

Oregon Occupational Safety and Health Division

Department of Consumer & Business Services
350 Winter Street, NE, Room 430
Salem, OR 97309-0405
(503) 378-3272

Puerto Rico Department of Labor and Human Resources

Occupational Safety and Health Office
505 Munoz Rivera Avenue
Hato Rey, PR 00918
(787) 754-2119

South Carolina Department of Labor, Licensing and Regulation
Synergy Business Park, Kingstree Building
110 Centerview Drive
PO Box 11329
Columbia, SC 29211
(803) 896-4300

Tennessee Department of Labor and Workforce Development
220 French Landing Drive
Nashville, TN 37243
(615) 741-2793

Utah Labor Commission
160 East 300 South
PO Box 146650
Salt Lake City, UT 84114-6650
(801) 530-6848

Vermont Department of Labor
Green Mountain Drive
PO Box 488
Montpelier, VT 05601-0488
(802) 828-4301

Virgin Islands Department of Labor
Division of Occupational Safety and Health
3012 Golden Rock
Christiansted
St. Croix, VI 00840
(340) 773-1994

Virginia Department of Labor and Industry
Occupational Safety and Health Cooperative Programs
Powers-Taylor Building
13 South 13th Street
Richmond, VA 23219
(804) 786-2377

Washington Department of Labor and Industry
WISHA Services Division
PO Box 44001
Olympia, WA 98504-4001
(360) 902-4200

Wyoming Department of Employment
Workers' Safety and Compensation Division
1510 East Pershing Boulevard - West Wing
Cheyenne, WY 82002
(307) 777-5110

REFERENCES

Best's Safety Directory. 2002. Vols. 1–2. Oldwick, NJ: A. M. Best.

Fischer, C. A., and Schwartz, C. A. (Eds.). 1996. *Encyclopedia of Associations: National Organizations of the U.S.*, 30th ed., Vol. 1. New York: Gale Research.

Thurn, L. (Ed.). 1966. *Encyclopedia of Associations; International Organizations*, 30th ed. New York: Gale Research.

Ulrich's International Periodicals Directory. 1994–1995. 33rd ed., Vols. 2–3. New Providence, NJ: R. R. Bowker.

Appendix A

29 CFR 1910—OSHA General Industry
Standards Summary and Checklist

TABLE OF CONTENTS

441

TITLE 29 CFR 1910—
OSHA GENERAL INDUSTRY STANDARDS

Introduction

In 1970, the U.S. Government passed the Occupational Health and Safety Act PL-91593. This act was designed to ensure safe and healthy working conditions for men and women in the United States and any other areas falling into the jurisdiction of the United States. The Secretary of Labor is in charge of OSHA. OSHA enacts and enforces laws that pertain to safety in industry. OSHA also has the power to inspect places of business and to levy fines for violations of the codes or laws. The *Code of Federal Regulations* (CFR) is long and arduous, and often confusing to the average person. Therefore, this is an overview of Title 29 CFR 1910, Occupational Safety and Health Standards for General Industry.

This document is not intended to be a replacement for 29 CFR 1910, but it can be used as a quick reference guide to 29 CFR 1910. Contractors or business persons look up a subpart in this appendix, get a brief overview of what it is about, and answer quick and easy questions. They can first go to this manual and decide which sections of the code to interpret. If they answer yes to a question in a subpart, then they know that they need to look that subpart up in 29 CFR 1910 and make sure that their company complies with that standard.

Purpose

This document is to be used as an aid for compliance with federal safety and health regulations. The main purpose is to assist employers in determining the applicability of standards for specific industries. This is to be used only as a supplement to 29 CFR 1910 and cannot, in any way, provide specific information independently on compliance or regulatory law.

How to Use This Document

Part 1910 is broken down into subparts and each subpart into sections. Under each section there is a question or series of questions to determine applicability. If the answer is yes to any question under the section, then the section applies and the individual needs to be referred to the appropriate subpart and section for clarification in 29 CFR 1910. If there is doubt about whether a company is in compliance on any portion of 29 CFR 1910, it is important to confirm that fact by referring to the standard.

Subpart A—General

This subpart explains the purpose and scope, definitions, petitions, amendments, and applicability of all the standards in 29 CFR 1910. Standards will be adopted as necessary by the Secretary to make a healthier and safer workplace. It explains each individual's role from the Secretary of Labor to that of the employee, as well as other terms used throughout 29 CFR 1910. Any interested person may petition in writing to the Assistant Secretary of Labor for the issuance, amendment, or repeal of a standard. It places the Assistant of the Secretary of Labor at the same level of authority as the Secretary of Labor. It lists all the territories that the standards apply to but excludes governmental agencies.

Subpart B—Adoption and Extension of Established Federal Standards

This subpart adopts and extends the applicability of established Federal standards to every employer, employee, and place of employment covered in the Act. Only standards relating to safety or health are adopted into this Act. This also pertains to any facility engaging in construction, alterations, or repair, including painting and decorating. Additionally, it pertains to facilities with employees engaged in ship repair or ship breaking, or that use craft as a means of transportation on water or a related employment. Companies employing people aboard a vessel on navigable waters of the United States and employees engaged in long shoring operations (i.e., the moving or handling of cargo, ship's stores, and so forth into, on, or out of any vessel) are also covered by the Act. Facilities that expose any employee to asbestos, tremolite, anthophyllite actinolite dust, vinyl chloride, inorganic arsenic, lead, ethylene oxide, acrylonitrile, formaldehyde, and cadmium are subject to the Act as well.

Subpart C—General Safety and Health Provisions

The purpose of this subpart is to provide employees and their designated representatives a right of access to relevant exposure and medical records. It also provides the representatives of the Assistant Secretary a right of access to these records in order to fulfill responsibilities under the Occupational Safety and Health Act.

Subpart D—Walking and Working Surfaces

The purpose of this subpart addresses the requirements for maintaining walking and working surfaces. Subpart D applies to all permanent places of employment, except where domestic, mining, or agricultural work only is performed. It contains regulations pertaining to housekeeping, aisles and passageways, ladders, scaffolding, railing, and working surfaces, including mobile surfaces. The section also contains information on guarding floor and wall openings and holes, fixed industrial stairs, portable wood ladders, portable metal ladders, and fixed ladders. Safety requirements for scaffolding, manually propelled mobile ladder stands and scaffolds (towers), and other working surfaces such as dock boards (bridge plates), forging machine areas, wooden platforms, veneer machinery space requirements, and walkways adjacent to large steam vats are presented in this subpart.

Subpart E—Means of Egress

This subpart deals specifically with Means of Egress, defined by the standard as "a continuous and unobstructed way of exit travel from any point in a building or structure to a public way and consists of three separate and distinct parts: the way of exit access, the exit, and the way of exit discharge." The standard addresses each of these items to include general requirements, the various items that make up an exit, specific physical requirements for an exit, and the number of exits required. This subpart also contains the general fundamental requirements essential to providing a safe means of egress from fire and like emergencies. The section has information on definitions, general requirements, means of egress, and employee emergency plans and fire prevention plans. It discusses the minimum requirements for the plans, alarm systems to be used, training requirements, and maintenance of equipment used under the plans.

Subpart F—Powered Platforms, Manlifts, and Vehicle-Mounted Work Platforms

This section covers powered platform installations permanently dedicated to interior or exterior building maintenance of a specific structure or group of structures. It does not apply to suspended scaffolds (swinging scaffolds) used to service buildings on a temporary basis and covered under subpart D of this part, or suspended scaffolds used for construction work and covered under subpart L of 29 CFR Part 1926. This section applies to all permanent installations completed after July 23, 1990, and contains information on powered platforms for building maintenance. Building maintenance includes, but is not limited to, such tasks as window cleaning, caulking, metal polishing, and reglazing. This section goes on to address such items as building owner and employer responsibilities associated with this equipment; specific make-up of the equipment, engineering, and the design of the equipment; and its inspection and maintenance requirements. It also addresses employee training, personal protective equipment requirements, and general housekeeping items.

In addition, this section specifically addresses the requirements for vehicle-mounted elevating and rotating work platforms. It specifically states, "Unless otherwise provided in this section, aerial devices (aerial lifts) acquired on or after July 1, 1975, shall be designed and constructed in conformance with the applicable requirements of the American National Standard for Vehicle-Mounted Elevating and Rotating Work Platforms, ANSI A92.2-1969, including appendix." The employee training, personal protective equipment required, boom operation around electrical hazards, and inspection and maintenance of equipment are also discussed.

Subpart G—Occupational Health and Environmental Control

The standards in subpart G deal with air quality, noise exposure exceeding 85 decibels, and radiation exposure in the workplace. This includes facilities that use abrasive blasting; facilities that have spray booths, spray rooms, or open surface tanks used for cleaning; facilities with grinding, polishing, and buffing operations; and facilities that have ionizing radiation such as alpha rays, beta rays, gamma rays, X rays, and other atomic particles, or radioactive materials such as nonionizing radiation that may be present in the facility.

Subpart H—Hazardous Materials

Subpart H contains information on compressed gases, acetylene, hydrogen, oxygen, nitrous oxide, flammable and combustible liquids, spray finishing using flammable and combustible materials, dip tanks using flammable or combustible liquids, explosive and blasting agents, storage and handling of liquid petroleum gases, and storage and handling of anhydrous ammonia. This section also covers process-safety management requirements of highly hazardous chemicals and hazardous-waste operations and emergency response.

Subpart I—Personal Protective Equipment

This subpart provides the general requirements for the minimum personal protective equipment to be used by the employees in the performance of their tasks for protection against recognized workplace hazards. The section specifically states, "Protective equipment including personal protective equipment for eyes, face, head, and extremities; protective clothing; respiratory devices; and protective shields and barriers; shall be provided, used, and maintained in a sanitary and reliable condition wherever it is necessary by reason of hazards of processes or environment, chemical hazards, radiological hazards, or mechanical irritants encountered in a manner capable of causing injury or impairment in the function of any part of the body through absorption, inhalation, or physical contact" (U.S. Department of Labor, 1993, p. 205). It goes on to address employee-owned equipment; design, hazard assessment, and equipment selection; employee training; and inspection and maintenance of personal protection, foot protection, and hand protection.

Subpart J—General Environmental Control

This section specifically applies to permanent places of employment. It addresses such items as toilet facilities, including numbers and type; washing facilities; sanitary food storage; and food handling. It addresses temporary labor camps, safety color-coding for marking physical hazards, and specifications for accident prevention signs and tags. Two additional items specifically addressed by this section and of considerable importance are permit-required confined spaces and the control of hazardous energy (lockout/tagout).

The permit-required confined spaces standard defines a "confined space"; lists specific employee training requirements, equipment, and personal protective equipment required prior to entry; and communications and emergency procedures. Appendix A to the 1910 standard also includes an informative Permit-Required Confined Space Decision Flow Chart, as well as an example of a Confined Space Entry Permit.

The standard for the control of hazardous energy (lockout/tagout) addresses "the servicing and maintenance of machines and equipment in which the unexpected energization or start-up of the machines or equipment or release of stored energy could cause injury to employees. This standard establishes minimum performance requirements for the control of such hazardous energy" (U.S. Department of Labor, 1993, p. 225). It further spells out the requirements of the standard including all aspects of training and personal protective equipment, as well as methods to meet the standards.

Subpart K—Medical and First Aid

The purpose of subpart K—Medical and First Aid is to provide the employee with readily available medical consultation on matters related to on-plant health. This service may be provided by on- or off-site personnel. First aid supplies shall also be readily available.

Subpart L—Fire Protection

This subpart contains requirements for fire brigades, all portable and fixed fire suppression equipment, fire detection systems, and fire or employee alarm systems installed to meet the fire protection requirements of 29 CFR 1910. It contains requirements for the organization, training, and personal protective equipment of fire brigades whenever they are established by an employer. These requirements apply to fire brigades, industrial fire departments, and private or contractual fire departments. Personal protective equipment requirements apply only to members of fire brigades, performing interior structural fire fighting. The requirements of this section do not apply to airport crash rescue or forest fire fighting operations. In addition, this subpart establishes the requirements for the placement, use, maintenance, and testing of portable fire extinguishers provided for use by employees, as well as the requirements for all automatic sprinkler systems installed to meet a particular OSHA standard.

Subpart M—Compressed-Gas and Compressed-Air Equipment

This section applies to compressed-air receivers and other equipment used in providing and utilizing compressed air for performing operations such as cleaning, drilling, hoisting, and chipping. However, this section does not deal with the special problems created by using compressed air to convey materials, or the problems created when work is performed in compressed-air environments such as in tunnels and caissons. This section is not intended to apply to compressed-air machinery and equipment used or transportation vehicles such as steam railroad cars, electric railway cars, and automotive equipment.

Subpart N—Materials Handling and Storage

This section is a general overview that covers the handling and storage of materials. The main concern of this subpart is the movement of materials on-site. Its four sections are devoted to material handling devices. These devices consist of industrial trucks, overhead and gantry cranes, crawler and truck cranes, derricks, and slings used to connect the loads. The use of helicopters to lift materials is covered, but more stringent FAA regulations exist that cover this in detail. The servicing of single- and multi-piece rims are also covered in this subpart.

The industrial trucks section covers the classifications of trucks and designated areas where a truck can be used. It also describes the required inspections and maintenance actions for those vehicles. Safe operation procedures are also covered in this section.

The same material is covered for cranes and derricks. Loads and proper lifting procedures are also covered. Required inspections are outlined and procedures to ensure that a non-functioning machine is marked and safe.

Procedures for keeping and using slings is also covered in this section. It describes the proper sizes for loads as well as safe hook-up procedures. Inspection requirements are stated and required markings are discussed.

The servicing of single-piece and multi-piece rim wheels is covered for those pieces of equipment that are not covered by the Construction Safety Standards, the Agriculture Standards, the Shipyard Standards, or the Longshoring Standards.

Subpart O—Machinery and Machine Guarding

Subpart O covers the machine guarding for any equipment that exposes the employee to a hazard during use due to exposed moving or rotating parts; generally speaking, this covers any device that has an exposed point of operation. This subpart covers guards for woodworking machinery, abrasive-wheel machinery, mills and calendars in the plastics industries, mechanical presses, forging machines, and mechanical power-transmission apparatus.

The woodworking section covers the parts that must be guarded and the type of guards that must be used, while the abrasive-wheel section describes the amount of wheel that can be exposed for the various types of abrasive-grinding equipment. Proper mounting is described for the various wheels. Charts of the minimum dimensions for abrasive wheels and their safety guards are included.

Mechanical-power presses are required to have switches and brakes that are operator proof. The size of some presses makes it imperative that they are switch-guarded as well as mechanically guarded. This section describes what actions must be taken to produce the most failsafe device possible. Minimum braking action and the proper brakes that must be installed on presses of differing specifications are outlined. Criteria for the use of presence-sensing devices and their proper utilization is described.

The safety criteria for a forging operation is the next section. Operation, inspection, and maintenance requirements are covered. The requirements for venting special processes and equipment that needs to be guarded are included.

The mechanical-power-transmission apparatus section covers all belts, pulleys, and conveyors that are used in industry. It describes which ones need to be guarded. For specific applications, guidance is given for preferred actions.

Subpart P—Hand and Portable Powered Tools and Other Hand-Held Equipment

This section describes the required guarding, inspections, and maintenance requirements for hand and portable tools. It includes tools employees may own and use on the job. Lawn mowers and other internal-combustion-engine-powered machines are included in this section. The proper procedures for using portable jacks are also covered in this section.

Subpart Q—Welding, Cutting, and Brazing

Subpart Q covers the use and installation of electric or gas welding, cutting, and brazing equipment. It covers the different types of welding and the specific safety needs of each. This subpart also regulates the use of oxygen-fuel gas welding and cutting, arc welding and cutting, and resistance welding. Basic fire precautions are included in this subpart, not detailed regulations, which are covered in NFPA standard 51B, 1962. Maintenance and safety requirements are included in this subpart, including welding in confined spaces and proper ventilation.

Subpart R—Special Industries

This section applies to special industries as defined by OSHA. These special industries include paper, pulp, and paperboard mills; textile mills; bakeries; laundries; and sawmills. Other industries covered are pulpwood logging; agricultural operations; telecommunications; electric-power generation, transmission, and distribution; and, finally, grain-handling facilities. These special industries are trades that perform specific tasks that are very distinct and, in turn, set standards for their own industries.

Subpart S—Electrical

This subpart addresses the electrical safety requirements for the safeguarding of employees in their workplace. It is divided into sections dealing with the general requirements of wiring design, protection of specific-purpose equipment, and installations. In addition, this subpart covers locations, wiring methods, wiring components, and training and equipment used.

Subpart T—Commercial Diving Operations

This subpart applies to every place of employment within the territory of the United States, or within any state, the District of Columbia, the Commonwealth of Puerto Rico, the Virgin Islands, American Samoa, Guam, the trust Territory of the Pacific Islands, Wake Island, Johnston Island, or the Canal Zone, or within the Outer Continental Shelf lands as defined in the Outer Continental Shelf Lands Act, where diving and related support operations are performed.

This standard applies to diving and related support operations conducted in connection with all types of work and employments including general industry, construction, ship repairing, shipbuilding, ship breaking, and long-shoring. It covers the personnel requirements, general operations procedures, specific operations procedures, equipment procedures and requirements, and recordkeeping for diving operations.

Subparts U–Y (Reserved)

Subpart Z—Toxic and Hazardous Substances

The purpose of this subpart is to protect employees from exposure to toxic and hazardous substances in the workplace. It covers the Permissible Exposure Limits (PEL) for all air contaminants, including all gases, vapors, and dusts. Some of the contaminants covered under this subpart include asbestos, coal tar pitch volatiles, vinyl chloride, inorganic arsenic, lead, cadmium, benzene, coke-oven emissions, blood-borne pathogens, cotton dust, ethylene oxide, and formaldehyde.

In addition, this subpart is Hazard Communication and Occupational Exposure to Hazardous Chemicals in the Laboratory. These regulations attempt to ensure worker awareness and knowledge of hazardous materials. Material Safety Data Sheets, container labeling, and employee-training requirements are specified in this subpart of the standards.

Subpart Z tables for the Permissible Exposure Limits of all air contaminants are published in this portion of the standards. Compliance requirements for hazardous materials found in these tables are presented in terms of eight-hour time weighted averages (TWA), maximum short-term exposure limits (STELs), and ceiling limits (C).

CHECKLIST FOR SUBPARTS A–Z

Subpart A—General

Purpose, scope, and definitions only in this subpart.

Subpart B—Adoption and Extension of Established Federal Standards

- Does this facility engage in construction, alterations, and/or repair, including painting and decorating?

- Does this facility have employees engaged in ship repair, ship breaking, or shipbuilding, or use craft as a means of transportation on water or a related employment?
- Does the company employ people aboard a vessel on the navigable waters of the United States?
- Does the company have employees engaged in longshoring operations—that is, the moving or handling of ship's cargo, stores, or gear into, on, or out of any vessel?
- Is the facility a marine terminal such as a pier, dock, or wharf?
- Does the facility expose any employee to asbestos, vinyl chloride, inorganic arsenic, lead, ethylene oxide, acrylonitrile, formaldehyde, or cadmium?

Subpart C—General Safety and Health Provisions

- Does the facility compile and retain employee medical records?

Subpart D—Walking and Working Surfaces

- Does the facility have passageways, storerooms, or service rooms?
- Does the facility have mechanical handling equipment that must be operated along aisles and passageways?
- Does the facility have open tanks, pits, vats, ditches, and so on which need covers and guardrails to protect employees?
- Is this facility a permanent place of employment?
- Does the facility have stairway, ladder way, hatchway, and chute floor openings?
- Does the facility have any wall openings, manhole floor openings, or temporary floor openings?
- Does the facility have fixed stairways inside or out; around machinery, tanks, and other equipment; and stairs leading to or from floors, platforms, or pits?
- Does the facility have ladders?
- Is fixed or portable scaffolding used at the facility?
- Are dock boards or bridge plates used at the facility?
- Does the facility have forging machines?
- Are steam vats used at the place of employment?

Subpart E—Means of Egress

- Does the facility have emergency exits?
- Does the facility have the required written emergency action plan and fire prevention plan?
- Has the facility properly trained a sufficient number of persons to assist in an orderly emergency evacuation?

Subpart F—Powered Platforms, Manlifts, and Vehicle-Mounted Work Platforms

- Are powered platforms of any type in use at the facility?
- Is training of employees in the operation and inspection of working platforms performed by a competent person?

Subpart G—Occupational Health and Environmental Control

- Does the facility use abrasive blasting—a solid substance used in blasting operations?
- Does the facility have spray booths, spray rooms, or open surface tanks used for cleaning?
- Does the facility have grinding, polishing, and buffing operations?
- Does the facility have noise levels exceeding 85 decibels?
- Is ionizing radiation such as alpha rays, beta rays, gamma rays, X rays, and other atomic particles present in the facility?
- Are radioactive materials present in the facility?
- Is nonionizing radiation such as electromagnetic radiation, which includes both the radio and microwave frequency regions, present in the facility?

Subpart H—Hazardous Materials

- Does the facility or work area have compressed-gas cylinders at the location?
- Is acetylene transferred through a piped system?
- Does the facility generate or fill cylinders with acetylene?
- Is hydrogen delivered, stored, or discharged at the facility?
- Does the facility have a gaseous hydrogen system in place?

- Does the facility have a liquefied hydrogen system in place?
- Does the facility have hydrogen containers?
- Does the facility have a bulk oxygen system?
- Is nitrous oxide used in the facility?
- Does the facility use flammable and combustible liquids in forms such as aerosol cans and atmospheric tanks?
- Does the facility have underground storage tanks for dispersement from fixed equipment into the fuel tanks of motor vehicles such as service stations?
- Are aerated solid powders used as a coating material used at the facility?
- Are spray booths, spray areas, water wash spray booth, or dry spray booth used at the facility?
- Does the facility have a tank, vat, or container of flammable or combustible liquid in which articles or materials are immersed for the purpose of coating, finishing, treating, or similar processes?
- Are any of the liquids in the dip tank heated?
- Are rags used or waste generated in connection with dipping operations?
- Does the facility have any vapor areas where dangerous quantities of flammable vapors can build up during operation or shutdown periods?
- Does the facility have or use class A, B, or C explosives and blasting agents?
- Are pyrotechnics, which includes fireworks, manufactured or stored at the facility?
- Does the company store, transport, or handle liquefied petroleum gases in containers such as tanks, cylinders, or drums?
- Does the company store, transport, or handle anhydrous ammonia (excluding manufacturing and plants where ammonia is used solely as a refrigerant), in containers such as tanks, cylinders, or drums?
- Does the operation have the potential for producing a catastrophic release of toxic, reactive, flammable, or explosive chemicals?
- Does the process involve a flammable liquid or gas on-site in one location in a quantity of 10,000 pounds or more?
- Does the company have employees who handle or may come in contact with hazardous waste or substances and chemicals?
- Do employees such as fire fighters or mutual aid groups respond to emergencies that are likely to result in an uncontrolled release of a hazardous substance?

- Are any employees designated by the employer as a Hazardous Materials Response Team (HAZMAT)?

Subpart I—Personal Protective Equipment

- Does the facility have any hazardous process, environmental hazards, chemical hazards, radiological hazards, or mechanical irritants which could cause injury to any part of the body by physical contact, breathing, or absorption?
- Does the facility require the employees to provide their own protective equipment?
- Are any employees exposed to eye or face hazards from flying particles, molten metal, liquid chemicals, acids, caustic liquids, chemical gases or vapors, or potentially injurious light radiation?
- Are employees exposed to hazards from flying objects?
- Do any employees wear prescription lenses while engaged in operations that involve eye hazards?
- Are employees exposed to any welding activities?
- Is there any exposure to occupational diseases caused by breathing contaminated air, whether in the form of dusts, fogs, fumes, mists, gases, smokes, sprays, or vapors?
- Are respirators provided when the equipment is necessary to protect the health of the employee?
- Do employees receive training in the use of respiratory protection?
- Are any employees working in areas where there is a potential for injury to the head from falling objects?
- Is there any possible exposure to electrical hazards?
- Are employees working in areas where there is a danger of foot injuries due to falling or rolling objects, objects piercing the sole, or where employees' feet are exposed to electrical hazards?
- Are employees exposed to electrical hazards?
- Are employees' hands exposed to hazards such as skin absorption, heat changes, or burns?

Subpart J—General Environmental Control

- Are there any physical hazards that will need to be marked with appropriate colors?

- Are there signs and symbols needed to indicate and define specific hazards?
- Are employees exposed to the hazards of entry into permit-required confined spaces?
- Does the facility have any machinery that could *unexpectedly* energize, start up, or release stored energy and cause injury to employees?
- Are employees required to remove or bypass a guard or other safety device for any reason?
- Are employees required to place any part of the body into an area on a machine where work is actually performed, or where an associated danger zone exists during machine operation?

Subpart K—Medical and First Aid

- Does the facility provide medical equipment and personnel for advice and consultation on medical matters?

Subpart L—Fire Protection

- Are fire protection methods in place at the facility?
- Does the facility have a fire brigade?
- Does the facility use industrial fire departments or private or contractual fire departments?
- Are fire extinguishing systems present at the facility?
- Does the facility have any automatic fire detection systems?
- Does the facility have any emergency employee alarms?

Subpart M—Compressed-Gas and Compressed-Air Equipment

- Does the facility have a compressed-air receiver?
- Does the facility have equipment that provides compressed air?

Subpart N—Materials Handling and Storage

- Does the facility store materials or have storage areas?
- Does the facility receive materials by rail or have a railway?

- Does the facility service single- or multi-piece rims for trucks, tractors, trailers, buses, and off-road machines? This does not apply to construction, agriculture, or longshoring.
- Does the facility utilize fork trucks, motorized hand trucks, or other specialized industrial trucks that are powered by electric motor or internal-combustion engines?
- Does the facility operate cranes of any type?
- Does the facility utilize derricks?
- Does the facility operate a helicopter to lift, move, or otherwise handle materials?
- Does the facility use any rope, chain, cable, or strap for slings that connect a load to a material-handling device for the purpose of hoisting or pulling?

Subpart O—Machinery and Machine Guarding

- Does the facility operate machinery that has any of the following characteristics including but not limited to point of operation area, ingoing nip points, rotating parts, or exposed or flying chips and sparks that create hazards to employees?
- Does the facility operate woodworking machinery?
- Does the facility operate any abrasive-wheel machinery?
- Is the facility in the rubber or plastics industries?
- If the above is true, does the facility operate mills and calendars?
- Does the facility operate mechanical-power presses, excluding the following machines: press brakes, hydraulic and pneumatic-power presses, forging presses and hammers, bulldozers, hot-bedding and hot-metal presses, riveting machines, and similar types of fastener applicators?
- Does the facility have a forging or die shop that uses lead casts or other uses of lead?
- Does the facility have power-transmission belts?

Subpart P—Hand and Portable Power Tools and Other Hand-Held Equipment

- Does the facility or its employees use hand tools that may pose a hazard to the employee if not maintained properly?
- Does the facility use compressed air for cleaning purposes?

- Does the facility use portable power tools that have an exposed point of operation that creates a hazard to the employee?
- Does the facility have any internal-combustion-engine-powered portable tools?
- Does the facility use portable jacks?
- Does the facility use portable blast cleaners?

Subpart Q—Welding, Cutting, and Brazing

- Does the work require the use of electric or gas welding and cutting equipment?
- Is the work to be done near a fire hazard?
- Is a mixture of fuel gas and air or oxygen to be used for welding or cutting?
- Are compressed-air or oxygen cylinders stored or transported?
- Are pressure generators utilized?
- Is arc welding and cutting equipment used?
- Is resistance welding utilized?

Subpart R—Special Industries

- Does the work involve the manufacturing and converting of pulp, paper, and paperboard?
- Are forklift trucks, hand/power trucks, cranes, ships, trucks, or railroad cars used in the moving of materials for manufacturing purposes?
- Does the work involve the design, installation, processes, operation, and maintenance of textile machinery (except synthetic fiber)?
- Is steam used in the processes?
- Does the work involve the design, installation, operation, and maintenance of machinery and equipment used within a bakery?
- Is flour-handling equipment used?
- Are steam pipes used?
- Does the laundry equipment utilize moving parts at the point of operation?
- Are chemicals or flammable liquids used in the processes throughout the facility?
- Are vehicles used to transport the material within the area?
- Are kilns utilized on the facility? Plywood, cooperage, and veneer manufacture are excluded from this standard.

- Does the work involve the normal operations included in the logging of pulpwood (excludes sawlogs)?
- Does the work involve the installation and processes performed at telecommunications centers and at telecommunications field installations?
- Is the work performed around energized power lines, batteries, or hazardous materials?
- Does the task require working in confined space or manhole worksites?
- Does the enterprise involve working with the operation and maintenance of electric-power generation, control transformation, transmission, and distribution lines and equipment?
- Does the work require the inspection of wood poles?
- Does the work require the trimming of trees around power lines?
- Does the enterprise require the operation of grain elevators, feed mills, flour mills, rice mills, dust-pelletizing plants, dry corn mills, soybean flaking operations, and dry grinding operations of soy cake?
- Is welding, cutting, or brazing to be performed?

Subpart S—Electrical

- Does the work include the installation and examination of electrical equipment used to provide electrical power and light for employee workplaces?
- Are all conductors and equipment approved?
- Are hoists or splices used during installation?
- Are there flammable or combustible hazards nearby?
- Is equipment used that requires markings?
- Are conductors utilized above ground that are near walkways, working areas, crossover roads or railways, near or over the tops of buildings, or near moving mechanisms such as cranes or booms?
- Are metal raceways, cable armor, and other metal enclosures used?
- Is temporary wiring done?
- Does wiring run through walls, floors, wood cross members, and the like?
- Does the work involve the installation or maintenance of electrical signs or directional lighting?
- Are cranes or hoists used to conduct the work?
- Are elevators, dumbwaiters, escalators, and moving parts used?
- Is the electrical work to be performed in an area that contains flammable vapors, liquids, or gases, or combustible dusts or fibers?

- Are there electrical systems over 600 volts?
- Is the electrical system for fixed installations or mobile equipment?
- Is the electrical system for emergency power supply?
- Is the employee to perform electrical work qualified or unqualified? The employee's risk of shock is not reduced by the electrical installation standards provided by the previous sections.
- Do the employees receive regular documented training?
- Is the work to be performed near energized or de-energized electrical equipment?
- Is corded portable electrical equipment used?
- Are employees working near areas where there is the potential for electrical hazard?

Subpart T—Commercial Diving Operations

- Is the diving work related to general industry, construction, shipbuilding, ship repairing, ship breaking, or longshoring?
- Are dive teams experienced and properly trained?
- Has the employer developed and maintained a safe-practices manual and made this manual available at the dive locations for all dive team members?
- Does the manual contain emergency and safety procedures, equipment procedures, and so on?
- Does the employer comply with the necessary safety and emergency requirements prior to each diving operation?
- Does the employer follow requirements that are applicable during all dives?
- Does the employer observe the required post-dive procedures?
- Are employees engaged in SCUBA diving?
- Does the employer engage surface-supplied air diving?
- Are employees engaged in mixed-gas diving?
- Are employees engaged in diving operations liveboating?

Subparts U–Y (Reserved)

Subpart Z—Toxic and Hazardous Substances

- Does the facility have any air contaminants listed in tables Z-1, Z-2, or Z-3?
- Does the facility contain any asbestos in any form?
- Does the facility have any of the following: 4-nitrobiphenyl, alpa-Napthylamine, Methyl Chloromethyl Ether, 3, 3'-Dichlorobenzidine, bis-Chlormethyl Ether, beta-Naphthylamine, Benzidine, 4-Aminodiphenyl, Ethyleneimine, beta-Propiolactone, 2-Acetylaminofluorene, 4-Dimethylaminoazobenzene, N-Nitrosodimethylamine, and Vinyl Chloride?
- Are employees at risk of exposure to inorganic arsenic, lead, cadmium or cadmium compounds, benzene, coke-oven emissions, occupational exposure to blood or other potentially infectious materials, or cotton dust?
- Does the facility manufacture yarn, perform slashing or weaving, or handle waste for textile operations?
- Are employees potentially exposed to 1, 2 dibromo-3-chloropropane, Acrylonitrile, Ethylene Oxide, Formaldehyde, or Methylenedianiline?
- Does the facility manufacture or import hazardous chemicals?
- Does this facility use hazardous chemicals in a laboratory?

REFERENCE

U.S. Department of Labor. 1993. *Occupational Safety and Health Standards for General Industry.* Washington, DC: U.S. Government Printing Office.

Appendix B

29 CFR 1926—OSHA Construction Standards Summary and Checklist

TABLE OF CONTENTS

29 CFR 1926—OSHA CONSTRUCTION STANDARDS

Introduction

The Code of Federal Regulations Title 29, part 1926, covers compliance safety issues for construction companies. This document is a consolidation of OSHA standards which are applicable to the construction industry. In publishing this standard, every effort was made to identify all the applicable General Industry standards from 29 CFR 1910. It should be noted that, under certain circumstances, other parts of 29 CFR 1910 may also be relevant. This is a short summary of the requirements needed to establish a safe work environment and should be used as a guide for compliance for companies, facilities, and other businesses.

The appendix is arranged so that 29 CFR 1926 standards, subparts A–X, appear first. This is followed by General Industry standards 29 CFR 1910 pertaining to construction, a subject index, and a listing of material approved for incorporation by reference. Note: Incorporation by reference is an established procedure whereby federal agencies meet publishing requirements by referring materials published elsewhere in the Federal Register

Subpart A—General

The first section of Subpart A is a general-purpose statement and an explanation of related policies. Subpart A is a general overview of the purpose, variances, inspections, and rules of practice used throughout the following subparts. In addition, the right of entry is addressed. This right of entry states that any contract subject to section 107 must allow the Secretary of Labor or a representative the right of entry. It is the right of the Secretary of Labor to conduct inspections or investigations to ensure compliance. The rules of practice for administrative adjudications for the enforcement of safety and health are also mentioned.

Subpart B—General Interpretations

Subpart B contains an interpretation and the applicability of construction health and safety standards to work performed under contract to government agencies. This subpart contains the general rules of the Secretary of Labor, which state that no contractor shall require any laborer to work in conditions that are unsanitary, hazardous, or dangerous to a person's safety and health. Section 107 of the Contract Work Hours and Safety Standards Act is addressed. As stated in section 107, a contract must be one which is entered into under statute that is subject to Reorganization Plan Number 14 and is for construction.

Further details on the Reorganization Plan Number 14 illustrates that appropriate regulations with respect to labor standards are met. There are 58 statutes listed in the Reorganization Plan Number 14. Rental Housing, Federal-Aided Highways, and Defense Housing are some statutes listed in the Plan. The intent of the subpart is to ensure a safe working environment at any federally assisted project. In addition, there is a discussion of the definition of contractor and subcontractor and their respective responsibilities. Other issues mentioned in this subpart were basic rules followed by contractors and how to use these in relation to construction. Issues like toilets or first aid are the responsibility of the prime contractor but can be delegated to other subcontractors. However, the prime contractor is never relieved of the responsibility. In either situation, all regulations must be followed under the contract.

Subpart C—General Safety and Health

Subpart C covers the contractor requirements for general safety and health. It states that no contractor or subcontractor may require a worker to perform hazardous work or perform in adverse conditions. A detailed discussion of contractor/employer responsibilities is broken into 11 broad areas that include:

- General Safety and Health Provisions
- Safety Training and Education
- Recording and Reporting of Injuries
- First Aid and Medical Attention
- Fire Protection and Prevention

- Illumination
- Sanitation
- Personal Protective Equipment (PPE)
- Acceptable Certification
- Shipbuilding and Ship Repair

In addition to the above referenced categories, there are other issues included in this subpart. For example, the contractor must also start and maintain accident prevention programs. Access to employee exposure and medical records is a key issue in this subpart.

Trade secrets are a big concern among employers and OSHA has allowed for them. Employers do not have to disclose secret formulas of dosages to inspectors or doctors in emergency situations if alternative information can be provided. However, under very special circumstances doctors may be authorized, following the special restraints of OSHA, to use the trade secrets in assisting them in their medical diagnoses. This procedure is governed by a confidentiality agreement of OSHA.

Further issues mentioned are means of exit. In general, all exits must be kept clear of any obstructions or impediments. Emergency escape procedures are also mandated by OSHA to ensure an effective and safe evacuation route. In light of emergency actions, alarm systems and detail planning routes should be utilized by all employers.

Subpart D—Occupational Health and Environmental Controls

In Subpart D, a diverse group of issues are discussed. The following is a list of some of the highlights mentioned in Subpart B.

- Medical services is the first mentioned. It states the employer shall ensure the availability of medical personnel for advice or consultation on the issues surrounding occupational health. For example, first-aid supplies shall be made available to employees.
- The second issue discussed is the number of toilet seats in a facility. This is designated by the number of employees at the site. Twenty or fewer people require at least one seat, twenty or more people have one seat and one urinal per forty employees, and two hundred or more people have one seat and one urinal for every fifty employees. Proper

washing facilities must also be present for those employees working with harmful substances.

- In the area of noise exposure, a chart lists the many combinations of time and noise levels. Ionizing and non-ionizing radiation requirements are stated, as well as all the different precautions needed.
- Approximately 500 different substances are listed in a Threshold Limit Value table showing the acceptable limits of airborne particles. There are many examples on the proper installation of exhaust fans on grinders and belt sanders. These exhaust fans help to disperse and filter the particles for compliance.
- Asbestos and its removal is another topic discussed in Subpart D. This section applies to nearly any work with or in close proximity to asbestos. An extensive program must be maintained while working with asbestos. Airborne particles have to be stringently monitored. Even the laundering of clothes used while working with asbestos must be controlled.
- Another main point illustrated in this subpart is Process Safety Management. This is a program used by those who frequently utilize very dangerous chemicals and other harmful substances. Process Safety Management is a system that helps to ensure a safe working environment for employees by using proper procedure and handling techniques with these materials.
- The final portion of this subpart deals with hazard communications. "This occupational safety and health standard is intended to address comprehensively the issue of evaluating the potential hazards of chemicals and communicating information concerning hazards and appropriate protective measures to employees" (1926.59(a)(2)).

Subpart E—Personal Protective and Lifesaving Equipment

This subpart deals with all the protective equipment that employers must supply to their employees for protection in the workplace. This includes personal protective equipment for the eyes, face, head, and extremities. In addition, protective clothing, respiratory devices, protective shields, and barriers shall be provided, used, and maintained in a sanitary and reliable condition wherever it is necessary because of hazards.

The employer is responsible for providing the employees with all the protective equipment that is needed for a particular job. They must also train the employees on how to properly use the equipment. Subpart E tells the employer what equipment is to be used for specific emergencies and where the equipment should be placed to ensure that all employees have knowledge of its location.

Hard hats are to be worn in areas where there is any possible danger of head injury from an impact of falling objects. Eye and face protection is to be used when machines or operations have the possibility of causing eye or face injury to the employee. This part also deals with the proper breathing apparatus needed in areas that are hazardous to the employees' health.

Subpart F—Fire Protection

Subpart F is concerned with fire protection, fire prevention, flammable and combustible liquids, liquefied petroleum, and temporary heating devices. Employers must have fire extinguishers in areas where fire is likely to occur. They must have the proper fire extinguisher for the hazard and the extinguishers must be approved by a nationally recognized testing laboratory. The facility must have either a temporary or permanent water supply that has enough pressure to put out a fire. A training course in proper fire extinguisher operation should be given to the employees to show how to operate fire extinguishers properly.

Flammable and combustible liquids should be stored in a container away from any activity which might cause an explosion. Most liquids must be stored in an approved room, and if they are stored outside of that room, it should be in quantities less than 25 gallons. In addition, this subpart informs the employer how close the employee must have the container for refilling operations and on moving containers from one area to another.

Tank storage is broken down into different areas such as welding tank storage and the different types of tanks. Specifications are also given in this subpart on how the tanks should be built. Each tank is required to have certain values that regulate the pressure in the tanks. Tanks are to be stamped or marked with the material the tank is holding. Finally, inspection of these tanks is required in order to prevent an accident from happening.

Fire detection systems are another way of preventing a fire from happening. The employer must inspect this equipment on a timely basis to make sure that it is working properly and is in compliance with the OSHA standard. The detection system must be located in the specified area and should be labeled to note its location.

Subpart G—Signs, Signals, and Barricades

This subpart details the signs required when hazards are present and the notifications required when an employee is near that hazard. There are six main signs—danger, caution, exit, safety instruction, directional, and traffic signs—that are used on a construction site. The subpart also states what color each sign must be and the size requirements for every sign.

Accident prevention tags are used to warn the employee of an existing hazard such as defective tools or equipment. Flagmen should be used in areas such as streets and highways to provide direction for vehicles to travel.

Subpart H—Materials Handling, Storage, Use, and Disposal

This subpart deals with the proper handling of materials. Stored materials must be stacked, racked, and secured to prevent sliding or falling. All materials must be stored away from aisles and passageways so employees can walk through these areas safely.

Rigging equipment, used for moving materials, is another factor associated with material handling. Every time material is moved, the equipment should be inspected prior to moving. This subpart breaks down the weight of materials that can be moved using a wire rope, which must also be inspected before use. There are several types of ropes that can be used for moving materials.

Proper disposal is another issue discussed in subpart H. When disposing of materials like scrap wood, such materials should be removed from the work area immediately to prevent employee injuries. Other materials such as oily rags and flammable liquids should be kept in covered containers until they can be removed from the worksite.

Subpart I—Hand and Power Tools

Subpart I of construction standard 29 CFR 1926 concentrates on the proper use and condition of both hand and power tools. Tool regulations on the jobsite apply to both employee-owned and employer-provided tools. This is important to note because so many times contractors and subcontractors are hired to work on the same jobsite.

Subpart I requires all power-operated tools that are designed with guards be equipped with those guards. Additionally, external protrusions such as wheels, pulleys, and shafts should be guarded according to the requirements presented by the American National Standards Institute (ANSI). Employees working near hazards or under hazardous conditions such as falling objects, dust, or harmful fumes will wear the appropriate personal protective equipment at all times.

Special hand tools shall be provided for the employee to prevent body parts from entering a danger zone of any machine. These tools should not be provided in lieu of using the appropriate guards. Large tools designed to operate in a fixed position must be securely anchored to the ground so that there is no potential for walking or tipping. According to Subpart I, hand tools should not be used if they are in any way damaged or unsafe. For example, chisels with mushroomed heads and wrenches with sprung jaws are unacceptable.

Proper use and handling of fuel-powered, pneumatic, and powder-actuated tools is also covered in Subpart I. Only employees who have been trained should operate powder-actuated tools. In addition, both power and hand tools should be maintained and serviced on a regular basis. Jacks are the last item covered under Subpart I. Proper blocking and securing of the lifted part is covered along with maintenance requirements. The standards also require the tagging of faulty or broken jacks so they will not be used by mistake.

Subpart J—Welding and Cutting

Subpart J of 29 CFR 1926 covers the methods and precautions associated with welding and cutting, arc welding, preventing fires, proper ventilation, and welding of materials. Subpart J covers the standards for most methods of welding and cutting and provides the guidelines for minimum worker safety when carrying out those tasks.

This subpart begins with gas welding and cutting. It addresses the proper storage and joist transportation of gas cylinders. In addition, Subpart J notes that damaged or deteriorated cylinders should not be used. Gas cylinders being used should be marked clearly with their names, using letters of at least one inch in height. Hoses used must be clearly distinguishable, and regulators and gauges should be in good working condition. All components should be inspected before being used.

Next, arc welding and cutting is covered. Proper grounding along with suitable cables and electrodes must be used when arc welding or cutting. Operators must be properly instructed on how to safely use the welding equipment and wear the appropriate personal protective equipment.

Fire protection and prevention is covered in the standards regulating welding and cutting. No welding is to be performed near flammable vapors and fumes, paints, or heavy dust. Fire extinguishing equipment must be available and in good working order at all times.

Finally, the subpart addresses acceptable methods of ventilation. The requirements for mechanical ventilation systems are listed in this subpart and it includes notes on confined spaces and toxic substance welding.

Subpart K—Electrical

Subpart K deals with the means of providing electrical power, both permanent and temporary, to a jobsite. Subpart K also states that the electrical standards do not apply to existing permanent installations that were on the site before construction began. There are four major divisions of this subpart, including installation safety requirements, safety-related work practices, maintenance and environmental considerations, and requirements for special equipment.

Installation Safety Requirements. Regulations require that all electrical parts be inspected for quality, durability, and appropriateness. Systems must be properly installed and grounded before use. Supports, holders, and equipment must also be grounded with only a few exceptions.

Safety-Related Work Practices. Subpart K describes acceptable and non-acceptable wiring methods for both temporary and permanent electrical installations. Temporary installations must be removed immediately when the work is finished. When applicable, the standards from the *National Electrical Code* (NEC) should be adhered to when performing any electrical wiring.

Maintenance and Environmental Conditions. Receptacles must be designed for their locations. For example, when one installs a receptacle in a damp or wet location, special grounding methods must be used to prevent accidental contact with water. Proper equipment should be used for those receptacles with special voltages such as air conditioners and clothes dryers.

Special Equipment. Power supplies for equipment such as elevators, escalators and moving walks are addressed under Subpart K. There are also requirements for electrical installations in hazardous environments such as where dust, fumes, and flammable vapor exist. Only equipment designed for the specific location should be used. Batteries and battery charging are regulated under this subpart. Proper handling of battery acids and charging equipment is addressed.

Subpart L—Scaffolding

Subpart L begins by noting that scaffolding will not be used unless it is constructed according to regulations. Footings and anchors must be sound and capable of carrying the intended load. No make-shift footings like bricks and concrete blocks can be used. Scaffolding should always be erected under the supervision of a competent person, which is defined in the subpart. Guardrail and toeboard regulations apply to all the open sides of scaffolding. Different types of scaffolding such as wood pole, tube, and outrigger scaffolds; proper construction and bracing; and anchoring requirements for scaffolding are covered in Subpart L.

After erection, all scaffolding must be capable of supporting its intended loads. This is also applicable to mobile scaffolding—scaffolding on wheels—which is addressed using 1910 standards. Employees should not work on scaffolding when high winds or stormy conditions pose a hazard. Scaffold work areas must be kept clean and free from the accumulation of tools and materials.

Scaffolds that exceed heights of 50 feet may only be erected by or under the supervision of the scaffold's manufacturer or designated agent unless the structure is approved by a licensed professional engineer. Ladders on scaffolding must be properly constructed and have slip-resistant treads. Scaffolding should never be unstable or of shoddy construction.

Subpart M—Floor and Wall Openings

This subpart applies to both temporary and emergency conditions where an employee may be susceptible to the danger of materials falling through roof, floor, or wall openings, or from runways or stairwells. In general, all floor openings will be guarded by a standard railing on all sides except at entrances to the stairway, and by toeboards or a cover. The height or placement of the wall opening shall be guarded by a standard or immediate rail. A standard toeboard or an enclosing screen should be used to protect a bottom-wall opening.

Types of floor holes and openings include skylights, ladderways, and temporary openings, and floor holes should be guarded by standard railings on all exposed sides and by standard toeboards. A ladderway opening should have an entrance that is blocked by a swinging gate or some other mechanism to prevent a person from falling into the entrance. Floor holes may use a cover in place of the standard railings. However, if the cover is not in use, the floor hole must be guarded by these standard railings.

Pits, trap-door openings, and manholes will be guarded by standard covers of normal construction strength. When the cover is not in place for pits or trap doors, the opening must be protected on all exposed sides by removable standard railings. If the manhole cover is not in place, then the opening will be protected by standard railings. Hatchways and chute floors must have standard railings with only one exposed side or removable railings with toe boards on two sides and fixed railings on the other side.

Standard railings consists of a top rail, intermediate rail, posts, and toeboards. The standard for toeboards includes a 4-inch minimum height from the top edge to the length of the floor, ramp, platform, or runway. Handrails shall have the same construction as a standard railing with the exception that the handrail is mounted onto some surface and there is no immediate rail involved. Floor openings such as manholes or trenches shall be capable of supporting the maximum intended load. Wall openings must withstand at least 200 pounds applied from any direction.

Subpart N—Cranes, Derricks, Hoists, Elevators, and Conveyors

Subpart N deals specifically with the operation of cranes, derricks, hoists, elevators, and conveyors in the construction industry. In this subpart, employees shall comply with the applicable manufacturer's standards for cranes and derricks. However, when there are no manufacturer's standards, the limits are assigned based on the determination of a qualified engineer. Warnings, special hazards, speed limits, load capacities, and additional instructions shall be posted on all equipment. Annual inspections shall be performed on the equipment by a competent person, government official, or by the U.S. Department of Labor.

This standard applies to such items as crawler, locomotive, truck, overhead, gantry, and tower cranes; helicopter operations; material and personnel hoists; elevators; and conveyors. Loads on the overhead cranes must be labeled and legible from the ground floor. The proper clearance must be maintained between moving and rotating structures on the crane and fixed objects. Employees on these cranes shall be protected from falling by guardrails or safety belts.

Mobile cranes may be mounted on barges but shall not exceed the original capacity set by the manufacturer. The use of a crane or derrick to hoist employees on a personnel platform is prohibited, except when the use or dismantling of a worksite would provide more danger because of the worksite's structural design. Load lines shall be capable of supporting seven times the maximum-intended load. Brakes on these cranes should only be used when the personnel platform is in a stationary position. Personnel platforms will be equipped with guardrails, toeboards, access gates, and headroom for employees to stand upright on the platform.

A trial lift must be performed with an unoccupied personnel platform at least once before the lift of the platform is taken from the ground floor. This trial run must be performed immediately before placing personnel on the platform and must be repeated whenever the cranes or derricks are moved or adjusted. A visual inspection of the crane, derricks, rigging, and personnel platform must be performed by a competent person immediately following the trial lift to determine if any defects or adverse effects have occurred.

Helicopter cranes must comply with the regulations set by the Federal Aviation Administration. The operator of the helicopter is responsible for determining if the load's weight, size, and manner that it is attached to the helicopter can be done safely. Static charge must be eliminated with a grounding device before personnel touch the load being suspended by the helicopter. Visibility is essential for the helicopter operator so he can keep in contact with the ground crew who is sending signals to the operator. When approaching or leaving a helicopter, all personnel must remain in full view of the pilot. Constant communication is essential between the pilot, the ground crew, and other employees to prevent accidents.

All hoists must comply with the manufacturer's specifications, but when they are not available, the limitations of the equipment shall be based on the determination of a professional engineer. Operating procedures should include a signal system, allowable line speeds, and a sign labeled "No Riders Allowed." Hoist cars shall be used in lifting personnel and will be permanently enclosed on all sides and at the top. Safety mechanisms should be capable of stopping or starting the car at anytime. Before placing them into service, hoists must be tested, inspected, and constantly maintained on a weekly basis. Employees using conveyors will sound an alarm before they start operation and must be capable of stopping it at anytime from the operator's station. In a place where employees come into contact with the conveyor, those individuals should be specially protected.

Subpart O—Motor Vehicles and Mechanized Equipment

All equipment not being used at night or near a highway or construction site must have either the appropriate lights, reflectors, or barricades to identify the location of the equipment. All controls on machinery must be in the neutral position unless work is being done with the machines. The neutral position involves the motor being off and the brakes being set, as if the machine were in a parked position. Safety glass will be used in all cabs and protection will be used when removing tires.

Motor vehicles include those vehicles that do not operate in public traffic, but within an off-highway jobsite. These vehicles will have a service, emergency, and parking brake system and shall be in operable condition. The vehicles will have two headlights and taillights, brake lights, a reverse-signal alarm system, and an audible warning system. Those vehicles with cabs will include windshields, power wipers, and a defogging system. Vehicles used to transport employees will have seats firmly secured in the cab with seat belts that meet Department of Transportation Standards.

Industrial trucks shall meet the following requirements: The capacity will be posted on the vehicle so it is clearly visible to the operator. No modifications should be made to the equipment without the consent of the manufacturer. Protection such as overhead guards or safety platforms should be secured to the lifting mechanism. Finally, all industrial trucks must meet all the ANSI standards for design, construction, stability, inspection, testing, maintenance, and operation.

All boilers and piping systems, either part of or on pile-driving equipment, will have overhead protection, guards, and stop blocks. A blocking device must support the weight of the hammer and will be provided for placement in the leads under the hammer while employees work. All employees shall follow the signals given by the designated signalmen and will be kept clear when piling is being hoisted into the leads. When piles are being placed into a pit, the pit shall be angled.

Marine operations include all loading, unloading, moving, handling, and so forth on or out of any vessel from a fixed structure onto a shore vessel. Ramps for the access vehicles between barges must be of adequate strength, protected by side boards, and properly secured. Unless employees can step safely across each barge, a safe walkway such as a "Jacob's Ladder" must be provided. This ladder will be of a flat-tread type, hang without slacking, and be maintained and properly secured.

Subpart P—Excavations

This subpart applies to all those open excavations, cuts, cavities, ortrenches made in the earth's surface. Details concerning employee exiting and hazardous atmospheres are of primary importance. In addition, instructions are given regarding underground utilities, necessary protective systems, and soil classifications for trenching, cutting, and excavating.

All underground utilities such as water, sewer, and gas must be identified prior to excavating. Any utility companies shall be contacted in order to advise the excavation crew on the location of these utilities and the proper means for encountering the utility lines. When the excavation is made, the utilities shall be protected or removed in order to safeguard employees while they are working. Structural ramps should be designed by a competent person and in accordance with the design of access and egress from the excavations. A means of egress must be located in trench excavations that are 4 feet deep, and no more than 25 feet deep in order for employees to exit and enter the excavation.

Employees must monitor for hazardous atmospheres to prevent the possible exposure to harmful levels and atmospheric contaminants. If oxygen levels are less than 19.5 percent, the atmosphere must be tested before employees can enter. Oxygen-deficiency exposure precautions should include proper ventilation, respiratory protection, and mandatory testing. If hazards arise while employees are present in the excavation, emergency rescue equipment such as a breathing apparatus, safety harnesses, and stretchers should be readily accessible.

Employees will not work in excavations where water has accumulated, unless measures have been taken to remove the water from the excavation. If excavation operations endanger the structure of buildings, support systems such as shoring, bracing, or underpinning shall be used to ensure the building's stability. Excavation should not be conducted under the base of any foundation because it can put employees in danger. Employees shall also be protected from loose soil or equipment that may fall into the excavation pit. Daily inspections of the excavation, its adjacent areas, and the protective systems shall be made by a competent person prior to the start of work as needed throughout the shift. Appendices A, C, and D cover excavation wall-shoring procedures.

Subpart Q—Concrete and Masonry Construction

This subpart sets forth the requirements that protect all construction employees from the hazards associated with concrete and masonry construction. In addition, this subpart includes the relevant general industry standards as they apply to this standard. Requirements for bull float, formwork, lift slab, precast concrete, re-shoring, shoring, vertical slip forms, and jacking operations are also discussed in this subpart.

In relation to structural integrity, all drawings and plans should be available at the worksite. Shoring and reshoring is discussed to ensure that maximum strength is achieved. Any design of the shoring should be done by an engineer or a qualified individual. All reinforcing steel should be adequately supported to make it stable. No framework should be removed until the employer determines the concrete is strong enough to support all loads. Plans, specifications, and tests are used to ensure the concrete has gained sufficient strength.

All precast wall units should be adequately braced to prevent overturning until other permanent structures are in place to support them. Any lifting points attached to the precast lift-up walls should be able to support two times the maximum-expected load. In addition, lifting points or attachments on pieces other than lift-up walls must be able to support four times the maximum-expected load. Hardware should be capable of lifting five times the maximum-expected load to be lifted.

All lift slab operations should be designed by an engineer, and plans should be included with the instructions on proper methods of construction. All jacks should be labeled with their rated capacities, and should be capable of lifting at least two-and-one-half times the load being lifted. Jacks should have a safety device that allows them to stop and support the weight if the load becomes too great to lift or if the jack malfunctions.

If any jacking point goes farther than one-half inch from level, jacking operations should stop and the problem fixed. There should be an adequate number of jacks present as specified in this section, but the number should never exceed 14 jacks at one time.

Subpart R—Steel Erection

The requirements for flooring, structural steel assembly, bolting, riveting, fitting-up, and plumbing-up are discussed in this subpart. The standard also addresses the Personal Protective Equipment (PPE) standards for steel erection.

Permanent floors should be installed as progress on the building continues. Information is given on the correct process for laying a temporary floor. These floors must be at least two inches thick or thick enough to support the load. They must also be bolted or otherwise tightly connected to the floor. When any beam or structure is being lifted, the hoisting line should not be loosened until there are at least two bolts at each joint that have been tightened with a wrench. Braces are needed on trusses spanning greater than 40 feet before the hoisting line is loosened. Guide lines should be attached to long beams being lifted in order to control them.

All materials, tools, rivets, bolts, and so on should be contained so they do not fall. All pneumatic tools should be disconnected from the power source before any adjustments are made to them. Turnbuckles used for plumbing-up should be properly secured and should not unwind under load. Connection points should also be easily accessible. The use of a safety harness is also stressed in this section.

Subpart S—Underground Construction, Caissons, Cofferdams, and Compressed Air

Subpart S applies to the construction of underground tunnels, shafts, chambers, and passageways. This includes the following conditions:

- Entrances and exits to underground/enclosed areas
- Air quality of underground/enclosed area
- Equipment used in these areas
- Flood controls used in these areas
- Environmental conditions including pressurization and flammability

Items such as check-in/check-out, safety instructions, communication requirements, rescue teams, and hazardous classifications for potentially gassy operations are covered.

There should be safe entrances and exits to tunnels and any unused tunnels or chutes are to be sealed. A check-in/check-out procedure should be used to keep a record of the number of people below ground.

Monitoring should be conducted to determine whether there is a possibility of a "gassy" situation below, that could cause an explosion. If the person who monitored for gassy situations determines that a hazardous situation exists, the employer is responsible for taking the proper action. If an individual determines that 20 percent of the lower explosive limit for a gas has been detected, then all personnel should be removed from the area except those necessary to resolve the problem.

All electrical devices should be turned off except for the necessary ventilation pumps. Local gas tests should be performed before and during any welding or other hot work. Exposure to toxic substances should be recorded and the proper fresh air should be supplied to all underground areas.

Open flames are prohibited below ground except for welding or other hot operations. All flammable materials that are permitted underground must be moved carefully and properly. Any leaks or spills should be cleaned up immediately. Proper fire extinguishers should be located below ground. Fire-retardant materials may be necessary around or within tunnels.

Proper bracing should be erected to support underground tunnels. When replacing damaged supports, the new ones should be in place before the damaged ones are removed. Any shaft deeper than 5 feet that workers enter must be encased by steel, concrete, or timber supports, unless the shaft is cut through rock. After any blasting operation, the shaft must be inspected by a competent person to make certain that it is safe.

When a hoist is in use to lift materials through a shaft, the materials should be secured so they will not fall. In some cases, a person may be needed at the bottom of the shaft to keep people clear while a load is being lowered to the bottom. Personnel should not ride the hoist unless it is specifically for human use or the person is inspecting the shaft. Wire rope used in the hoist must be capable of lifting five times the expected load, or the factor recommended by the rope manufacturer, whichever is greater. A record must be kept of the periodic inspections of lifts. Cages for transporting people must be specifically designed for this purpose and must not exceed certain speeds.

Subpart T—Demolition

A survey should be done by a qualified individual of any structure that is to be demolished. The survey is designed to check the integrity of the framework in order to prevent unplanned collapses. In a building damaged by fire, flood, or explosion, supports should be erected for walls and floors. All gas, sewer, steam, and electrical lines should be cut off prior to demolition. If hazards exist from glass or any other material, then the material should be removed. When material is dropped through holes without a chute, there should be a wall surrounding the landing area that meets all specifications. Any holes not used for dropping debris should be substantially covered, and employee entrances to the building should be protected from falling debris.

Only stairs, passageways, and ladders approved for use in the demolition process are to be used in these buildings. These stairs or ladders are to be periodically inspected. Any stairwell being used in a multilevel building must be properly illuminated and have protective roof coverings where necessary.

A chute with an angle steeper than 45 degrees must be completely covered. The bottom of chutes should be closed off when work is not in progress. Top chute openings should be protected with guard rails and any space between the edge of the chute and the side of the building must be covered. Chutes must be strong enough to handle the falling debris.

In addition, all holes cut in the floor must not exceed a certain size and should not compromise the structural integrity of the rest of the floor. In steel and masonry constructions, loose masonry must be removed before continuing down to the next level. Walkways or ladders must be available for access to scaffolds or walls. Walls which serve as retaining walls for debris must be capable of retaining the debris. Once floors have been removed, walkways at least 18 inches wide must be installed. These walkways must be substantial and planks should overlap at least one foot over solid beams.

Floors must be strong enough to support the equipment being used. Barriers should be erected to keep equipment from falling over the side or through holes. The stored material should not exceed the capacity of the floor. Storage space in which material is to be dumped or dropped is to be concealed.

Except for designated people, no one is to be near the area where mechanical demolition is taking place. Any crane and ball must meet certain requirements. All structural steel beams should be cut before using the demolition ball. Continued monitoring should be conducted throughout the demolition process to check structural integrity of other parts of the building.

Subpart U—Blasting and the Use of Explosives

Only authorized and qualified persons can handle and use explosives. No person handling explosives should be under the influence of alcohol or drugs. Explosives should be accounted for at all times and never be abandoned. Fires should not be fought if there is an imminent danger of contact with explosives. Above-ground blasting should be done between sunrise and sunset. Precautions should be taken to prevent accidental discharge of electric blasting caps from other sources of current. Operators or owners should be notified when blasting will be done near power lines. Loading and firing should be done only by competent individuals. Blasters should be competent, in good physical condition, qualified, and be required to provide evidence of their credentials. They should also be knowledgeable in the use of each type of blasting method. No smoking or flame-producing devices are allowed in the area.

Transportation of explosives shall meet the provisions of the Department of Transportation. The electrical system of trucks should be checked weekly. Auxiliary lights are prohibited. Explosive equipment and supplies should be conveyed singly. They should not be transported on locomotives and the materials should be carried in separate containers. The powder car or conveyance should have a reflectorized sign with the word "Explosives."

Explosives and blasting agents should be stored in approved IRS. Materials should be stored separately from explosives and blasting agents. In addition, smoking is not allowed within 50 feet of explosives. Two modes of exit must be identified if explosives are stored underground.

A warning signal should be given prior to the blast. Flagmen should be stationed on highways, and the blaster fixes the time of blasting. Blasting signals should be posted in a place for employees to study.

Detonators and explosives should not be stored in tunnels, shafts, or caissons. Detonators and explosives are carried separately into the working chambers. The blaster or powder man is responsible for the management aspects.

Subpart V—Power Transmission and Distribution

General requirements relating to power transmission and distribution should identify the proper procedures and actions for initial inspections, tests, and determinations. Gloves or gloves with sleeves should be worn for insulation from the energized parts. Employees should maintain the minimum working distances for various voltage ranges. Lines or equipment should be deenergized before repairing, and guards or barriers should be erected as necessary. Employers should provide training for workers so they all are familiar with emergency procedures and first aid.

Rubber equipment should be visually inspected and gloves should have an air test done before use. Protective hats and climbing equipment should be worn. Safety lines should be used in emergency rescue situations. Metal or conductive ladders should not be used near energized lines or other electrical equipment.

Mechanical equipment should be visually inspected daily. Aerial lift trucks should be grounded or barricaded when working near energized lines or they should be insulated for the work being performed. For material handling, materials should be examined before unloading can begin. When hauling poles, the load should be secure and have a red flag attached on the back. Materials should not be stored near energized equipment if it can be stored elsewhere. Tag lines should be used to control loads handled by hoisting equipment. Grounding for protection of employees is also very important. New construction cannot begin unless the equipment is grounded or other means have been implemented to prevent contact with energized lines.

Guarding and ventilating street openings to underground lines should be done. Oxygen levels should be monitored for the presence of fumes or gases. Underground facilities such as gas or telephone lines should be protected from damage when exposed to energized parts. Energized facilities should be identified and the protective equipment needed should be specified. Deenergizing of equipment or barricades should be used for the particular situation. It is important for body belts, safety straps, and lanyards to be used by all personnel. The hardware should be forged or pressed steel and tensile tests should be performed to prove reliability. Testing of safety straps, body belts, and lanyards should be performed regularly.

Subpart W—Rollover Protective Structures and Overhead Protection

Rollover protective structures as well as overhead protection are included in this section's regulations. This includes not only heavy construction equipment but also farming equipment, mostly tractors, used by the general public. There are rollover protective structures (ROPS) for material handling equipment. If a ROPS is removed, it should be remounted with one of equal or better quality. Labeling should include the manufacturer's name, the model number, and the machine make or model. Machines should meet the requirements of the State of California, the U.S. Army Corps of Engineers, or the Bureau of Reclamation of the U.S. Department of the Interior.

There are minimum performance criteria for ROPS on designated scrapers, loaders, dozers, graders, and crawler tractors. There should be material, equipment, and tie-down means adequate for the vehicle frame to absorb the applied energy. A testing procedure that includes the energy absorbing capabilities of ROPS, the support capability, and the low-temperature impact strength of the material used in the ROPS should be performed.

There are protective-frame (ROPS) test procedures and performance requirements for wheel-type agricultural and industrial tractors used in construction. The protective frame is a structure mounted to the tractor that extends above the operator's seat. It is also used to minimize the chance of injury resulting from normal operations. A field upset test, either a static or a dynamic test, is done to determine ROPS performance. The frame, overhead weather shield, fenders, or other parts may be deformed but should not shatter or leave sharp edges.

Overhead protection for operators of agricultural and industrial tractors is also required under Subpart W. The purpose of overhead protection is to minimize the possibility of operator injury from falling objects or from upset of the cover itself. Overhead protection may be constructed of a solid material and should not be installed in a way that causes a hazard. A drop test should be performed following either the static or dynamic test. The same frame should then be subjected to a crush test.

Subpart X—Stairways and Ladders

This subpart concerns stairways and ladders used in construction, alteration, repair, and demolition of workplaces. This section deals with the stairways and ladders used during one of the following events:

- Demolition of buildings and site
- Repair of building and site
- Construction of worksites
- Alteration of buildings and site

It also is applied when ladders and stairways are required to be provided for certain circumstances. A stairway or a ladder is required when there is a break in elevation of 19 inches or more and no ramp, runway, sloped embankment, or personnel hoist has been provided. Employers are required to provide and install stairwells, ladders, and fall-protection systems before employees can begin work.

Stairs should be installed between 30 degrees and 50 degrees from the horizontal. Variations in riser height and tread depth should not be over one-fourth inch in any stairway system. All stairways must be free of hazardous projections and slip-resistant. Stairways having four or more risers or a rising of more than 30 inches, whichever is less, are required to have at least one handrail and one stair rail system along each unprotected edge. Handrails and the top rails of stairwell systems should be capable of withstanding at least 200 pounds at any point along the top edge. Unprotected sides and edges of stairway landings should be provided with guardrail systems.

Ladders should be capable of supporting specified loads without failure. Ladder rungs, cleats, and steps should be parallel, level, and uniformly spaced. The rungs of individual-rung step ladders should be shaped so that employees' feet can't slide off. They should be coated with a skid-resistant material or treated to minimize slipping. Ladders should not be tied together to provide longer sections unless they are designed forth at purpose, and ladders should be surfaced in a way to prevent injury caused by punctures or lacerations.

When the total length of a climb equals or exceeds 24 feet, fixed ladders must have one of the following: a ladder safety device, a self-retracting life line and rest platform or a cage or well, and multiple ladder sections. Cages and wells should conform to various specifications. When using a ladder to gain access to an upper-landing service, the ladder side rails must extend at least 3 feet above the upper-landing service, or the ladder should be secured at its top to a rigid support that will not deflect. Ladders should be maintained free of oil, grease, and other slipping hazards and should not be loaded beyond their maximum-designed load. They should be used on stable and level surfaces unless secured to prevent an accident. When an employee uses a ladder, she should always be facing it. The area around the top and bottom of a ladder should be kept clear and ladders should never be moved or extended when occupied. The top panel of a stepladder should never be used as a step. Ladders should be inspected for defects on a periodic basis and after any accident. Repairs should restore the ladder to its original condition before it is returned to use. Single-rail ladders are prohibited for use.

If necessary, employers should provide a training program on the use of ladders and stairways. The employer should ensure that each employee is trained by a competent individual. Employees should be able to understand the nature of fall hazards, the correct procedures for constructing and maintaining fall-protection systems, the correct use of stairways and ladders, and the maximum-intended carrying capacities of ladders. Retraining should be provided for each employee if necessary.

CHECKLIST FOR SUBPARTS A–X

Subpart A—General

- Does your company have to comply with the construction standards?
- Does your company need to be granted a variance from the safety and health standards?
- Does your company make special accommodations for inspectors?
- Is your company aware of the secretary's right of entry procedure?

Subpart B—General Interpretations

- Is your contract one of the 58 statutes listed in the Reorganization Plan No. 14?
- Does your company perform work under federal contracts?
- Does your contract entail the use of subcontractors?
- Does your company delegate responsibility to subcontractors in contracts?

Subpart C—General Safety and Health

- Does your company have accident-prevention programs in place?
- Does your company promote education for prevention of unsafe working conditions?
- Does your company have an elaborate fire protection and prevention program?
- Does your company keep detailed medical records of employees; if so, how long?
- Does your process incorporate any trade secrets that should not be disclosed to doctors?
- Are all of your exits unlocked and free from impediments?

Subpart D—Occupational Health and Environmental Controls

- Does your company have a written HAZCOM program?
- Does your company offer medical assistance or advice?
- Does your company have a sufficient potable water system?
- Do you have any areas of high noise exposure? How loud?
- Does your facility contain any asbestos?
- Does your facility have the proper illumination devices?
- Have you incorporated a Process Safety Management System?
- Do you have the proper personal safety equipment on hand?

Subpart E—Personal Protective and Lifesaving Equipment

- Do the employees at your facility use PPE or lifesaving equipment?
- Are your employees trained on the proper use of PPE?
- Does the facility have the protective equipment for its employees?
- Does the facility have the proper safety equipment needed?
- Does the facility provide safety classes for its employees?
- Does the facility provide the proper respiratory protective equipment?

Subpart F—Fire Protection

- Does your company have a fire protection and prevention program in place on-site?
- Does the facility have the proper fire extinguisher in areas needed?
- Does the facility provide training to the employees on extinguishers?
- Does the facility have a sufficient water supply to put out a fire?
- Does the facility have storage area for flammable liquids?
- Does the facility detection system meet OSHA standards?

Subpart G—Signs, Signals, and Barricades

- Does the facility have safety signs in place?
- Does the facility have the proper coloring on the signs?
- Do the employees have knowledge of the signs?
- Does the facility have the signs in the proper areas?

Subpart H—Materials Handling, Storage, Use, and Disposal

• Do the employees of your facility use proper techniques and equipment when handling materials on-site?
• Does the facility have materials stored in proper areas not blocking aisles?
• Does the facility have proper ropes and wires when moving materials?
• Does the facility dispose of the materials correctly?

Subpart I—Hand and Power Tools

• Is there, on a permanent basis, any power tools or hand tools for use by the company on-site?
• Is there machinery that must be in a fixed position requiring anchors?
• Are there fans with blades less than 7 feet above the floor?
• Is there equipment requiring a special hand tool or tools to keep fingers, hands, or any other body part from being harmed?

Subpart J—Welding and Cutting

• Do you employ any full- or part-time welders to weld or cut materials on the site?
• Do you store or use gases commonly used in welding operations?
• Is there any welding required on the job for more than just minor repairs?
• Do you have repairs requiring major welding or cutting?

Subpart K—Electrical

• Does the establishment accommodate elevators, escalators, or powered walks?
• Is there any construction on the site using power tools or any temporary power installation?
• Are there circuit breakers on the premises?
• Does the environment consist of hazardous surroundings such as extremely wet, dusty, or vapor-rich conditions?

Subpart L—Scaffolding

- Is there any construction on the grounds requiring scaffolding?
- Is mobile scaffolding used in the operation?
- Are you at any time required to construct any sort of scaffolding to perform certain tasks?
- Are there any renovations or remodeling that would require scaffolding to be erected on the premises?

Subpart M—Floor and Wall Openings

- Does the facility have floor openings or floor holes?
- If there are wall openings in a facility, do they consist of more than a drop of 4 feet?
- Have you checked the guidelines pertaining to the width of your roof in Appendix A of Subpart M of the Construction Standards in 29 CFR 1926?

Subpart N—Cranes, Derricks, Hoists, Elevators, and Conveyors

- Are cranes of any type being used in your facility?
- Does your facility use hoists for materials, personnel, or overhead objects that are hard to reach?
- Are there conveyor systems located anywhere in your facility and, if so, do they send a warning signal before starting their operation?

Subpart O—Motor Vehicles and Mechanized Equipment

- Are there any motorized vehicles being used at your facility?
- Is any material-handling, earth-moving, or pile-driving equipment used at your company?
- Are operations occurring between moored vessels and your company's personnel or equipment?
- If motorized vehicles are present on-site, are they equipped with seat belts and adequate safety protection?
- Is your facility near a body of water, and, if so, are there ongoing operations between the land vessel and the marine vessel?

Subpart P—Excavations

- Will there be any forms of excavations at your facility?
- If excavations are present at your place of business, have you considered the location of underground utilities, means of access and egress, and the possibility of hazardous atmospheres prior to excavating on-site?
- What type of support and protective systems are going to be utilized when dealing with the excavations, safety mechanisms, and protection from cave-ins?
- Has the soil at the desired location for the excavation been classified?

Subpart Q—Concrete and Masonry Construction

- Does your company ever perform work such as lift slab, precastconcrete, shoring, jacking, or similar operations?

Subpart R—Steel Erection

- Does your company erect steel buildings or perform other steel erection operations?

Subpart S—Underground Construction, Caissons, Cofferdams, and Compressed Air

- Does your company construct underground tunnels, shafts, chambers, or passageways?

Subpart T—Demolition

- Does your company demolish existing structures via mechanical or explosive means?

Subpart U—Blasting and the Use of Explosives

- Does your company use explosives for blasting or other purposes?
- Is the blaster qualified by training, knowledge, or experience in the field of explosives?

- In transporting explosives, does the vehicle have placards on all four sides?
- Are explosives and blasting agents being stored separately?
- Are safety fuses being used in areas where extraneous electricity is present?
- When blasting underwater, are vessels at least 1,500 feet away?

Subpart V—Power Transmission and Distribution

- Does your company erect or improve electrical transmission lines and equipment?
- Are gloves being worn for insulation and are employees maintaining minimum working distances for different voltages?
- Has the equipment been grounded or has other means been implemented to prevent contact with energized lines?
- Have the oxygen levels and the presence of fumes or gases been tested before working on underground lines?
- Are the equipment and rigging being inspected regularly?

Subpart W—Rollover Protective Structures and Overhead Protection

- Is your company's heavy equipment in good condition and equipped with roll cages?
- Is the equipment certified?
- Are the material, equipment, and tie-down means adequate for the vehicle to absorb the applied energy?
- Does the labeling include the manufacturer's name, the model number, and the machine make or model?
- Have a drop test and a crush test been done?

Subpart X—Stairways and Ladders

- Are stairways and ladders used within your facility?
- Are stairways and ladders not being used in your facility when they should be in use?

- Is there a break in elevation of 19 inches, or is there no ramp, runway, sloped embankment, or personnel hoist provided?
- Has the employer installed a fall-protection system?
- Is the stairway or ladder free of hazardous objects and treated to minimize slipping?
- Are the unprotected sides and edges of stairways provided with guardrail systems?
- Are ladders being used on stable, level surfaces, or have they been secured?

Appendix C

Anthropometric Data

TABLE OF CONTENTS

493

Table C1.   Standing Body Dimensions (From MIL-STD-1472D)

Percentile Values in Centimeters

	5th Percentile			95th Percentile		
	Ground Troops	Aviators	Women	Ground Troops	Aviators	Women
Weight (kg)	55.5	60.4	46.4	91.6	96.0	74.5
Standing Body Dimensions						
1. Stature	000.8	000.2	46.4	000.6	000.7	000.1
2. Eye Height (standing)	000.1	000.1	000.4	000.3	000.2	000.2
3. Shoulder (acromiale) Height	000.6	000.3	000.9	000.2	000.8	000.7
4. Chest (nipple) Height	000.9	000.8	000.0	000.5	000.5	000.8
5. Elbow (radiale) Height	000.0	000.8	94.9	000.8	000.0	000.7
6. Fingertip (dactylion) Height		61.5			73.2	
7. Waist Height	96.6	97.6	93.1	000.2	000.1	000.3
8. Crotch Height	76.3	74.7	68.1	91.8	92.0	83.9
9. Gluteal Furrow Height	73.3	74.6	66.4	87.7	88.1	81.0
10. Kneecap Height	47.5	46.8	43.8	58.6	57.8	52.5
11. Calf Height	31.1	30.9	29.0	40.6	39.3	36.6
12. Functional Reach	72.6	73.1	64.0	90.9	87.0	80.4
13. Functional Reach, Extended	84.2	82.3	73.5	000.2	97.3	92.7

Percentile Values in Inches

	5th Percentile			95th Percentile		
	Ground Troops	Aviators	Women	Ground Troops	Aviators	Women
Weight (kg)	000.4	000.1	000.3	000.9	000.6	000.3
Standing Body Dimensions						
1. Stature	64.1	64.6	60.0	73.1	73.9	68.5
2. Eye Height (standing)	59.5	59.9	55.5	68.2	69.0	63.9
3. Shoulder (acromiale) Height	52.6	52.5	48.4	60.7	60.9	56.6
4. Chest (nipple) Height	46.4	47.5	43.0	53.7	54.5	50.3
5. Elbow (radiale) Height	39.8	41.3	37.4	46.4	47.2	43.6
6. Fingertip (dactylion) Height		24.2			28.8	
7. Waist Height	38.0	38.4	36.6	45.3	45.3	43.4
8. Crotch Height	30.0	29.4	26.8	36.1	36.2	33.0
9. Gluteal Furrow Height	28.8	29.4	26.2	34.5	34.7	31.9
10. Kneecap Height	18.7	18.4	17.2	23.1	22.8	20.7
11. Calf Height	12.2	12.2	11.4	16.0	15.5	14.4
12. Functional Reach	28.6	28.8	25.2	35.8	34.3	31.7
13. Functional Reach, Extended	33.2	32.4	28.9	39.8	38.3	36.5

Table C2. Seated Body Dimensions (From MIL-STD-1472D)

Percentile Values in Centimeters

	5th Percentile			95th Percentile		
	Ground Troops	Aviators	Women	Ground Troops	Aviators	Women
Standing Body Dimensions						
14. Vertical Arm Reach, Sitting	000.6	000.0	000.4	000.8	000.2	000.4
15. Sitting Height, Erect	83.5	85.7	79.0	96.9	98.6	90.9
16. Sitting Height, Relaxed	81.5	83.6	77.5	94.8	96.5	89.7
17. Eye Height, Sitting Erect	72.0	73.6	67.7	84.6	86.1	79.1
18. Eye Height, Sitting Relaxed	70.0	71.6	66.2	82.5	84.0	77.9
19. Mid-shoulder Height	56.6	58.3	53.7	67.7	69.2	62.5
20. Shoulder Height, Sitting	54.2	54.6	49.9	65.4	65.9	60.3
21. Shoulder-Elbow Length	33.3	33.2	30.8	40.2	39.7	36.6
22. Elbow-Grip Length	31.7	32.6	29.6	38.3	37.9	35.4
23. Elbow-Fingertip Length	43.8	44.7	40.0	52.0	51.7	47.5
24. Elbow Rest Height	17.5	18.7	16.1	28.0	29.5	26.9
25. Thigh Clearance Height		12.4	10.4		18.8	17.5
26. Knee Height, Sitting	49.7	48.9	46.9	60.2	59.9	55.5
27. Popliteal Height	39.7	38.4	38.0	50.0	47.7	45.7
28. Buttock-Knee Length	54.9	55.9	53.1	65.8	65.5	63.2
29. Buttock-Popliteal Length	45.8	44.9	43.4	54.5	54.6	52.6
30. Buttock-Heel Length		46.7			56.4	
31. Functional Leg Length	000.6	000.6	99.6	000.7	000.4	000.6

(*continued*)

Table C2. *(continued)*

Percentile Values in Inches

	5th Percentile			95th Percentile		
	Ground Troops	Aviators	Women	Ground Troops	Aviators	Women
Standing Body Dimensions						
14. Vertical Arm Reach, Sitting	50.6	52.8	46.2	58.2	60.3	54.9
15. Sitting Height, Erect	32.9	33.7	31.1	38.2	38.8	35.8
16. Sitting Height, Relaxed	32.1	32.9	30.5	37.3	38.0	35.3
17. Eye Height, Sitting Erect	28.3	30.0	26.6	33.3	33.9	31.2
18. Eye Height, Sitting Relaxed	27.6	28.2	26.1	32.5	33.1	30.7
19. Mid-Shoulder Height	22.3	23.0	21.2	26.7	27.3	24.6
20. Shoulder Height, Sitting	21.3	21.5	19.6	25.7	25.9	23.7
21. Shoulder-Elbow Length	13.1	13.1	12.1	15.8	15.6	14.4
22. Elbow-Grip Length	12.5	12.8	11.6	15.1	14.9	14.0
23. Elbow-Fingertip Length	17.3	17.6	15.7	20.5	20.4	18.7
24. Elbow Rest Height	6.9	7.4	6.4	11.0	11.6	10.6
25. Thigh Clearance Height		4.9	4.1		7.4	6.9
26. Knee Height, Sitting	19.6	19.3	18.5	23.7	23.6	21.8
27. Popliteal Height	15.6	15.1	15.0	19.7	18.8	18.0
28. Buttock-Knee Length	21.6	22.0	20.9	25.9	25.8	24.9
29. Buttock-Popliteal Length	17.9	17.7	17.1	21.5	21.5	20.7
30. Buttock-Heel Length		18.4			22.2	
31. Functional Leg Length	43.5	40.9	39.2	50.3	47.4	46.7

Table C3. Body Depth and Breadth Dimensions (From MIL-STD-1472D)

Percentile Values in Centimeters

	5th Percentile			95th Percentile		
	Ground Troops	Aviators	Women	Ground Troops	Aviators	Women
Depth and Breadth Dimensions						
32. Chest Depth	18.9	20.4	19.6	26.7	27.8	27.2
33. Buttock Depth		20.7	18.4		27.4	24.3
34. Chest Breadth	27.3	29.5	25.1	34.4	38.5	31.4
35. Hip Breadth, Standing	30.2	31.7	31.5	36.7	38.8	39.5
36. Shoulder (bideltoid) Breadth	41.5	43.2	38.2	49.8	52.6	45.8
37. Forearm - Forearm Breadth	39.8	43.2	33.0	53.6	60.7	44.9
38. Hip Breadth, Sitting	30.7	33.3	33.0	38.4	42.4	43.9
39. Knee-to-Knee Breadth		19.1			25.5	

Percentile Values In Inches

	5th Percentile			95th Percentile		
	Ground Troops	Aviators	Women	Ground Troops	Aviators	Women
Depth and Breadth Dimensions						
32. Chest Depth	7.5	8.0	7.7	10.5	11.0	10.7
33. Buttock Depth		8.2	7.2		10.8	9.6
34. Chest Breadth	10.8	11.6	9.9	13.5	15.1	12.4
35. Hip Breadth, Standing	11.9	12.5	12.4	14.5	15.3	15.6
36. Shoulder (bideltoid) Breadth	16.3	17.0	15.0	19.6	20.7	18.0
37. Forearm - Forearm Breadth	15.7	17.0	13.0	21.1	23.9	17.7
38. Hip Breadth, Sitting	12.1	13.1	13.0	15.1	16.7	17.3
39. Knee-to-Knee Breadth		7.5			10.0	

Table C4. Body Circumference and Surface Dimensions (From MIL-STD-1472)

Percentile Values in Centimeters

	5th Percentile			95th Percentile		
	Ground Troops	Aviators	Women	Ground Troops	Aviators	Women
Circumferences						
40. Neck Circumference	34.2	34.6	29.9	41.0	41.6	36.7
41. Chest Circumference	83.8	87.5	78.4	000.9	000.9	000.2
42. Waist Circumference	68.4	73.5	59.5	95.9	000.7	83.5
43. Hip Circumference	85.1	87.1	85.5	000.9	000.4	000.1
44. Hip Circumference, Sitting		97.0	87.7		000.3	000.8
45. Vertical Trunk Circumference, Standing	000.6	000.3	000.2	000.6	000.9	000.3
46. Vertical Trunk Circumference, Sitting		000.4	000.8		000.0	000.0
47. Arm Scye Circumference	39.6	39.9	33.6	50.3	53.0	41.7
48. Biceps Circumference, Flexed	27.0	27.8	23.2	37.0	36.9	30.8
49. Elbow Circumference, Flexed		28.5	23.5		34.2	30.0
50. Forearm Circumference, Flexed	26.1	26.3	22.2	33.1	33.1	27.5
51. Wrist Circumference	15.7	15.3	13.6	18.6	19.2	16.2
52. Upper Thigh Circumference	48.1	49.6	48.7	63.9	66.9	64.5
53. Calf Circumference	31.6	33.3	30.6	41.2	41.3	39.2
54. Ankle Circumference	19.3	20.0	18.7	25.2	24.8	23.3
55. Waist Back Length	39.2	42.4	36.7	50.8	50.9	45.4
56. Waist Front Length	36.1	35.7	30.5	46.2	44.2	41.4

Table C4. (*continued*)

Percentile Values in Inches

	5th Percentile			95th Percentile		
	Ground Troops	*Aviators*	*Women*	*Ground Troops*	*Aviators*	*Women*
			Circumferences			
40. Neck Circumference	13.5	13.6	11.8	16.1	16.4	14.4
41. Chest Circumference	33.0	34.4	30.8	41.7	43.3	39.5
42. Waist Circumference	26.9	28.9	23.4	37.8	40.0	32.9
43. Hip Circumference	33.5	34.3	33.7	42.1	42.7	41.8
44. Hip Circumference, Sitting		38.2	34.5		47.0	43.6
45. Vertical Trunk Circumference, Standing	59.3	61.6	56.0	70.3	71.6	65.5
46. Vertical Trunk Circumference, Sitting		59.2	53.1		68.9	63.4
47. Arm Scye Circumference	15.6	15.7	13.2	19.8	20.9	16.4
48. Biceps Circumference, Flexed	10.6	11.0	9.1	14.6	14.5	12.1
49. Elbow Circumference, Flexed		11.2	9.2		13.5	11.8
50. Forearm Circumference, Flexed	10.3	10.4	8.7	13.0	13.0	10.8
51. Wrist Circumference	6.2	6.0	5.4	7.3	7.6	6.4
52. Upper Thigh Circumference	18.9	19.5	19.2	25.1	26.3	25.4
53. Calf Circumference	12.4	13.1	12.0	16.2	16.3	15.4
54. Ankle Circumference	7.6	7.9	7.4	9.9	9.7	9.2
55. Waist Back Length	15.4	16.7	14.4	20.0	20.0	17.9
56. Waist Front Length	14.2	14.1	12.0	18.2	17.4	16.3

Table C5.   Hand and Foot Dimensions (From MIL-STD-1472)

Percentile Values in Centimeters

	5th Percentile			*95th Percentile*		
	Ground Troops	*Aviators*	*Women*	*Ground Troops*	*Aviators*	*Women*
Hand Dimensions						
57. Hand Length	17.4	17.7	16.1	20.7	20.7	20.0
58. Palm Length	9.6	10.0	9.0	11.7	11.9	10.8
59. Hand Breadth	8.1	8.2	6.9	9.7	9.7	8.5
60. Hand Circumference	19.5	19.6	16.8	23.6	23.1	19.9
61. Hand Thickness		2.4			3.5	
Foot Dimensions						
62. Foot Length	24.5	24.4	22.2	29.0	29.0	26.5
63. Instep Length	17.7	17.5	16.3	21.7	21.4	19.6
64. Foot Breadth	9.0	9.0	8.0	10.9	11.6	9.8
65. Foot Circumference	22.5	22.6	20.8	27.4	27.7	24.5
66. Heel-Ankle Circumference	31.3	30.7	28.5	37.0	36.3	33.3

Percentile Values in Inches

	5th Percentile			*95th Percentile*		
	Ground Troops	*Aviators*	*Women*	*Ground Troops*	*Aviators*	*Women*
Hand Dimensions						
57. Hand Length	6.85	6.98	6.32	8.13	8.14	7.89
58. Palm Length	3.77	3.92	3.56	4.61	4.69	4.24
59. Hand Breadth	3.20	3.22	2.72	3.83	3.80	3.33
60. Hand Circumference	7.68	7.71	6.62	9.28	9.11	7.82
61. Hand Thickness		0.95			1.37	
Foot Dimensions						
62. Foot Length	9.65	9.62	8.74	11.41	11.42	10.42
63. Instep Length	6.97	6.88	6.41	8.54	8.42	7.70
64. Foot Breadth	3.53	3.54	3.16	4.29	4.58	3.84
65. Foot Circumference	8.86	8.91	8.17	10.79	10.62	9.65
66. Heel-Ankle Circumference	12.32	12.08	11.21	14.57	14.30	13.11

Index

Note: Page numbers in *italics* indicate photographs, tables, or figures

501

About the Authors

Dr. Mark A. Friend, CSP, is professor in the Department of Doctoral Studies in the College of Aviation at Embry-Riddle Aeronautical University, Daytona Beach, Florida. Dr. Friend is the former director of East Carolina University's Occupational Safety and Health Consortium and of the Center for Applied Technology and former chair of the Occupational Safety and Health Department at Murray State University. Dr. Friend is also past president of the National Occupational Safety and Health Educators' Association and a member of the American Society of Safety Engineers.

At the time of his death in 1999, **James P. Kohn** was associate professor of Industrial Technology at East Carolina University and president of OccuSafe Service Corporation. He had formerly been vice president for professional development of the American Society of Safety Engineers and director of the Occupational Safety Management program at Indiana State University. He held a doctorate in education, and was a certified safety professional, certified industrial hygienist, and Certified Professional Ergonomist. Dr. Kohn authored, co-authored, or edited seven books on safety and industrial hygiene topics. He was a member of the American Society of Safety Engineers, American Industrial Hygiene Association, and the American Conference of Governmental Industrial Hygienists.